ADVANCED CIRCUIT ANALYSIS AND DESIGN

H. Michael Thomas

CONTENTS

Preface v

Acknowledgements vii

Chapter 1 Introduction 1

   1-1. Overview 1
   1-2. The Need for a More Advanced Circuit Analysis Method 1
   1-3. The Frequency Domain Analysis Method 4

Chapter 2 Function Definitions 7

   2-1. Introduction 7
   2-2. The t^n Family of Functions 7
   2-3. The Exponential Function 11
   2-4. The Sinusoidal Function 11
   2-5. Time Shifting 12
   2-6. Complex Functions 13

Chapter 3 Circuit Equations 21

   3-1. Introduction 21
   3-2. The Resistor 21
   3-3. The Capacitor 22
   3-4. The Inductor 23
   3-5. Determining the Circuit Equations 24

Chapter 4 The Laplace Transform 31

   4-1. Introduction 31
   4-2. Periodic Waveforms and the Fourier Series 32
   4-3. Non-Periodic Waveforms and the Fourier and Laplace
         Transforms 33
   4-4. Laplace Transform of Waveforms 34
   4-5. Laplace Transform of Time Shifted Waveforms 38
   4-6. Laplace Transform of the Derivative of a Waveform ... 38
   4-7. Laplace Transform of the Integral of a Waveform 39
   4-8. Laplace Transform of Integral-Differential Equations ... 41
   4-9. The Inverse Laplace Transform 42
   4-10. The Laplace Transform of a Circuit 45

Chapter 5 System Analysis 55

    5-1. Introduction 55
    5-2. System Stability 55
    5-3. Frequency Domain Analysis 60

Chapter 6 Introduction to Filters 81

    6-1. Why Filters 81
    6-2. Filter Types 81
    6-3. Filter Specification 82
    6-4. The Butterworth Approximation 85
    6-5. The Chebyshev Approximation 92
    6-6. Low Pass to High Pass Transformation 101
    6-7. Low Pass to Band Pass Transformation 109
    6-8. Low Pass to Notch Transformation 117

Chapter 7 Passive Filter Realizations 129

    7-1. Introduction 129
    7-2. Low Pass Filter Realization 129
    7-3. Frequency and Impedance Translation 135
    7-4. Low Pass to High Pass Transformation 142
    7-5. Low Pass to Band Pass Transformation 146
    7-6. Low Pass to Notch Transformation 150

Chapter 8 Active Filter Realizations 157

    8-1. Introduction 157
    8-2. The Operational Amplifier 157
    8-3. Low Pass Filter Realization 160
    8-4. High Pass Filter Realization 175
    8-5. Band Pass Filter Realization 186
    8-6. Notch Filter Realization 197
    8-7. State Variable Filter Realization 206

Appendix A The Nodal Analysis Method 229

Appendix B Laplace Transform Tables 239

Appendix C Review of Complex Numbers 247

Appendix D Partial Fraction Expansion 251

Problem Solutions 257

Index 395

PREFACE

This book is intended to be a follow on to a basic circuit analysis text that can be offered in an upper level term. It could also be used by students as supplementary material for self study and as an additional source of information. Problem solutions are provided for all the problems in the book in order to provide the student with an extensive source of worked examples. The book covers advanced circuit analysis using the Laplace transform, system analysis in the frequency domain using Bode plots, and the design of filter circuits.

The student is expected to have strong skills in algebra, trigonometry, and basic circuit analysis as well as a good understanding of differential and integral calculus. These are typical skills for electronics students in their upper terms.

Another problem that students are confronted with is the high cost of technical textbooks. I chose to self publish this book in paperback form in order to reduce the cost to students for textbooks or additional self study material.

Chapter 1 establishes the need for a more advanced analysis method to analyze non steady state sinusoidal circuits. The frequency domain concept of analysis is introduced using an optical analog approach.

Chapter 2 introduces the various non sinusoidal functions used in advanced circuit analysis and develops the mathematical representations for the functions. Chapter 3 defines the differential and integral relationships between voltage and current for resistors, capacitors, and inductors.

Chapter 4 introduces the Laplace transform and inverse Laplace transform as an advanced analysis method for circuit analysis. Circuits are transformed into the Laplace domain, analyzed, and inverse transformed back to the time domain.

Chapter 5 introduces the concept of systems analysis. Circuit performance is analyzed in the frequency domain through the use of Bode plots. The concept of system stability is also discussed.

Chapters 6, 7, and 8 apply the analysis methods of the first five chapters to the design of filter circuits. Chapter 6 introduces filters and the basic types along with the filter transfer functions. Chapter 7 develops methods to design passive (resistors, capacitors, and inductors) filter circuits. Chapter 8 develops methods to design active (resistors, capacitors, and operational amplifiers) filter circuits.

H. Michael Thomas

ACKNOWLEDGEMENTS

I would like to thank my wife, Bonnie, for encouraging me to publish the book and for her endless patience while I spent countless hours researching, writing, and publishing the book.

CHAPTER 1
INTRODUCTION

1-1. Overview

Basic circuit analysis is limited to steady state analysis of resistive, inductive, and capacitive circuits with DC and sinusoidal AC sources. This book extends those basic circuit analysis methods to include transient as well as steady state analysis and include non-sinusoidal AC sources as well. A good understanding of basic circuit analysis methods including Ohm's law, Kirchhoff's voltage and current laws, voltage and current divider principles, series/parallel circuits, superposition, Thévenin equivalent circuits, Norton equivalent circuits, and nodal voltage analysis is assumed.

In addition, a good understanding of algebra, trigonometry, complex numbers, and differential and integral calculus is assumed. The Laplace transform will be developed as the primary tool to extend the analysis concepts. Once the advanced circuit analysis concepts are mastered, they will be applied to the design of filter circuits.

1-2. The Need for a More Advanced Circuit Analysis Method

DC and AC circuit analysis procedures previously studied were limited to a steady-state analysis of circuitry with either a DC or a sinusoidal AC source applied. Non-sinusoidal sources such as ramps, square waves, and pulses were conveniently ignored as well as transient response.

Consider the simple RC circuit of Figure 1-1. Analyzing this circuit to determine the voltage $\left(v_{N1}(t)\right)$ across the capacitor (C) is simple when the input source $\left(v_S(t)\right)$ is a sinusoid. If the input source is non-sinusoidal, it is necessary to result to calculus to analyze the circuit. This circuit will be analyzed using

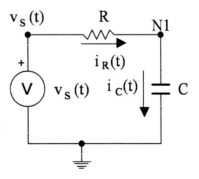

Figure 1-1 Simple RC Circuit

1

nodal voltage analysis. Appendix A is a review of nodal voltage analysis.

The reference node for the circuit of Figure 1-1 is the ground node. The node connecting $v_S(t)$ and R is a grounded voltage source node and is equal to $v_S(t)$. The node connecting R and C is N1, the only unknown node for the circuit. Branch currents $i_R(t)$ and $i_C(t)$ have been assigned based on expected current direction. Applying Kirchhoff's current law (the sum of the currents entering a node is equal to the sum of the currents leaving a node) to node N1 yields Equation (1-1).

$$i_R(t) = i_C(t) \tag{1-1}$$

Converting $i_R(t)$ to node voltages using Ohm's law yields Equation (1-2).

$$i_R(t) = \frac{v_S(t) - v_{N1}(t)}{R} \tag{1-2}$$

Converting $i_C(t)$ to node voltages requires calculus and yields Equation (1-3). A more detailed discussion of this equation can be found in Chapter 3.

$$i_C(t) = C \cdot \frac{d(\text{voltage across C})}{dt} = C \cdot \frac{d(v_{N1}(t) - 0)}{dt}$$
$$= C \cdot \frac{d(v_{N1}(t))}{dt} \tag{1-3}$$

Substituting Equations (1-2) and (1-3) into Equation (1-1) yields Equation (1-4).

$$\frac{v_S(t) - v_{N1}(t)}{R} = C \cdot \frac{d(v_{N1}(t))}{dt} \tag{1-4}$$

Rearranging Equation (1-4) yields Equation (1-5).

$$\frac{d(v_{N1}(t))}{dt} = \frac{v_S(t)}{R \cdot C} - \frac{v_{N1}(t)}{R \cdot C} = -\frac{1}{R \cdot C} \cdot (v_{N1}(t) - v_S(t)) \tag{1-5}$$

Rearranging Equation (1-5) yields Equation (1-6).

$$\frac{d(v_{N1}(t))}{v_{N1}(t) - v_S(t)} = -\frac{1}{R \cdot C} \cdot dt \tag{1-6}$$

In order to solve this differential equation, it will be necessary to integrate both sides. The limits of integration for the left side will be from

$v_{N1}(t)$ at $t = 0$ $\left(v_{N1}(0)\right)$ to $v_{N1}(t)$ at any arbitrary time t $\left(v_{N1}(t)\right)$. The limits of integration for the right side will be from $t = 0$ to t. To simplify the math, assume that $v_S(t)$ is 0 for $t < 0$ and a constant V for $t \geq 0$. Integrating both sides of Equation (1-6) yields Equation (1-7).

$$\int_{V_{N1}(0)}^{V_{N1}(t)} \left(\frac{1}{V_{N1}(t) - V} \right) d\left(V_{N1}(t)\right) = \int_0^t \left(-\frac{1}{R \cdot C} \right) dt \qquad (1\text{-}7)$$

Performing the integration of Equation (1-7) yields Equation (1-8).

$$\ln\left(V_{N1}(t) - V\right) - \ln\left(V_{N1}(0) - V\right) = \left(-\frac{1}{R \cdot C} \right) \cdot t + 0$$

$$\ln\left(\frac{V_{N1}(t) - V}{V_{N1}(0) - V} \right) = \left(-\frac{1}{R \cdot C} \right) \cdot t \qquad (1\text{-}8)$$

$v_{N1}(0)$ is the initial voltage across the capacitor C which will be renamed V_0. Exponentiating both sides of Equation (1-8) yields Equation (1-9).

$$e^{\ln\left(\frac{V_{N1}(t) - V}{V_{N1}(0) - V} \right)} = e^{\left(-\frac{1}{R \cdot C} \right) \cdot t}$$

$$\frac{V_{N1}(t) - V}{V_{N1}(0) - V} = e^{-\left(\frac{1}{R \cdot C} \right) \cdot t} \qquad (1\text{-}9)$$

Rearranging Equation (1-9) yields Equation (1-10).

$$V_{N1}(t) = \left(V_0 - V\right) \cdot e^{-\left(\frac{t}{R \cdot C} \right)} + V$$

$$= V_0 \cdot e^{-\left(\frac{t}{R \cdot C} \right)} + V \cdot \left(1 - e^{-\left(\frac{t}{R \cdot C} \right)} \right) \qquad (1\text{-}10)$$

Summarizing Equation (1-10): when $v_S(t)$ (a constant voltage V) is applied to the circuit, the voltage across the capacitor $\left(v_{N1}(t)\right)$ starts at the initial voltage across the capacitor V_0 which decays to zero at an exponential rate determined by R and C while also charging up to the constant voltage V at the same exponential rate. Thus, Equation (1-10) yields not only the steady state solution ($v_{N1}(t) = V$ at t equal to infinity) but also the transient solution (how the capacitor voltage got to the steady state value).

This was a simple circuit and yet the analysis was extremely complex. Consider the increase in complexity of the analysis if the circuit had more elements and the input source was a function of time rather than a constant. This illustration definitely demonstrates the need for a simpler analysis process for more advanced circuit analysis of non-sinusoidal applications and transient as well as steady state analysis.

1-3. The Frequency Domain Analysis Method

The key to an easier analysis method involves a transformation of the problem from the time domain (which has been the focus of previous circuit analysis methods) to the frequency domain. The frequency domain is used when the frequency response of a sinusoidal AC circuit is analyzed.

A sinusoidal input is applied to a circuit and the frequency is varied while the output magnitude and phase are determined for each value of the input frequency. This frequency response data represents the circuit characteristics in the frequency domain. As will be seen later, it is also possible to process signals and analyze circuits in the frequency domain.

An analogous example to the frequency domain for circuit analysis is that of transforming light into its spectrum of colors using a prism. Figure 1-2 illustrates white light passing through a prism and being transformed into the visible spectrum.

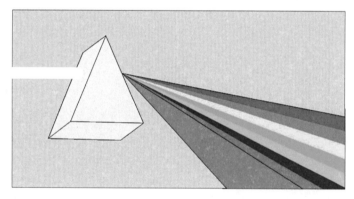

Figure 1-2 Optical Analog of Time/Frequency Domain

This process is reversible as illustrated in Figure 1-3 which uses one prism to transform white light into the visible spectrum and a second prism transforms the visible spectrum back into white light.

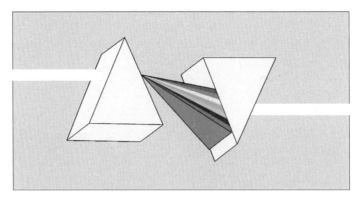

Figure 1-3 Optical Analog
with Both Forward and Inverse Transformations

Since the latter part of this book is to apply advanced circuit analysis methods to the design of filter circuits, this optical analog will also be extended to an optical filter. Figure 1-4 illustrates an optical equivalent to a time domain analog filter. White light is passed through colored glass and only the light that is the color of the glass is emitted from the other side of the colored glass.

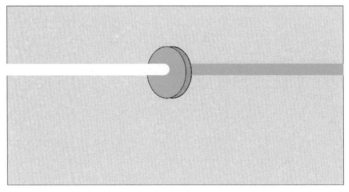

Figure 1-4 Optical Analog of Time Domain Filter

The filtering process can be performed in the frequency domain by using a process like the one illustrated in Figure 1-5. Here the white light is passed through a prism, transforming it to the visible or frequency spectrum. An opaque object with a narrow slit is used to block all of the colors but the desired color that passes through the slit. The single color light is then passed through a second prism and exits the filter as the filtered light.

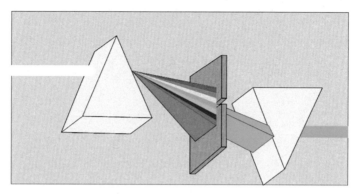

Figure 1-5 Optical Analog of Frequency Domain Filter

In this illustration the filtering process is accomplished in the frequency domain analogous to the way filter circuits will be designed later in this book.

Before developing the mathematical frequency domain analysis methods required, it will be necessary to establish some understanding of the basics. The next three chapters will establish the basis for the new mathematical methods that will be used for analyzing circuits in the frequency domain, which will then be applied to the design of filter circuits.

CHAPTER 2
FUNCTION DEFINITIONS

2-1. Introduction

Basic circuit analysis was limited to steady state analysis of circuits with sources that were either DC or sinusoidal AC. Advanced circuit analysis will extend that scope to include transient analysis as well as other sources. Sinusoidal waves as well as triangle waves, square waves, rectangular waves, and saw tooth waves are all what is termed periodic functions. Periodic functions repeat at regular intervals of time in both positive and negative time directions to positive and negative infinity respectively.

Since these waveforms don't have a starting point, there can be no transient solution but only a steady state solution. Therefore, a totally new set of functions need to be defined that have a starting point which is usually at time equal to zero. That doesn't mean that functions such as sinusoids can't be used in advanced circuit analysis, but that new functions need to be defined that have a starting point and may or may not repeat at regular intervals of time from that point forward. This chapter will define a set of functions for advanced circuit analysis use and will develop mathematical representations for those functions.

2-2. The t^n Family of Functions

The first group of functions that will be defined is the t^n family of functions. This group of functions can be defined mathematically by Equation (2-1).

$$f_n(t) = \begin{cases} 0 & \text{for } t < 0 \\ \dfrac{K \cdot t^n}{n!} & \text{for } t \geq 0 \end{cases} \qquad (2\text{-}1)$$

Where n is an integer 0, 1, 2, etc. and K is a constant

The entire family is defined as zero for negative time which gives a starting point for the functions. The starting point is a requirement for transient analysis solutions. Each function is designated a number of n which is the power of t for that particular function. The power of t portion of the func-

7

tion starts at t equal to zero and proceeds for positive time. A significant property of the t^n family of functions is defined by Equation (2-2).

$$\frac{d\left(f_n(t)\right)}{dt} = f_{(n-1)}(t) \qquad (2\text{-}2)$$

First consider the $f_0(t)$ function for n equal to 0 in Equation (2-1). From Equation (2-1), the $f_0(t)$ function is just K for $t \geq 0$. This function is plotted in Figure 2-1. Here, the $u(t)$ operator has been used as a short hand notation for $f(t) = 0$ for $t < 0$ and $f(t) = 1$ for $t \geq 0$. The $u(t)$ operator is

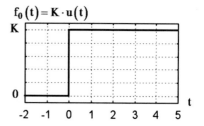

Figure 2-1 The Step Function

referred to as the unit step function and the $f_0(t)$ function pictured in Figure 2-1 is referred to as the step function. The unit step function $u(t)$ is also called the Heavyside step function named after Oliver Heavyside an English electrical engineer, mathematician, and physicist who introduced the function in 1880.

The $u(t)$ function is useful in mathematically describing an instantaneous action such as applying a constant voltage to a circuit as was done in Chapter 1. The unit step function is also useful as a multiplier operator on common functions to force a zero value for negative time as was done to change the constant K to a step function.

Example 2-1

Determine the mathematical representation for a voltage source of a constant 5 volts applied at t equal to zero.

$$v(t) = 5 \cdot u(t)$$

Next consider the $f_1(t)$ function for n equal to 1 in Equation (2-1). From Equation (2-1), the $f_1(t)$ function is $K \cdot t$ for $t \geq 0$. Figure 2-2 is a plot of the waveform $K \cdot t$. But notice that for negative time the function is not zero which is required by Equation (2-1).

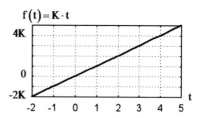

Figure 2-2 $K \cdot t$ Waveform

By multiplying $K \cdot t$ by $u(t)$, the negative time portion can be forced to zero. Figure 2-3 shows the desired $f_1(t)$ function which has been accomplished using the unit step operator as a multiplier. This $f_1(t)$ function is called the ramp function.

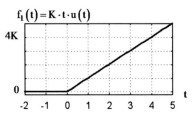

$$f_1(t) = K \cdot t \cdot u(t)$$

Figure 2-3 Ramp Function

Example 2-2

Determine the mathematical representation for a ramp voltage source with a slope of 5 volts per second applied at t equal to zero.

$$v(t) = 5 \cdot t \cdot u(t)$$

This process can be continued for the $f_2(t)$ square function $\frac{K}{2} \cdot t^2 \cdot u(t)$ (n = 2 in Equation (2-1)), the $f_3(t)$ cubic function $\frac{K}{6} \cdot t^3 \cdot u(t)$ (n = 3 in Equation (2-1)), and so on but these functions are rarely used.

Applying Equation (2-2) to these $f_n(t)$ functions yields Equations (2-3).

$$\frac{d(f_3(t))}{dt} = \frac{d\left(\frac{K}{6} \cdot t^3 \cdot u(t)\right)}{dt} = \frac{3 \cdot K}{6} \cdot t^2 \cdot u(t) = f_2(t)$$

$$\frac{d(f_2(t))}{dt} = \frac{d\left(\frac{K}{2} \cdot t^2 \cdot u(t)\right)}{dt} = \frac{2 \cdot K}{2} \cdot t \cdot u(t) = f_1(t) \qquad (2\text{-}3)$$

$$\frac{d(f_1(t))}{dt} = \frac{d(K \cdot t \cdot u(t))}{dt} = K \cdot u(t) = f_0(t)$$

What if Equation (2-2) is applied to $f_0(t)$? The derivative of $f_0(t)$ for $t > 0$ and $t < 0$ are both zero but the derivative of $f_0(t)$ at $t = 0$ is undefined because the step function is discontinuous at $t = 0$. The solution to this problem can be approached by assuming a non-ideal step function that instead of jumping instantaneously to a value of K at zero ramps up to that value over a short time interval. This function and its derivative are shown in Figure 2-4. The step function ramps up to K over a 2 second interval.

9

The derivative is a pulse of width 2 seconds and amplitude of K/2. Note that the area under the derivative curve is K (2 times K/2).

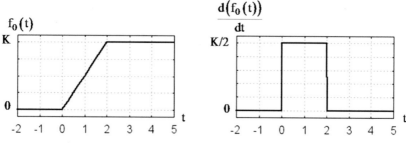

Figure 2-4 Non-Ideal Step Function and Derivative

If the step function ramps up to K in 1 second instead of 2 seconds, the derivative pulse has a width of 1 second and an amplitude of K. Note that the area under the derivative curve is still K. Now as the time interval of the ramp (and the width of the derivative pulse) is decreased further, the amplitude of the derivative curve gets larger and the pulse width gets narrower but the area under the derivative curve remains constant at K. If the time interval of the ramp (and the width of the derivative pulse) approach zero, the amplitude of the derivative pulse approaches infinity and the area under the derivative curve remains K.

This function is called the impulse function and has the characteristics of infinite amplitude, zero width, and an area of K. The accepted representation for the impulse function is $K \cdot \delta(t)$. The accepted format for drawing an impulse function is shown in Figure 2-5. The arrow indicates an infinite amplitude and the K next to the arrow represents the area of the impulse.

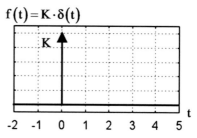

Figure 2-5 Impulse Function

The $\delta(t)$ function is referred to as the unit impulse function the same as $u(t)$ is the unit step function. The unit impulse function $\delta(t)$ is also called the Dirac delta function named after Paul Dirac an English physicist who introduced the function in 1927.

Example 2-3

Determine the mathematical representation for an impulse voltage source with an area of 5 applied at t equal to zero.

$$v(t) = 5 \cdot \delta(t)$$

10

2-3. The Exponential Function

The exponential function frequently arises in the solution of differential equations as was demonstrated in Chapter 1. Since the exponential function may be an output of a circuit, it can also be an input to a subsequent circuit. The two forms most often encountered are shown in Equations (2-4).

$$f(t) = K \cdot e^{-(\alpha \cdot t)} \cdot u(t)$$

(where α = damping factor and K is a constant)

or   (2-4)

$$f(t) = K \cdot e^{-\left(\frac{t}{\tau}\right)} \cdot u(t)$$

(where τ = time constant and K is a constant)

Note that the unit step function multiplier $u(t)$ has been used as it was with the t^n functions to force the exponential function to zero for negative time. Figure 2-6 is a plot of the exponential function with α or τ equal to 1. The exponential function is zero for negative time and starts at K for t equal to 0

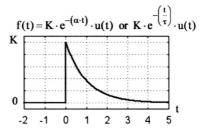

Figure 2-6 Exponential Function

($K \cdot e^0 = K$), then decays exponentially towards zero. The exponential decay reaches zero at t equal to infinity but the accepted practice is that when $\alpha \cdot t$ or t/τ equals 5 , the exponential function has decayed to less than 1% of its initial value $\left(e^{-5} = 0.00674\right)$ and can be considered to be zero.

Example 2-4

Determine the mathematical representation for an exponential voltage source starting at 5 volts with a damping factor of 7 applied at t equal to zero.

$$v(t) = 5 \cdot e^{-(7 \cdot t)} \cdot u(t)$$

2-4. The Sinusoidal Function

As was discussed in Section 2-1, advanced circuit analysis is concerned with transient as well as steady state analyses. Since the sinusoidal waveform is periodic, it has no starting point and therefore can't be used for transient analysis. By multiplying the sinusoid by the unit step function, the negative time portion of the sinusoid can be forced to zero as was done for the t^n and exponential functions.

Figure 2-7 is a plot of the sinusoidal function that can be used for transient analysis. For the plot $\omega = 2 \cdot \pi$ (the frequency is 1 hertz) and θ is zero. Note that the sinusoid is zero for negative time and repeats every 1 second for positive time. This form of the sinusoidal function will work for transient analysis since it has a

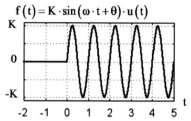

$$f(t) = K \cdot \sin(\omega \cdot t + \theta) \cdot u(t)$$

Figure 2-7 Sinusoidal Function

starting point at t equal to zero. Non-zero values of θ will result in a jump at t equal to zero to the value of $K \cdot \sin(\theta)$.

Example 2-5

Determine the mathematical representation for a 60 hertz sinusoidal voltage source with an amplitude of 5 volts and zero phase shift applied at t equal to zero.

$$\omega = 2 \cdot \pi \cdot f = 2 \cdot \pi \cdot 60 = 120 \cdot \pi$$
$$v(t) = 5 \cdot \sin(120 \cdot \pi \cdot t) \cdot u(t)$$

2-5. Time Shifting

What if the function needs to start at some time value other than zero? Recalling from algebra that translation of a function in a positive direction along the x axis is equivalent to replacing x with (x - a) where "a" is a constant equal to the desired translation amount. This concept can also be applied to the functions. Since most functions use the unit step to force all negative time values to zero, a time shifted function and a time shifted unit step can be used to force all values of the function less than the time shift value to zero. Using the ramp function as an example, $f_1(t) = K \cdot t \cdot u(t)$ is the original function and the time shifted function is:

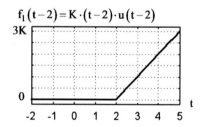

$$f_1(t-2) = K \cdot (t-2) \cdot u(t-2)$$

$$f_1(t-a) = K \cdot (t-a) \cdot u(t-a)$$

Figure 2-8 is a plot of the time shifted ramp function with a equal to 2.

Figure 2-8 Time Shifted Ramp Function

Example 2-6

Determine the mathematical representation for an exponential voltage source starting at 5 volts with a damping factor of 7 applied at t equal to 3 seconds.

$$v(t) = 5 \cdot e^{-(7(t-3))} \cdot u(t-3)$$

2-6. Complex Functions

So far only relatively simple non-periodic functions have been defined. How can more complex functions be created? Any of the t^n, exponential, and sinusoidal functions can be combined by addition, subtraction, multiplication, and time shifting to form more complex functions. Even functions that repeat at regular time intervals but with a starting point can be created from a single cycle description of the repeating function using a summation notation and iterative application of the time shifted single cycle description. Following are several examples of how more complex functions can be created.

Example 2-7

Determine the mathematical representation for the time shifted pulse function of Figure 2-9.

This function can be created using two time shifted step functions. The first is a unit step function (the pulse amplitude is 1) shifted by 1 and the second is a unit step function shifted by 3 that is subtracted from the first step function. The resulting equation is:

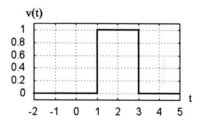

Figure 2-9 Time Shifted Pulse

$$v(t) = u(t-1) - u(t-3)$$

Example 2-8

Determine the mathematical representation for the triangle pulse function of Figure 2-10.

This function can be created from three ramp functions. The first is a ramp function with a slope of 1 starting at t equal to 0. The second ramp function must start at t = 2 and must cancel out the slope

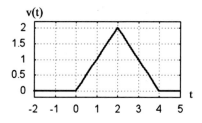

Figure 2-10 Triangle Pulse

of the first ramp plus provide a negative slope ramp of -1. This is accomplished by subtracting a time shifted ramp function (starting at t = 2) with a slope of 2 (1 to cancel out the slope of the first ramp and 1 to give the slope of -1). The third ramp function must start at t = 4 and cancel out the negative slope resulting from the second ramp function. This is accomplished by adding a time shifted ramp function (starting at t = 4) with a slope of 1. The resulting equation is:

$$v(t) = t \cdot u(t) - 2 \cdot (t-2) \cdot u(t-2) + (t-4) \cdot u(t-4)$$

Example 2-9

Determine the mathematical representation for the trapezoid pulse function of Figure 2-11.

This function can be created from four ramp functions. The first is a ramp function with a slope of 1 starting at t = 0. The second ramp function must also have a slope of 1, start at t = 1, and is subtracted from the first ramp

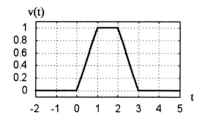

Figure 2-11 Trapezoid Pulse

function in order to cancel out the slope of the first ramp function (creating the flat top). The third ramp function must also have a slope of 1, start at t = 2, and is subtracted from the second ramp function in order to cancel out the flat top and create a negative ramp with a slope of -1. The fourth ramp function must also have a slope of 1, start at t = 3, and is added to the third ramp function to cancel out the negative slope and create the flat area for the remaining positive time values. The resulting equation is:

$$v(t) = t \cdot u(t) - (t-1) \cdot u(t-1) - (t-2) \cdot u(t-2)$$
$$+ (t-3) \cdot u(t-3)$$

Example 2-10

Determine the mathematical representation for the sinusoid pulse function of Figure 2-12.

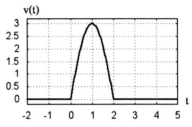

v(t)

This function is the first half period of a sinusoidal waveform. Since half the period is 2 seconds, the period of the sinusoid is 4 seconds and the frequency is 1/4 hertz. The radian frequency is then:

Figure 2-12 Sinusoid Pulse

$$\omega = 2 \cdot \pi \cdot f = 2 \cdot \pi \cdot 0.25 = 0.5 \cdot \pi$$

The amplitude of the sinusoid is 3 and the sinusoid representation is:

$$3 \cdot \sin(0.5 \cdot \pi \cdot t) \cdot u(t)$$

This, however, is a repeating sine wave not just the first half cycle. To cancel out all but the first half cycle, a second time shifted sinusoid of the same amplitude and frequency starting at t = 2 must be added to the first sinusoid. The resulting equation is:

$$v(t) = 3 \cdot \sin(0.5 \cdot \pi \cdot t) \cdot u(t)$$
$$+ 3 \cdot \sin[0.5 \cdot \pi \cdot (t-2)] \cdot u(t-2)$$

Example 2-11

Determine the mathematical representation for the decaying sinusoid function of Figure 2-13.

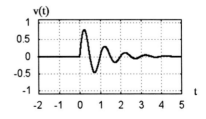

v(t)

This function can be created by multiplying the sinusoidal function times the exponential function. Since both functions have the u(t) multiplier, it need only be used once.

Figure 2-13 Decaying Sinusoid

The period of the sine wave of Figure 2-13 is 1 second, therefore, the frequency is 1 hertz and the resulting equation is:

$$v(t) = e^{-t} \cdot \sin(2 \cdot \pi \cdot t) \cdot u(t)$$

The damping factor of the exponential function and the frequency of the sine wave are interrelated in determining how many sine wave cycles will occur before the exponential function gets to less than 1% of its initial value and can be considered zero. For Figure 2-13 a damping factor of 1 was used. A higher sine wave frequency would result in more sine wave cycles and a lower sine wave frequency would result in fewer sine wave cycles. Also, a lower damping factor would result in more sine wave cycles and a higher damping factor would result in fewer sine wave cycles.

Example 2-12

Determine the mathematical representation for the repeating pulse function of Figure 2-14.

This function can be created by generating a single pulse at time equal to 0 and then successively adding time shifted versions at 3 second intervals. The single pulse is created with two step functions. The first step function has

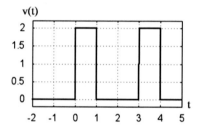

Figure 2-14 Repeating Pulse

an amplitude of 2 and starts at t equal to 0. The second step function also has an amplitude of 2 and is time shifted by 1 second and is subtracted from the first step function in order to cancel out its amplitude. The resulting equation for the single pulse is:

$$2 \cdot u(t) - 2 \cdot u(t-1)$$

The resulting equation for the first time shifted pulse is:

$$2 \cdot u(t-3) - 2 \cdot u(t-4)$$

This is the first pulse delayed by 3 seconds. The equation for the repeating pulse is then:

$$v(t) = 2 \cdot u(t) - 2 \cdot u(t-1) + 2 \cdot u(t-3) - 2 \cdot u(t-4)$$
$$+ 2 \cdot u(t-6) - 2 \cdot u(t-7) \cdots$$

This equation can be simplified by using a summation format for the equation:

$$v(t) = \sum_{n=0}^{n=\infty} \left(2 \cdot u(t-3\cdot n) - 2 \cdot u(t-1-3\cdot n)\right)$$

PROBLEMS

2-1. Determine the mathematical representation for $v(t)$ that is a DC voltage of 12 volts switched on at t equal to 0.

2-2. A voltage described by the equation $2 \cdot t + 3 \cdot t^2$ is applied to a circuit at t equal to 0. Determine the mathematical representation for the applied voltage $v(t)$.

2-3. A ramp voltage with a slope of 4 volts per second is applied to a circuit at t equal to 0. Determine the mathematical representation for the applied voltage $v(t)$.

2-4. Plot the voltage $v(t) = -0.5 \cdot t \cdot u(t)$.

2-5. Determine the damping factor and time constant for:
$$v(t) = 12 \cdot e^{-(10 \cdot t)} \cdot u(t)$$

2-6. A voltage $v(t)$ applied at t equal to 0 starts at 8 volts and decays exponentially with a time constant of 0.5 seconds. Determine the mathematical representation for the applied voltage.

2-7. Plot the voltage $v(t) = 10 \cdot \sin\left(1000 \cdot t + 30^\circ\right) \cdot u(t)$ and determine the amplitude, period, frequency, and radian frequency.

2-8. Plot the voltage $v(t) = 10 \cdot e^{-2t} \cdot \sin\left(20 \cdot \pi \cdot t\right) \cdot u(t)$.

2-9. Determine the mathematical representation for $v(t)$ that is a DC voltage of 5 volts switched on at t equal to 3 seconds.

2-10. A voltage described by the equation $3 \cdot t + 4$ is applied to a circuit at t equal to 2 seconds. Determine the mathematical representation for the applied voltage $v(t)$.

2-11. A ramp voltage with a slope of 5 volts per second is applied to a circuit at t equal to 2 seconds. Determine the mathematical representation for the applied voltage $v(t)$.

2-12. Determine the mathematical representation for the voltage shown below:

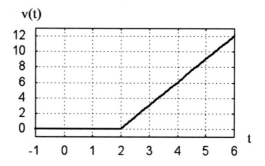

2-13. Determine the mathematical representation for the voltage shown below:

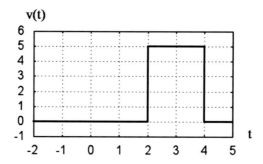

2-14. Determine the mathematical representation for the voltage shown below:

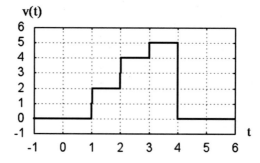

2-15. A sinusoidal source with an amplitude of 2 and a frequency of 5 is turned on at t equal to 0.2 seconds. Determine its mathematical representation.

2-16. Plot the voltage $v(t) = 5 \cdot (t-2) \cdot u(t-2)$.

2-17. Plot the voltage $v(t) = 3 \cdot u(t-2) - 3 \cdot u(t-4)$.

2-18. Plot the voltage:

$$v(t) = 5 \cdot t \cdot u(t) - 5 \cdot (t-1) \cdot u(t-1) + 5 \cdot (t-2) \cdot u(t-2)$$
$$-5 \cdot (t-3) \cdot u(t-3)$$

2-19. Determine the mathematical representation for an impulse function with an area of 3 that starts at t equal to 2.

2-20. Determine the mathematical representation in summation format for the repeating voltage that starts at t equal to 0 shown below:

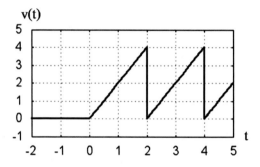

2-21. Determine the mathematical representation in summation format for the repeating half-wave rectified sine wave voltage that starts at t equal to 0 shown below:

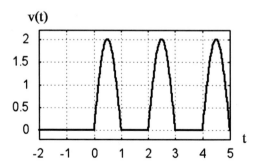

CHAPTER 3
CIRCUIT EQUATIONS

3-1. Introduction

In steady state basic circuit analysis, relationships for voltage and current relative to each of the circuit element types (resistors, capacitors, and inductors) were developed based on the source type (DC or AC sinusoidal). Advanced circuit analysis will require relationships for voltage and current relative to the circuit element type to be independent of the source type. As was seen in Chapter 1 these relationships involve calculus.

This chapter will develop the required relationships to allow the determination of the equations necessary to analyze the circuits for transient and steady state solutions for any source type. The equations will not be solved because the Laplace transform discussed in Chapter 4 will be needed to make the solution practical.

3-2. The Resistor

Steady state DC circuit analysis developed Ohm's law to define voltage and current relationships for resistive circuits. When the analysis was extended to steady state AC sinusoidal analysis of resistive circuits, the relationships were unchanged. Ohm's law still defined the voltage and current relationships for resistive circuits. This is because the resistor is a constant element independent of the source being applied.

In 1827 Georg Ohm developed the relationship between voltage, current and resistance known as Ohm's law that applies for both transient and steady state analysis of circuits with any type source. Ohm's law states that the voltage across a resistor, given the current flowing through the resistor, can be determined from Equation (3-1).

$$v(t) = R \cdot i(t) \quad \text{where R is the resistance of the resistor} \qquad (3-1)$$

The voltage and current are shown as functions of time since any source type can be present in the circuit. Equation (3-1) can be rearranged as shown in Equation (3-2) to define the current through a resistor given the voltage across the resistor.

$$i(t) = \frac{v(t)}{R} \tag{3-2}$$

These equations can be used when analyzing any circuit which contains resistors. Equation (3-2) was used in the analysis of the RC circuit in Chapter 1.

3-3. The Capacitor

For steady state sinusoidal AC analysis, the voltage and current relationships were combined to define a capacitive impedance. This method is only valid for steady state sinusoidal AC analysis and the capacitor must be reevaluated to determine the voltage and current relationships that will apply for any analysis.

When a capacitor is in a circuit, the circuit will establish currents and voltages such that the capacitor voltage satisfies Kirchhoff's voltage law. If the capacitor voltage is either too low or too high, circuit current will flow into one capacitor plate and out of the other capacitor plate until the capacitor voltage is satisfied. The current, however, does not flow through the capacitor since it is two conductive plates separated by an insulating material and the current can't flow through the insulator. Since the plates are conductors, the circuit current causes electrons flow out of one plate leaving it positively charged and into the other plate leaving it negatively charged thus creating a voltage across the capacitor.

The voltage across the capacitor creates an electric field between the plates which stores a charge on the capacitor. This charge is proportional to the voltage across the capacitor. The proportionality constant is called the capacitance and the charge (q) is determined by Equation (3-3).

$$q(t) = C \cdot v(t) \tag{3-3}$$

The capacitor charge and voltage are both functions of time in order for the equation to apply for all types of circuit sources. Conventional current flow in a circuit is actually electrons flowing in the opposite direction. Electrons are charged particles and $2 \cdot \pi \cdot 10^{18}$ electrons constitute a coulomb which is the unit of charge. Current is defined as the rate of flow of charge as indicated by Equation (3-4).

$$i(t) = \frac{d(q(t))}{dt} \tag{3-4}$$

Taking the derivative of Equation (3-3) and recalling that C is a constant yields Equation (3-5).

$$\frac{d(q(t))}{dt} = C \cdot \frac{d(v(t))}{dt} \qquad (3\text{-}5)$$

Combining Equations (3-4) and (3-5) yields Equation (3-6), the expression for the capacitor current given the capacitor voltage.

$$i(t) = C \cdot \frac{d(v(t))}{dt} \qquad (3\text{-}6)$$

Remember that this current is the current that flows in the circuit due to the capacitor and the current does not actually flow through the capacitor as was explained above. Equation (3-6) was used in Chapter 1 in the solution of the simple RC circuit.

By integrating Equation (3-6), an expression for the capacitor voltage given the capacitor current can be obtained as shown in Equation (3-7).

$$\int_0^t (i(t)) \, dt + V_0 = \int_0^t \left(C \cdot \frac{d(v(t))}{dt} \right) dt$$

or

$$V(t) = \frac{1}{C} \cdot \int_0^t (i(t)) \, dt + V_0 \qquad (3\text{-}7)$$

V_0 is the constant of integration and is the voltage across the capacitor at t equal to zero. Equations (3-6) and (3-7) can be used in advanced circuit analysis involving transient as well as steady state analysis with any type source.

3-4. The Inductor

As with the capacitor, for steady state sinusoidal AC analysis, the voltage and current relationships were combined to define an inductive impedance. This method is only valid for steady state sinusoidal AC analysis and the inductor must be reevaluated to determine the voltage and current relationships that will apply for any analysis.

When an inductor is in a circuit, the circuit will establish currents and voltages such that the inductor voltage satisfies Kirchhoff's voltage law. If the inductor voltage is either too low or too high, circuit current will flow through the inductor until the inductor voltage is satisfied.

Current flowing through an inductor creates a magnetic field which stores energy in the inductor. This magnetic field is represented by the magnetic flux, Φ. This magnetic flux is proportional to the current flowing through the inductor. The proportionality constant is called the inductance and the magnetic flux is determined by Equation (3-8).

$$\Phi(t) = L \cdot i(t) \tag{3-8}$$

The inductor magnetic flux and current are both functions of time in order for the equation to apply for all types of circuit sources. A changing magnetic flux creates a voltage across the inductor that is determined by Equation (3-9).

$$v(t) = \frac{d(\Phi(t))}{dt} \tag{3-9}$$

Taking the derivative of Equation (3-8) and recalling that L is a constant yields Equation (3-10).

$$\frac{d(\Phi(t))}{dt} = L \cdot \frac{d(i(t))}{dt} \tag{3-10}$$

Combining Equations (3-9) and (3-10) yields Equation (3-11), the expression for the inductor voltage given the inductor current.

$$v(t) = L \cdot \frac{d(i(t))}{dt} \tag{3-11}$$

By integrating Equation (3-11), an expression for the inductor current given the inductor voltage can be obtained as shown in Equation (3-12).

$$\int_0^t (v(t)) \, dt + I_0 = \int_0^t \left(L \cdot \frac{d(i(t))}{dt} \right) dt$$

or

$$i(t) = \frac{1}{L} \cdot \int_0^t (v(t)) \, dt + I_0 \tag{3-12}$$

I_0 is the constant of integration and is the current flowing in the inductor at t equal to zero. Equations (3-11) and (3-12) can be used in advanced circuit analysis involving transient as well as steady state analysis with any type source.

3-5. Determining the Circuit Equations

This section will use Equations (3-1), (3-2), (3-6), (3-7), (3-11), and (3-12) along with basic circuit analysis methods to determine the integral-differential equations that are needed to analyze a circuit with any type sources. Chapter 1 used Equations (3-2) and (3-6) to analyze a simple RC circuit. This was a complete solution but in this section the analysis will be stopped at the determination of the integral-differential equations. The solution to these type equations will be discussed in Chapter 4.

Figure 3-1 is a simple series RLC circuit. Kirchhoff's voltage law along with Equations (3-1), (3-7), and (3-11) will be used to determine the circuit equation. Kirchhoff's voltage law states that the sum of the voltages around a closed loop must equal zero. For Figure 3-1, the voltage source, R, L, and C form a closed loop. Then, summing the voltages around the closed loop, starting with the ground point and progressing clockwise, yields Equation (3-13). Note that the

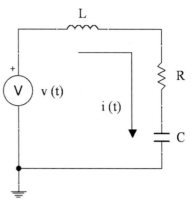

Figure 3-1 Series RLC Circuit

voltage source is a voltage rise while the R, L, and C voltages are voltage drops based on the assumed current direction.

$$v(t) - v_L(t) - v_R(t) - v_C(t) = 0 \qquad (3\text{-}13)$$

Substituting Equation (3-1) for the resistor voltage, Equation (3-7) for the capacitor voltage, and Equation (3-11) for the inductor voltage yields Equation (3-14) which is the integral-differential equation for Figure 3-1.

$$v(t) - L \cdot \frac{d(i(t))}{dt} - R \cdot i(t) - \frac{1}{C} \cdot \int_0^t (i(t)) \, dt - V_{C0} = 0 \qquad (3\text{-}14)$$

V_{C0} is the voltage across the capacitor at t equal to zero.

Figure 3-2 is a simple parallel RLC circuit. Kirchhoff's current law along with Equations (3-2), (3-6), and (3-12) will be used to determine the circuit equation. Kirchhoff's current law states that the sum of the currents entering a node equals the sum of the currents leaving a node. The

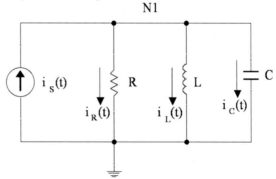

Figure 3-2 Parallel RLC Circuit

circuit of Figure 3-2 has a single node, N1. Applying Kirchhoff's current law to node N1 yields Equation (3-15).

$$i_S(t) = i_R(t) + i_L(t) + i_C(t) \qquad (3\text{-}15)$$

The voltage across R, L, and C is $v_{N1}(t)$. Substituting Equation (3-2) for the resistor current, Equation (3-6) for the capacitor current, and Equation (3-12) for the inductor current yields Equation (3-16) which is the integral-differential equation for Figure 3-2.

$$i_S(t) = \frac{v_{N1}(t)}{R} + \frac{1}{L} \cdot \int_0^t (v_{N1}(t)) \, dt + I_{L0} + C \cdot \frac{d(v_{N1}(t))}{dt} \tag{3-16}$$

I_{L0} is the current flowing in the inductor at t equal to zero.

Figure 3-3 is a more complex series/parallel circuit. For this circuit assume that at t equal to zero, inductor currents and capacitor voltages are zero. Nodal voltage analysis along with Equations (3-2), (3-6), and (3-12) will be used to determine the integral-differential equations. Appendix A is a review of nodal voltage analysis.

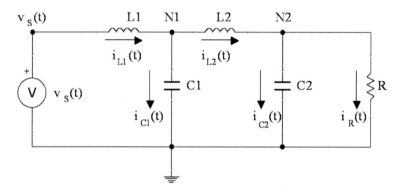

Figure 3-3 Series/Parallel RLC Circuit

The reference node for the circuit of Figure 3-3 is the ground node. The node connecting $v_S(t)$ and L1 is a grounded voltage source node and is equal to $v_S(t)$. The node connecting L1, L2, and C1 is node N1 and the node connecting L2, C2, and R is node N2. These are the two unknown nodes which means there will be two integral-differential equations. Branch currents $i_{L1}(t)$, $i_{L2}(t)$, $i_{C1}(t)$, $i_{C2}(t)$, and $i_R(t)$ have been assigned based on expected current direction.

Applying Kirchhoff's current law to node N1 yields Equation (3-17).

$$i_{L1}(t) = i_{C1}(t) + i_{L2}(t) \tag{3-17}$$

Applying Kirchhoff's current law to node N2 yields Equation (3-18).

$$i_{L2}(t) = i_{C2}(t) + i_R(t) \tag{3-18}$$

The node voltage for node N1 is $v_{N1}(t)$ and the node voltage for node N2 is $v_{N2}(t)$. Substituting Equation (3-12) for $i_{L1}(t)$ and $i_{L2}(t)$, and Equation (3-6) for $i_{C1}(t)$ into Equation (3-17) yields Equation (3-19).

$$\frac{1}{L1}\cdot\int_0^t \left(v_S(t) - v_{N1}(t)\right)dt = C1\cdot\frac{d\left(v_{N1}(t)\right)}{dt}$$

$$+\frac{1}{L2}\cdot\int_0^t\left(v_{N1}(t) - v_{N2}(t)\right) \tag{3-19}$$

Substituting Equation (3-12) for $i_{L2}(t)$, Equation (3-6) for $i_{C2}(t)$, and Equation (3-2) for $i_R(t)$ into Equation (3-18) yields Equation (3-20).

$$\frac{1}{L2}\cdot\int_0^t\left(v_{N1}(t) - v_{N2}(t)\right) = C2\cdot\frac{d\left(v_{N2}(t)\right)}{dt} + \frac{v_{N2}(t)}{R} \tag{3-20}$$

Equations (3-19) and (3-20) are the integral-differential equations for the circuit of Figure 3-3. As can be seen, the solution to these equations has become extremely difficult. Chapter 4 will discuss a method using Laplace transforms that will greatly simplify the solution of the integral-differential equations.

PROBLEMS

3-1. What is the expression for the current through a resistor given the voltage across the resistor?

3-2. What is the expression for the capacitor current given the voltage across the capacitor?

3-3. What is the expression for the current through an inductor given the voltage across the inductor?

3-4. What is the expression for the voltage across a resistor given the current through the resistor?

3-5. What is the expression for the voltage across a capacitor given the capacitor current?

3-6. What is the expression for the voltage across an inductor given the current through the inductor?

3-7. Use nodal voltage analysis to determine the integral-differential equation for the following circuit:

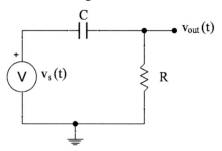

3-8. Use nodal voltage analysis to determine the integral-differential equation for the following circuit:

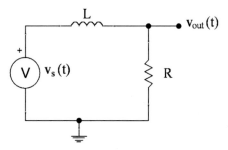

3-9. Use nodal voltage analysis to determine the integral-differential equation for the following circuit:

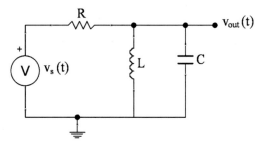

3-10. Use nodal voltage analysis to determine the integral-differential equation for the following circuit:

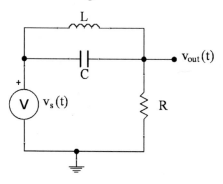

CHAPTER 4
THE LAPLACE TRANSFORM

4-1. Introduction

In Chapter 1, the concept of solving integral-differential equations by transformation to the frequency domain was discussed. The method presented was an optical method that is analogous to the mathematical methods that will be discussed in this chapter.

Chapter 3 discussed circuit integral-differential equations which involved current and voltage waveforms and their integrals and derivatives. Chapter 2 discussed various current and voltage waveforms and their mathematical definitions. This chapter will discuss how to transform those mathematical definitions as well as their integrals and derivatives to the frequency domain, how to generate a solution in the frequency domain, and how to transform the results back to the time domain.

There are two classifications of current and voltage waveforms. One is continuous time (also referred to as analog) and the other is discrete time (also referred to as digital). Continuous time waveforms have a value for any value of time. Discrete time waveforms have a value only at specific repeating intervals of time. This book will only be concerned with continuous time waveforms.

Continuous time waveforms can be either periodic or non-periodic. Periodic waveforms repeat at regular time intervals in both negative and positive time directions. Non-periodic waveforms don't repeat. The Fourier series is generally used to transform periodic waveforms to the frequency domain. The Fourier transform or the Laplace transform are generally used to transform non-periodic waveforms to the frequency domain.

Joseph Fourier, a French mathematician, developed a mathematical treatment of heat theory in 1822 that solved integral-differential equations by using an infinite series of trigonometric functions. The Fourier series and Fourier transforms are attributed to him. Pierre Simon De Laplace, a French astronomer and mathematician, developed the theory of motion of the solar system during the period from 1799 to 1825. This also involved the solution of integral-differential equations but used a method different from but similar to Fourier's. The Laplace transform is attributed to him. Oliver Heaviside, an English physicist and electrical engineer, is credited with adapting Fourier and Laplace solutions of integral-differential equations to electronics in the late 1800's and early 1900's.

4-2. Periodic Waveforms and the Fourier Series

The Fourier series actually is the first introduction to transforming signals to the frequency domain. The approach is usually simplified to a series of sines and cosines so as not to complicate the math. The more familiar representation of the Fourier series is shown in Equation (4-1).

$$F(m \cdot \omega) = A_0 + \sum_{m=1}^{\infty} \left(A_m \cdot \cos(m \cdot \omega \cdot t) + B_m \cdot \sin(m \cdot \omega \cdot t) \right) \quad (4\text{-}1)$$

where :

m is an integer between 1 and ∞

$$\omega = \frac{2 \cdot \pi}{t_p} \quad \text{and } t_p \text{ is the period of f(t)}$$

$$A_0 = \frac{1}{t_p} \int_0^{t_p} f(t) \, dt$$

$$A_m = \frac{2}{t_p} \int_0^{t_p} \left(f(t) \cdot \cos(m \cdot \omega \cdot t) \right) dt$$

$$B_m = \frac{2}{t_p} \int_0^{t_p} \left(f(t) \cdot \sin(m \cdot \omega \cdot t) \right) dt$$

This form which has both a sine and cosine function at each frequency does not lend itself to calculation in the frequency domain. If Euler's exponential representation for the sine and cosine, shown in Equations (4-2), is substituted into Equation (4-1), Equation (4-3) results.

$$\sin(m \cdot \omega \cdot t) = \frac{e^{(j \cdot m \cdot \omega \cdot t)} - e^{(-j \cdot m \cdot \omega \cdot t)}}{2 \cdot j}$$

$$\cos(m \cdot \omega \cdot t) = \frac{e^{(j \cdot m \cdot \omega \cdot t)} + e^{(-j \cdot m \cdot \omega \cdot t)}}{2} \quad (4\text{-}2)$$

$$F(m \cdot \omega) = A_0 + \sum_{m=1}^{\infty} \left(A_m \cdot \frac{e^{j \cdot m \cdot \omega \cdot t} + e^{-j \cdot m \cdot \omega \cdot t}}{2} + B_m \cdot \frac{e^{j \cdot m \cdot \omega \cdot t} - e^{-j \cdot m \cdot \omega \cdot t}}{2 \cdot j} \right) \quad (4\text{-}3)$$

Equation (4-3) can be rearranged to collect like exponential terms and change the summation limits from 1 to ∞ to $-\infty$ to $+\infty$ in order to make the negative exponentials positive (the negative m's yield the negative exponentials). This also brings the A_0 term inside the summation since $m = 0$ is

32

now one of the summation values. The Fourier series expression now becomes Equation (4-4).

$$F(m \cdot \omega) = \sum_{m=-\infty}^{+\infty} \left(C_m \cdot e^{j \cdot m \cdot \omega \cdot t} \right) \qquad (4\text{-}4)$$

where:

m is an integer between $-\infty$ and $+\infty$

$$C_m = \frac{A_m - j \cdot B_m}{2}$$

$$= \frac{1}{t_p} \cdot \int_{t_p} \left(f(t) \cdot e^{-j \cdot m \cdot \omega \cdot t} \right) dt$$

Periodic waveforms can be transformed from the time domain to the frequency domain using Equation (4-4). The result is a discrete frequency spectrum of the frequencies present in the waveform. Note that the frequencies are integer multiples of the frequency of the time domain waveform.

4-3. Non-Periodic Waveforms and the Fourier and Laplace Transforms

The Fourier transform extends the Fourier series to transform to the frequency domain non-periodic continuous time waveforms. This is the method that Fourier and Laplace both researched and came up with different but similar approaches. Fourier's approach assumes that time is continuous and infinite in both the positive and negative direction and thus his transformation ranges from $-\infty$ to $+\infty$. Equation (4-5) is the definition of the Fourier Transform.

$$F(\omega) = \int_{-\infty}^{+\infty} \left(f(t) \cdot e^{-j \cdot \omega \cdot t} \right) dt \qquad (4\text{-}5)$$

Non-periodic waveforms can be transformed from the time domain to the frequency domain using Equation (4-5). The result is a continuous frequency spectrum unlike the discrete frequency spectrum obtained for periodic waveforms.

Laplace made the assumption that every real process started at some point in time and defined this point as $t = 0$. He then assumed that everything prior to $t = 0$ (or negative time) was zero. This made his transformation range from 0 to $+\infty$. Laplace was further concerned about the existence of the transform for a large range of functions which meant that the integral must converge for all values of the function. To improve this process he added a second negative real exponential function of time to the transform. Equation (4-6) is the result.

$$F(\omega) = \int_0^{+\infty} \left(f(t) \cdot e^{-\sigma t} \cdot e^{-j\omega \cdot t} \right) dt$$

$$= \int_0^{+\infty} \left(f(t) \cdot e^{-(\sigma + j\omega)t} \right) dt \tag{4-6}$$

Laplace then renamed $(\sigma + j \cdot \omega)$ s and called the new domain the Laplace domain since it is similar to but not identical to the frequency domain. His final transform expression is shown in Equation (4-7).

$$F(s) = \int_0^{+\infty} \left(f(t) \cdot e^{-st} \right) dt \tag{4-7}$$

Non-periodic waveforms can be transformed from the time domain to the Laplace domain using Equation (4-7). The result is a continuous frequency spectrum similar to that of the Fourier transform.

The standard practice is to use the Fourier series and the Fourier transform when analyzing the frequency spectrum of waveforms and to use the Laplace transform when analyzing circuits or systems. Since this book is concerned with the analysis of circuits and systems, the Laplace transform is used.

4-4. Laplace Transform of Waveforms

A new convention was used in the above equations. Functions in the time domain are represented with lowercase letters and functions in the Laplace or frequency domain are represented with uppercase letters. For example: $f(t)$, $F(s)$, or $F(\omega)$. When a function is transformed from the time domain to the Laplace domain, the definition of Equation (4-8) is used.

$$\mathscr{L}\left(f(t)\right) = F(s) = \int_0^{\infty} \left(f(t) \cdot e^{(-st)} \right) dt \tag{4-8}$$

Before applying Equation (4-8) to some of the waveforms of Chapter 2, there is an important property of the Laplace transform that must be considered. That is the property of linearity. The linearity property has two components. The first is "the Laplace transform of a constant times a function of time is equal to the constant times the Laplace transform of the original function of time". The second is "the Laplace transform of the sum (or difference) of several functions of time is equal to the sum of the individual Laplace transforms of the functions of time".

Equation (4-9) uses the Laplace transform definition of Equation (4-8) and the linearity property of integration to demonstrate the Laplace transform linearity property.

$$\mathcal{L}\left(K_1 \cdot f_1(t) \pm K_2 \cdot f_2(t) \pm \cdots\right) = \int_0^\infty \left(\left(K_1 \cdot f_1(t) \pm K_2 \cdot f_2(t) \pm \cdots\right) \cdot e^{-st}\right) dt$$

$$= \int_0^\infty \left(\left(K_1 \cdot f_1(t)\right) \cdot e^{-st}\right) dt \pm \int_0^\infty \left(\left(K_2 \cdot f_2(t)\right) \cdot e^{-st}\right) dt \pm \cdots$$

$$= K_1 \cdot \int_0^\infty \left(\left(f_1(t)\right) \cdot e^{-st}\right) dt \pm K_2 \cdot \int_0^\infty \left(\left(f_2(t)\right) \cdot e^{-st}\right) dt \pm \cdots$$

$$= K_1 \cdot F_1(s) \pm K_2 \cdot F_2(s) \pm \cdots \tag{4-9}$$

It is important to note that the Laplace linearity property only applies to functions that are added or subtracted not multiplied as indicated by the inequality in Equation (4-10).

$$\mathcal{L}\left(f_1(t) \cdot f_2(t)\right) \neq F_1(s) \cdot F_2(s) \tag{4-10}$$

The first waveform discussed in Chapter 2 was the step function. Applying Equation (4-8) to the step waveform yields Equation (4-11).

$$\mathcal{L}\left(K \cdot u(t)\right) = F(s) = \int_0^\infty \left(\left(K \cdot u(t)\right) \cdot e^{-st}\right) dt \tag{4-11}$$

Using the Laplace transform linearity property and the definition of $u(t)$, which is 1 for t between 0 and ∞, on Equation (4-11) yields Equation (4–12).

$$F(s) = K \cdot \int_0^\infty e^{-st} \, dt = K \cdot \left.\frac{e^{-st}}{-s}\right|_{t=0}^{\infty} = K \cdot \left(\frac{0}{-s} - \frac{1}{-s}\right) = \frac{K}{s} \tag{4-12}$$

Another waveform discussed in Chapter 2 was the ramp waveform. Applying Equation (4-8) to the ramp waveform yields Equation (4-13).

$$\mathcal{L}\left(K \cdot t \cdot u(t)\right) = F(s) = \int_0^\infty \left(\left(K \cdot t \cdot u(t)\right) \cdot e^{-st}\right) dt \tag{4-13}$$

Using the Laplace transform linearity property and the definition of $u(t)$, which is 1 for t between 0 and ∞, on Equation (4-13) yields Equation (4–14).

$$F(s) = K \cdot \int_0^\infty \left(t \cdot e^{-st}\right) dt \tag{4-14}$$

Using integration by parts with $u = t$, and $dv = e^{-st}dt$ in Equation (4-14), then $\int u \, dv = u \cdot v - \int v \, du$ where $du = dt$ and $v = \int e^{-st} \, dt = -\frac{e^{-st}}{s}$. Equation (4-14) then becomes Equation (4-15).

$$F(s) = K \cdot \left(\left(-\frac{t \cdot e^{-s \cdot t}}{s} \right) \Big|_0^\infty - \int_0^\infty \left(-\frac{e^{-s \cdot t}}{s} \right) dt \right)$$

$$= K \cdot \left((0) - \left(\frac{e^{-s \cdot t}}{s^2} \right) \Big|_0^\infty \right) = \frac{K}{s^2} \tag{4-15}$$

Another waveform discussed in Chapter 2 was the impulse waveform. Applying Equation (4-8) to the impulse waveform yields Equation (4-16).

$$\mathcal{L}(K \cdot \delta(t)) = F(s) = \int_0^\infty \left((K \cdot \delta(t)) \cdot e^{-s \cdot t} \right) dt \tag{4-16}$$

Using the Laplace transform linearity property Equation (4-16) becomes Equation (4-17).

$$F(s) = K \cdot \int_0^\infty \left((\delta(t)) \cdot e^{-s \cdot t} \right) dt \tag{4-17}$$

The $\delta(t)$ function is zero everywhere except at t equal to 0 and at t equal to 0 the $e^{-s \cdot t}$ is 1. $F(s)$ is then K times the integral of the $\delta(t)$ function. Since the integral is the area under the curve and the area of the $\delta(t)$ function is 1, $F(s)$ is K.

Another waveform discussed in Chapter 2 was the exponential waveform. Applying Equation (4-8) to the exponential waveform yields Equation (4-18).

$$\mathcal{L}(K \cdot e^{-\alpha \cdot t} \cdot u(t)) = F(s) = \int_0^\infty \left((K \cdot e^{-\alpha \cdot t} \cdot u(t)) \cdot e^{-s \cdot t} \right) dt \tag{4-18}$$

Using the Laplace transform linearity property and the definition of $u(t)$, which is 1 for t between 0 and ∞, on Equation (4-18) yields Equation (4–19).

$$F(s) = K \cdot \int_0^\infty \left(e^{-\alpha \cdot t} \cdot e^{-s \cdot t} \right) dt = K \cdot \int_0^\infty e^{-(s+\alpha) \cdot t} \, dt$$

$$= K \cdot \left(\frac{e^{-(s+\alpha) \cdot t}}{-(s+\alpha)} \right) \Big|_0^\infty = 0 - \frac{K}{-(s+\alpha)} = \frac{K}{s+\alpha} \tag{4-19}$$

An extension of the exponential waveform is the Laplace complex translation property. The Laplace transform of any function multiplied by $e^{-\alpha \cdot t}$ will result in a translation of $+\alpha$ $(s = s + \alpha)$ in the Laplace domain. Equation (4-20) applies Equation (4-8) to this property.

$$\mathcal{L}(e^{-\alpha \cdot t} \cdot f(t) \cdot u(t)) = \int_0^\infty \left((e^{-\alpha \cdot t} \cdot f(t) \cdot u(t)) \cdot e^{-s \cdot t} \right) dt \tag{4-20}$$

Using the definition of $u(t)$, which is 1 for t between 0 and ∞, on Equation (4-20) yields Equation (4-21).

$$\mathcal{L}\left(e^{-\alpha t} \cdot f(t) \cdot u(t)\right) = \int_0^\infty \left(f(t) \cdot e^{-(s+\alpha)t}\right) dt = F(s+\alpha) \tag{4-21}$$

Equation (4-21) states that the Laplace transform of $f(t)$ is replaced with $s = (s+\alpha)$ which is a translation by α in the Laplace domain.

Another waveform discussed in Chapter 2 was the sinusoidal waveform. Using the Laplace transform linearity property, the definition of $u(t)$ equal to 1, and Euler's exponential representation for the sine function, yields Equation (4-22).

$$\mathcal{L}\left(K \cdot \sin(\omega \cdot t) \cdot u(t)\right) = K \cdot \mathcal{L}\left(\frac{e^{j\omega t} - e^{-j\omega t}}{2 \cdot j}\right)$$

$$= \frac{K}{2 \cdot j} \cdot \left(\mathcal{L}(e^{j\omega t}) - \mathcal{L}(e^{-j\omega t})\right) \tag{4-22}$$

Applying the Laplace transform of the exponential function from Equation (4-19) to Equation (4-22) yields Equation (4-23).

$$F(s) = \frac{K}{2 \cdot j} \cdot \left(\frac{1}{s - j \cdot \omega} - \frac{1}{s + j \cdot \omega}\right)$$

$$= \frac{K}{2 \cdot j} \cdot \left(\frac{(s + j \cdot \omega) - (s - j \cdot \omega)}{s^2 + \omega^2}\right)$$

$$= \frac{K}{2 \cdot j} \cdot \left(\frac{2 \cdot j \cdot \omega}{s^2 + \omega^2}\right) = \frac{K \cdot \omega}{s^2 + \omega^2} \tag{4-23}$$

Another waveform discussed in Chapter 2 was the decaying sinusoid waveform. Using the Laplace transform linearity property, the definition of $u(t)$ equal to 1, the Laplace complex translation property and the Laplace transform of the sinusoidal function of Equation (4-23), yields Equation (4-24).

$$\mathcal{L}\left(K \cdot e^{-\alpha t} \cdot \sin(\omega \cdot t) \cdot u(t)\right) = \mathcal{L}\left(K \cdot \sin(\omega \cdot t) \cdot u(t)\right)\Big|_{s=(s+\alpha)}$$

$$= \frac{K \cdot \omega}{(s+\alpha)^2 + \omega^2} \tag{4-24}$$

4-5. Laplace Transform of Time Shifted Waveforms

What about time shifted waveforms? Equation (4-25) is the definition of the Laplace transform of a function of x times the unit step function.

$$\int_0^\infty \left(f(x) \cdot u(x) \cdot e^{-sx} \right) dx = F(s) \tag{4-25}$$

Change the variable to t by letting $x = t - a$. Substituting into Equation (4-25) yields Equation (4-26).

$$\int_0^\infty \left(f(t-a) \cdot u(t-a) \cdot e^{-s(t-a)} \right) d(t-a) = F(s)$$

$$\int_0^\infty \left(f(t-a) \cdot u(t-a) \cdot e^{a \cdot s} \cdot e^{-st} \right) dt = F(s)$$

$$e^{a \cdot s} \cdot \int_0^\infty \left(f(t-a) \cdot u(t-a) \cdot e^{-st} \right) dt = F(s)$$

$$\int_0^\infty \left(f(t-a) \cdot u(t-a) \cdot e^{-st} \right) dt = e^{-a \cdot s} \cdot F(s) \tag{4-26}$$

The left side of Equation (4-26) is the definition of the Laplace transform of a time shifted function. The right side of Equation (4-26) defines the Laplace transform of a function time shifted by a as e^{-as} times the Laplace transform of the non-shifted function. This operation is the Laplace real translation property. Note the similarity to the complex translation property of Equation (4-21). Both translations result in multiplication by an exponential function in the other domain.

Determine the Laplace transform of a rectangular pulse of amplitude K and width a. Equation (4-27) is the Laplace transform of the pulse.

$$F(s) = \mathcal{L}\left(K \cdot u(t) - K \cdot u(t-a) \right)$$

$$= \mathcal{L}\left(K \cdot u(t) \right) - \mathcal{L}\left(K \cdot u(t-a) \right)$$

$$= \frac{K}{s} - e^{-a \cdot s} \cdot \frac{K}{s} \tag{4-27}$$

4-6. Laplace Transform of the Derivative of a Waveform

In Chapter 3 integral-differential equations were determined for some typical circuits. These equations contained the derivative of voltage and current waveforms in the circuit. How is the Laplace transform of the derivative of functions determined? Equation (4-28) is the definition of the Laplace transform of the derivative of a function of time.

$$\mathcal{L}\left(\frac{df(t)}{dt} \right) = \int_0^\infty \left(\frac{df(t)}{dt} \cdot e^{-st} \right) dt \tag{4-28}$$

Using integration by parts with $u = e^{-st}$, and $dv = \dfrac{df(t)}{dt}dt$ in Equation (4-28), then $\int u\,dv = u \cdot v - \int v\,du$ where $du = -s \cdot e^{-st}\,dt$ and $v = f(t)$. Equation (4-28) then becomes Equation (4-29).

$$\mathcal{L}\left(\frac{df(t)}{dt}\right) = e^{-st} \cdot f(t)\Big|_0^\infty - \int_0^\infty \left(f(t) \cdot \left(-s \cdot e^{-st}\right)\right) dt$$

$$= \left(0 - f(0)\right) + s \cdot \int_0^\infty \left(f(t) \cdot e^{-st}\right) dt$$

$$= s \cdot F(s) - f(0) \tag{4-29}$$

This is the Laplace transform real differentiation property which states that the Laplace transform of the derivative of a function is s times the Laplace transform of the un-differentiated function minus the value of the function at t equal to zero.

Just like the real and complex translation properties there is also a complex differentiation property dealing with differentiation in the Laplace domain. Equation (4-30) is the definition of the Laplace transform.

$$F(s) = \int_0^\infty \left(f(t) \cdot e^{-st}\right) dt \tag{4-30}$$

Differentiate both sides of Equation (4-30) with respect to s to arrive at Equation (4-31).

$$\frac{dF(s)}{ds} = \frac{d\left(\int_0^\infty \left(f(t) \cdot e^{-st}\right) dt\right)}{ds} = \int_0^\infty \left(f(t) \cdot \frac{d\left(e^{-st}\right)}{ds}\right) dt$$

$$= \int_0^\infty \left(-t \cdot f(t) \cdot e^{-st}\right) dt \tag{4-31}$$

Thus, differentiation in the Laplace domain is multiplication by −t in the time domain. Again, note the similarity of differentiation in the time domain results in multiplication by s (the Laplace domain variable) in the Laplace domain while differentiation in the Laplace domain results in multiplication by −t (the time domain variable) in the time domain.

4-7. Laplace Transform of the Integral of a Waveform

The Chapter 3 integral-differential equations also involved Laplace transforming the integral of voltage and current waveforms in a circuit. How is the Laplace transform of the integral of functions determined? Equation (4-32) is the definition of the Laplace transform of the integral of a function of time.

$$\mathcal{L}\left(\int_0^t f(t)\, dt\right) = \int_0^\infty \left(\int_0^t f(t)\, dt\right) \cdot e^{-st}\, dt = F(s) \tag{4-32}$$

Using integration by parts with $u = \int_0^t f(t)\, dt$, and $dv = e^{-st}dt$ in Equation (4-32), then $\int u\, dv = u \cdot v - \int v\, du$ where $du = f(t)dt$ and $v = -\dfrac{e^{-st}}{s}$. Equation (4-32) then becomes Equation (4-33).

$$\left(-\frac{e^{-st}}{s} \cdot \int_0^t f(t)\, dt\right)\Bigg|_0^\infty - \int_0^\infty \left(-\frac{e^{-st}}{s} \cdot f(t)\right) dt$$

$$= 0 + \frac{1}{s} \cdot \int_0^\infty f(t) \cdot e^{-st}\, dt = \frac{F(s)}{s} \tag{4-33}$$

This is the Laplace transform real integration property which states that the Laplace transform of the integral of a function is the Laplace transform of the un-integrated function divided by s. Note that real differentiation results in multiplication by s while real integration results in division by s in the Laplace domain.

Just like the real and complex differentiation properties there is also a complex integration property dealing with integration in the Laplace domain. Equation (4-34) is the definition of the Laplace transform.

$$F(s) = \int_0^\infty \left(f(t) \cdot e^{-st}\right) dt \tag{4-34}$$

Integrate both sides of Equation (4-34) with respect to s to arrive at Equation (4-35).

$$\int_s^\infty F(s)\, ds = \int_s^\infty \left(\int_0^\infty \left(f(t) \cdot e^{-st}\right) dt\right) ds$$

$$= \int_0^\infty \left(f(t) \cdot \left(\int_s^\infty e^{-st}\, ds\right)\right) dt$$

$$= \int_0^\infty \left(f(t) \cdot \frac{e^{-st}}{t}\right) dt$$

$$\int_s^\infty F(s)\, ds = \int_0^\infty \left(\frac{f(t)}{t} \cdot e^{-st}\right) dt = \mathcal{L}\left(\frac{f(t)}{t}\right) \tag{4-35}$$

Thus, integration in the Laplace domain is division by t in the time domain. Again, note the similarity of integration in the time domain results in division by s (the Laplace domain variable) in the Laplace domain while integration in the Laplace domain results in division by t (the time domain variable) in the time domain.

4-8. Laplace Transform of Integral-Differential Equations

At this point all of the Laplace operations necessary to transform integral-differential equations into the Laplace domain have been discussed. The series RLC circuit of Chapter 3 will be used as an example. Figure 4-1 is the series RLC circuit of Chapter 3. The integral-differential equation was determined using Kirchhoff's voltage law on the loop formed by $v(t)$, L, R, and C. The expressions for L, R, and C voltages given the current $i(t)$ were then substituted into the

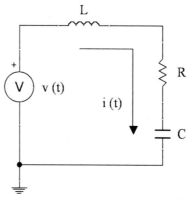

Figure 4-1 Series RLC Circuit

Kirchhoff's voltage law equation yielding the integral-differential equation of Equation (4-36).

$$v(t) - L \cdot \frac{d(i(t))}{dt} - R \cdot i(t) - \frac{1}{C} \cdot \int_0^t (i(t)) \, dt - V_{C0} = 0 \qquad (4\text{-}36)$$

For this transformation assume that $v(t)$ is a step function of magnitude V. In Equation (4-36) $i(t)$ is the unknown and therefore, its Laplace transform is not known. However, it will be assumed to be $I(s)$. Then, applying the Laplace transform operations and properties to Equation (4-36) yields Equation (4-37).

$$\frac{V}{s} - L(s \cdot I(s) - I_{L0}) - R \cdot I(s) - \frac{1}{C} \cdot \frac{I(s)}{s} - \frac{V_{C0}}{s} = 0 \qquad (4\text{-}37)$$

I_{L0} and V_{C0} are the inductor current and capacitor voltage at t equal to zero and are constants which is why V_{C0} transforms as V_{C0}/s. Now that the integral-differential equation has been transformed into the Laplace domain, it is an algebraic equation without any derivatives or integrals. This allows the equation to be solved for the unknown $I(s)$ using algebraic methods as shown in Equation (4-38).

$$\left(s \cdot L + R + \frac{1}{s \cdot C} \right) \cdot I(s) = \frac{V}{s} - L \cdot I_{L0} - \frac{V_{C0}}{s}$$

$$\left(\frac{L \cdot C \cdot s^2 + R \cdot C \cdot S + 1}{s \cdot C} \right) \cdot I(s) = \frac{V}{s} - L \cdot I_{L0} - \frac{V_{C0}}{s}$$

$$I(s) = \left(\frac{s \cdot C}{L \cdot C \cdot s^2 + R \cdot C \cdot S + 1} \right) \cdot \left(\frac{V}{s} - L \cdot I_{L0} - \frac{V_{C0}}{s} \right) \qquad (4\text{-}38)$$

This is the solution for $I(s)$ in the Laplace domain but what is needed is the solution for $i(t)$ in the time domain. To accomplish this an inverse Laplace transform method is needed.

4-9. The Inverse Laplace Transform

The inverse Laplace transform is defined by Equation (4-39).

$$f(t) = \mathcal{L}^{-1}(F(s)) = \frac{1}{2 \cdot \pi \cdot j} \cdot \int_{\gamma - j \cdot \infty}^{\gamma + j \cdot \infty} (F(s) \cdot e^{-s \cdot t}) ds \qquad (4\text{-}39)$$

This integration over a complex variable is very difficult and not a practical solution to the inverse Laplace transform. If a table of Laplace transforms were constructed with $f(t)$ in one column and the corresponding $F(s)$ in the other column, then the inverse Laplace transform could be determined by locating the function in the $F(s)$ column and determining the inverse Laplace transform from the entry in the $f(t)$ column.

Appendix B is a table of Laplace transforms that actually consists of 3 tables. The first table is a table of the Laplace transform properties. The second table is a table of Laplace transforms of some typical circuit waveforms used in advanced circuit analysis. The third table is a table of Laplace transform responses to typical circuit waveforms. The third table is intended for use in determining the inverse Laplace transform and is even arranged with the $F(s)$ functions in the left column.

A more modern approach to table lookup for Laplace transforms and inverse Laplace transforms is to use computer software. There are several programs available that determine these transforms. There is even downloadable software for programmable scientific calculators that works well.

Going back to the example of Figure 4-1, determine the inverse Laplace transform of Equation (4-38). Since there are 3 sources, V, I_{L0}, and V_{C0}, there will be 3 components to $i(t)$. The V component is shown in Equation (4-40).

$$I_1(s) = \left(\frac{s \cdot C}{L \cdot C \cdot s^2 + R \cdot C \cdot S + 1} \right) \cdot \frac{V}{s}$$

$$I(s) = \frac{\dfrac{V}{L}}{s^2 + \dfrac{R}{L} \cdot s + \dfrac{1}{L \cdot C}} \qquad (4\text{-}40)$$

Next, find a transform pair from the third table of Appendix B that resembles Equation (4-40). Select transform pair number 20 which is one pair of the same form and is shown in Equation (4-41).

$$\mathcal{L}^{-1}\left(\frac{K}{(s+a)^2 + b^2}\right) = \frac{K \cdot e^{-a \cdot t} \cdot \sin(b \cdot t)}{b} \cdot u(t) \qquad (4\text{-}41)$$

Equation (4-41) can be rearranged to Equation (4-42).

$$\mathcal{L}^{-1}\left(\frac{K}{s^2 + 2 \cdot a \cdot s + a^2 + b^2}\right) = \frac{K \cdot e^{-a \cdot t} \cdot \sin(b \cdot t)}{b} \cdot u(t) \qquad (4\text{-}42)$$

Comparing Equation (4-42) with Equation (4-40) yields Equation (4-43).

$$i_1(t) = \mathcal{L}^{-1}\left(I_1(s)\right) = \frac{K \cdot e^{-a \cdot t} \cdot \sin(b \cdot t)}{b} \cdot u(t) \qquad (4\text{-}43)$$

where :

$$K = \frac{V}{L} \qquad 2 \cdot a = \frac{R}{L} \quad \text{or} \quad a = \frac{R}{2 \cdot L}$$

$$a^2 + b^2 = \frac{1}{L \cdot C} \quad \text{or} \quad b = \sqrt{\frac{1}{L \cdot C} - \left(\frac{R}{2 \cdot L}\right)^2}$$

The V_{C0} component is identical to the V component with the exception of the value for K. Equation (4-44) shows the inverse transform for the V_{C0} component.

$$i_3(t) = \mathcal{L}^{-1}\left(I_3(s)\right) = \frac{K \cdot e^{-a \cdot t} \cdot \sin(b \cdot t)}{b} \cdot u(t) \qquad (4\text{-}44)$$

where :

$$K = \frac{-V_{C0}}{L} \qquad 2 \cdot a = \frac{R}{L} \quad \text{or} \quad a = \frac{R}{2 \cdot L}$$

$$a^2 + b^2 = \frac{1}{L \cdot C} \quad \text{or} \quad b = \sqrt{\frac{1}{L \cdot C} - \left(\frac{R}{2 \cdot L}\right)^2}$$

The I_{L0} component is shown in Equation (4-45).

$$I_2(s) = \left(\frac{s \cdot C}{L \cdot C \cdot s^2 + R \cdot C \cdot S + 1}\right) \cdot (-L \cdot I_{L0})$$

$$= \frac{-I_{L0} \cdot s}{s^2 + \dfrac{R}{L} \cdot s + \dfrac{1}{L \cdot C}} \tag{4-45}$$

Next, find a transform pair from the third table of Appendix B that resembles Equation (4-45). Select transform pair number 25 which is one pair of the same form and is shown in Equation (4-46).

$$\mathcal{L}^{-1}\left(\frac{K \cdot s}{(s+a)^2 + b^2}\right) = \frac{K \cdot \sqrt{a^2 + b^2} \cdot e^{-a \cdot t} \cdot \sin(b \cdot t + \theta)}{b} \cdot u(t) \tag{4-46}$$

where: $\theta = \text{angle}(-a, b)$

Equation (4-46) can be rearranged to Equation (4-47).

$$\mathcal{L}^{-1}\left(\frac{K \cdot s}{s^2 + 2 \cdot a \cdot s + a^2 + b^2}\right) = \frac{K \cdot \sqrt{a^2 + b^2} \cdot e^{-a \cdot t} \cdot \sin(b \cdot t + \theta)}{b} \cdot u(t) \tag{4-47}$$

where: $\theta = \text{angle}(-a, b)$

Comparing Equation (4-47) with Equation (4-45) yields Equation (4-48).

$$i_2(t) = \mathcal{L}^{-1}(I_2(s)) = \frac{K \cdot \sqrt{a^2 + b^2} \cdot e^{-a \cdot t} \cdot \sin(b \cdot t + \theta)}{b} \cdot u(t) \tag{4-48}$$

where:

$$K = -I_{L0} \qquad 2 \cdot a = \frac{R}{L} \quad \text{or} \quad a = \frac{R}{2 \cdot L}$$

$$a^2 + b^2 = \frac{1}{L \cdot C} \quad \text{or} \quad b = \sqrt{\frac{1}{L \cdot C} - \left(\frac{R}{2 \cdot L}\right)^2}$$

$$\theta = \text{angle}(-a, b) = \tan^{-1}\left(\frac{b}{-a}\right)$$

The time domain solution is shown in Equation (4-49).

$$i(t) = i_1(t) + i_2(t) + i_3(t) \tag{4-49}$$

All three of the components of $i(t)$ are decaying sinusoids and result in $i(t) = 0$ for large values of t. This is expected since the source of Figure 4-1 is a DC source applied at t equal to 0, there is a series capacitor, and the steady state DC current for a capacitor is zero. The solution for $i(t)$ is a transient solution with the steady state solution equal to zero.

The above selection of transform pairs from Appendix B assumed that b was not zero or imaginary $\left(\dfrac{1}{L \cdot C} > \left(\dfrac{R}{2 \cdot L}\right)^2\right)$. If b is zero or imaginary, transform pairs 9 and 14 would apply rather than transform pairs 20 and 25. This would make the solution decaying exponentials rather than decaying sinusoids.

The $i_2(t)$ solution of Equation (4-48) contains a $\sin(b \cdot t + \theta)$ expression. If a different Laplace table or a computer software program is used for the inverse Laplace transform, this expression may be of the form $A \cdot \sin(b \cdot t) + B \cdot \cos(b \cdot t)$. Some authors prefer this form but when circuit analysis is involved the phase shifted sine form is preferred. Since circuits can only change the magnitude and phase of sinusoids, the phase shifted sine form presents the magnitude and phase change directly.

If the Inverse transform is in the sin/cos form, it can be converted to the phase shifted sine form using the complex number representation method of Appendix C. The cos function is converted to a sin function by adding a 90 degree phase shift. The sinusoid magnitudes are the complex number magnitudes and the sinusoid phases are the complex number angles. The complex numbers are then combined to yield an equivalent single magnitude and an equivalent single angle. This magnitude is then the magnitude of the combined sinusoid and the angle is the combined sinusoid phase. Note, this process can only be used for sinusoids of the same frequency.

4-10. The Laplace Transform of a Circuit

When analyzing mechanical systems, thermal systems, motion dynamics, etc., it is more practical to determine the integral-differential equations and transform them into the Laplace domain to generate a solution which can then be inverse Laplace transformed. When analyzing circuits there is a simpler method. The circuit can be transformed to the Laplace domain and then all of the standard circuit analysis methods can be used to generate a solution in the Laplace domain prior to inverse transforming back to the time domain.

In order to transform a circuit to the Laplace domain all of the sources are transformed to the Laplace domain and the resistors, capacitors and inductors are converted to a Laplace impedance. To determine the Laplace impedance, transform the voltage and current relationships of Chapter 3 to the Laplace domain and determine $V(s)/I(s)$. The capacitor voltage and inductor current at t equal to zero will be assumed zero in determining the

Laplace impedance but will be considered later. The equation for the resistor voltage given the resistor current is shown in Equation (4-50).

$$v(t) = R \cdot i(t) \tag{4-50}$$

Converting Equation (4-50) to the Laplace domain yields Equation (4-51).

$$V(s) = R \cdot I(s) \tag{4-51}$$

Converting Equation (4-51) to impedance form yields Equation (4-52).

$$ZR(s) = \frac{V(s)}{I(s)} = R \tag{4-52}$$

Thus, the resistor Laplace impedance is R.

The equation for the capacitor current given the capacitor voltage is shown in Equation (4-53).

$$i(t) = C \cdot \frac{dv(t)}{dt} \tag{4-53}$$

Converting Equation (4-53) to the Laplace domain yields Equation (4-54).

$$I(s) = C \cdot s \cdot V(s) \tag{4-54}$$

Converting Equation (4-54) to impedance form yields Equation (4-55).

$$ZC(s) = \frac{V(s)}{I(s)} = \frac{1}{s \cdot C} \tag{4-55}$$

Thus, the capacitor Laplace impedance is $\dfrac{1}{s \cdot C}$.

The equation for the inductor voltage given the inductor current is shown in Equation (4-56).

$$v(t) = L \cdot \frac{di(t)}{dt} \tag{4-56}$$

Converting Equation (4-56) to the Laplace domain yields Equation (4-57).

$$V(s) = L \cdot s \cdot I(s) \tag{4-57}$$

Converting Equation (4-57) to impedance form yields Equation (4-58).

$$ZL(s) = \frac{V(s)}{I(s)} = s \cdot L \tag{4-58}$$

Thus, the inductor Laplace impedance is $s \cdot L$.

The capacitor voltage at t equal to zero can be represented as a DC voltage source in series with the capacitor. Since this is a DC source, its Laplace transform is V_0/s where V_0 is the capacitor voltage at t equal to zero. V_0 is also referred to as the capacitor initial voltage.

The inductor current at t equal to zero can be represented as a DC current source in parallel with the inductor. Since this is a DC source, its Laplace transform is I_0/s where I_0 is the inductor current at t equal to zero. I_0 is also referred to as the inductor initial current. Figure 4-2 summarizes the Laplace transform of a circuit.

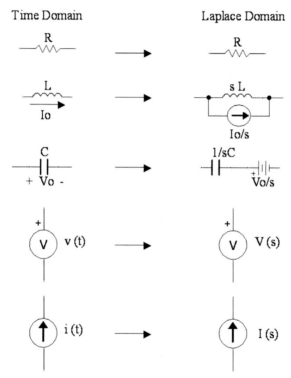

Figure 4-2 Circuit Laplace Transform Operations

In Chapter 1 a simple RC Circuit was analyzed using standard calculus methods. That analysis proved to be very difficult and led to the Laplace transform circuit analysis method discussed in this chapter. Now, go back to that simple RC circuit of Chapter 1 and apply the Laplace transform circuit analysis method.

Figure 4-3a shows the original simple RC circuit and Figure 4-3b shows the circuit transformed to the Laplace domain. The source is a step function of magnitude V as it was in Chapter 1. The capacitor initial voltage is V_0.

Figure 4-2 was used to determine the transform of the resistor and capacitor to the Laplace domain.

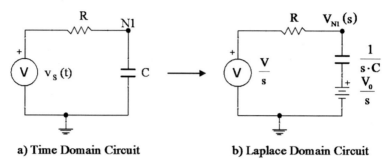

a) Time Domain Circuit   **b) Laplace Domain Circuit**

Figure 4-3 Simple RC Circuit

The analysis task is to determine the voltage across the capacitor. Since the time domain capacitor transformed to the Laplace domain as a capacitor in series with its initial voltage source, the voltage across the capacitor is the voltage across the Laplace domain capacitor plus the initial voltage source. Equation (4-59) uses the voltage divider rule to determine $V_{N1}(s)$.

$$V_{N1}(s) = \left(\frac{V}{s} - \frac{V_0}{s}\right) \cdot \left(\frac{\frac{1}{s \cdot C}}{R + \frac{1}{s \cdot C}}\right) + \frac{V_0}{s}$$

$$= \frac{V - V_0}{s \cdot (R \cdot C \cdot s + 1)} + \frac{V_0}{s}$$

$$= \frac{V}{s \cdot (R \cdot C \cdot s + 1)} + \frac{V_0 \cdot R \cdot C}{(R \cdot C \cdot s + 1)}$$

$$= \frac{\frac{V}{R \cdot C}}{s \cdot \left(s + \frac{1}{R \cdot C}\right)} + \frac{V_0}{s + \frac{1}{R \cdot C}} \tag{4-59}$$

$v_{N1}(t)$ is then the inverse Laplace transform of Equation (4-59). Appendix B Laplace transform pairs 1 and 2 give the inverse Laplace transform of Equation (4-59) as shown in Equation (4-60).

$$v_{N1}(t) = \mathcal{L}^{-1}(V_{N1}(s)) = \frac{K_1 \cdot (1 - e^{-a \cdot t})}{a} \cdot u(t) + K_2 \cdot e^{-a \cdot t} \cdot u(t) \tag{4-60}$$

$$\text{where: } K_1 = \frac{V}{R \cdot C} \quad K_2 = V_0 \quad \text{and } a = \frac{1}{R \cdot C}$$

Substituting circuit values into Equation (4-60) and dropping $u(t)$ yields Equation (4-61) which agrees with the solution determined in Chapter 1.

$$V_{NI}(t) = V_0 \cdot e^{\frac{-t}{R \cdot C}} + V \cdot \left(1 - e^{\frac{-t}{R \cdot C}}\right) \tag{4-61}$$

This was simpler to solve but was also a very simple circuit. The circuit of Figure 4-4a is a more complex circuit. The circuit will be analyzed to determine the voltage across R. Again, rather than transforming the integral-differential equations, the circuit will be transformed to the Laplace domain prior to analyzing the circuit. Figure 4-4b is the transformed circuit.

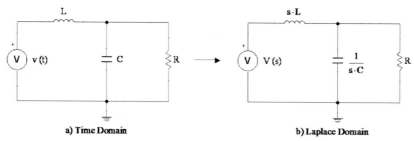

a) Time Domain   b) Laplace Domain

Figure 4-4 RLC Circuit

The source voltage $v(t)$ will be $\sin(\omega \cdot t) \cdot u(t)$ where ω is 10,000. Also, L will be 100 millihenry, C will be 0.01 microfarad, R will be 1000 ohms, and the capacitor initial voltage and inductor initial current will be zero. The voltage divider rule will be used to determine the voltage across R. First, the resistor and capacitor must be combined in parallel to form $Z(s)$ as shown in Equation (4-62).

$$Z(s) = \frac{R \cdot \dfrac{1}{s \cdot C}}{R + \dfrac{1}{s \cdot C}} = \frac{R}{s \cdot R \cdot C + 1} \tag{4-62}$$

Since the voltage across R is the same as the voltage across $Z(s)$, $V_R(s)$ can be determined using the voltage divider rule as shown in Equation (4-63).

$$V_R(s) = V(s) \cdot \frac{\dfrac{R}{s \cdot R \cdot C + 1}}{s \cdot L + \dfrac{R}{s \cdot R \cdot C + 1}}$$

$$V_R(s) = V(s) \cdot \frac{R}{s^2 \cdot R \cdot L \cdot C + s \cdot L + R}$$

$$= V(s) \cdot \frac{\dfrac{1}{L \cdot C}}{s^2 + s \cdot \dfrac{1}{R \cdot C} + \dfrac{1}{L \cdot C}} \tag{4-63}$$

Substituting the values for R, L, C, and ω along with the Laplace transform of $v(t)$ into Equation (4-63) yields Equation (4-64).

$$V_R(s) = \left(\frac{10000}{s^2 + 10000^2}\right) \cdot \left(\frac{1 \cdot 10^9}{s^2 + 1 \cdot 10^5 \cdot s + 1 \cdot 10^9}\right)$$

$$= \frac{1 \cdot 10^{13}}{(s^2 + 10000^2) \cdot (s + 11270) \cdot (s + 88730)} \tag{4-64}$$

Equation (4-64) is of the form of Laplace transform pair 12a in Appendix B. Equation (4-65) is the expression for $v_R(t)$.

$$v_R(t) = \mathcal{L}^{-1}\left(\frac{K}{(s^2 + \omega^2) \cdot (s+a) \cdot (s+b)}\right)$$

$$= \left(\frac{K \cdot e^{-a \cdot t}}{(b-a) \cdot (a^2 + \omega^2)} + \frac{K \cdot e^{-b \cdot t}}{(a-b) \cdot (b^2 + \omega^2)}\right.$$

$$\left. + \frac{K \cdot \sin(\omega \cdot t + \theta)}{\omega \cdot \sqrt{(a^2 + \omega^2) \cdot (b^2 + \omega^2)}}\right) \cdot u(t)$$

where $\theta = $ angle (x,y)

and $x = a \cdot b - \omega^2$; $y = -\omega \cdot (a+b)$ \qquad (4-65)

Comparing Equations (4-64) and (4-65) yields:

$$K = 1 \cdot 10^{13} \quad \omega = 10000 \quad a = 11270 \quad b = 88730$$

Substituting these values into Equation (4-65) yields the final expression for $v_R(t)$ as shown in Equation (4-66).

$$v_R(t) = (0.5687 \cdot e^{-11270 \cdot t} - 0.01619 \cdot e^{-88730 \cdot t}$$

$$+ 0.7433 \cdot \sin(10000 \cdot t - 48.01^\circ)) \cdot u(t) \tag{4-66}$$

Equation (4-66) contains both transient and steady state solutions. The two decaying exponential terms are the transient solution and the sinusoidal term is the steady state solution. Figure 4-5 is a plot of Equation (4-66) versus time.

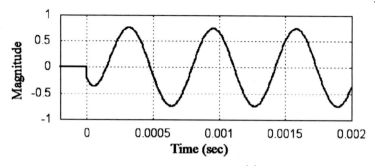

Figure 4-5 Plot of $v_R(t)$

The period of the sinusoid is 0.628 milliseconds $(2 \cdot \pi/\omega)$ and the first exponential term decays to less than 1% in 0.444 milliseconds $(5/a)$. The second exponential term has a much smaller magnitude and decays much faster than the first and has negligible effect in Figure 4-5. The transient terms only have an effect for the first 0.444 milliseconds as seen in Figure 4-5 and after that only the sinusoid term is observable.

The inverse Laplace transform tables of Appendix B only cover lower order denominator factors. When higher order denominator factors exist, the partial fraction expansion method of Appendix D can be used to expand the higher order functions into a sum of lower order functions. Appendix B can then be used for each of the lower order functions using the Laplace linearity property.

PROBLEMS

4-1. What are the two classifications of current and voltage waveforms?

4-2. What method is used to transform periodic waveforms from the time domain to the frequency domain?

4-3. What two methods are used to transform non-periodic waveforms from the time domain to the frequency domain?

4-4. What are the differences between the Fourier and Laplace transforms?

4-5. What is the linearity property of the Laplace transform?

4-6. What is the Laplace transform of a step function of magnitude 5?

4-7. What is the Laplace transform of a ramp function of slope 4?

4-8. What is the Laplace transform of an impulse function of area 3?

4-9. What is the Laplace transform of $6 \cdot e^{-7t} \cdot u(t)$?

4-10. What is the complex translation property of the Laplace transform?

4-11. What is the Laplace transform of $3 \cdot \sin(5 \cdot t) \cdot u(t)$?

4-12. What is the Laplace transform of $4 \cdot e^{-2t} \cdot \sin(3 \cdot t) \cdot u(t)$?

4-13. What is the Laplace transform of $6 \cdot \sin(3 \cdot t + 60°) \cdot u(t)$?

4-14. What is the Laplace transform of $6 \cdot e^{-2t} \cdot \sin(3 \cdot t + 60°) \cdot u(t)$?

4-15. What is the real translation property of the Laplace transform?

4-16. What is the Laplace transform of $5 \cdot (t - 2) \cdot u(t - 2)$?

4-17. What is the Laplace transform of a rectangular pulse with an amplitude of 3 volts and a width of 0.5 seconds that starts at t equal to 0?

4-18. What is the real differentiation property of the Laplace transform?

4-19. What is the complex differentiation property of the Laplace transform?

4-20. What is the real integration property of the Laplace transform?

4-21. What is the complex integration property of the Laplace transform?

4-22. The integral-differential equation for the circuit of Chapter 3 problem 7 is (see problem solutions chapter 3 problem 7 for details):

$$C \cdot \frac{d\left(v_s(t) - v_{out}(t)\right)}{dt} = \frac{v_{out}(t)}{R}$$

Assume zero initial values, convert the equation to the Laplace domain, and solve for $V_{out}(s)$.

4-23. The integral-differential equation for the circuit of Chapter 3 problem 8 is (see problem solutions chapter 3 problem 8 for details):

$$\frac{1}{L} \cdot \int_0^t \left(v_s(t) - v_{out}(t)\right) dt + I_{L0} = \frac{v_{out}(t)}{R}$$

Assume zero initial values, convert the equation to the Laplace domain, and solve for $V_{out}(s)$.

4-24. The integral-differential equation for the circuit of Chapter 3 problem 9 is (see problem solutions chapter 3 problem 9 for details):

$$\frac{v_S(t) - v_{out}(t)}{R} = \frac{1}{L} \cdot \int_0^t (v_{out}(t)) \, dt + I_{L0} + C \cdot \frac{d(v_{out}(t))}{dt}$$

Assume zero initial values, convert the equation to the Laplace domain, and solve for $V_{out}(s)$.

4-25. The integral-differential equation for the circuit of Chapter 3 problem 10 is (see problem solutions chapter 3 problem 10 for details):

$$\frac{1}{L} \cdot \int_0^t (v_S(t) - v_{out}(t)) \, dt + I_{L0} + C \cdot \frac{d(v_S(t) - v_{out}(t))}{dt} = \frac{v_{out}(t)}{R}$$

Assume zero initial values, convert the equation to the Laplace domain, and solve for $V_{out}(s)$.

4-26. What is the inverse Laplace transform of $\dfrac{8}{s+3}$?

4-27. What is the inverse Laplace transform of $\dfrac{32}{s^2 + 64}$?

4-28. What is the inverse Laplace transform of $\dfrac{6 \cdot s + 15}{s^2 + 9}$?

4-29. What is the inverse Laplace transform of $\dfrac{4 \cdot s - 22}{s^2 + 4 \cdot s + 29}$?

4-30. What is the inverse Laplace transform of $\dfrac{s+2}{s+1}$?

4-31. Determine the voltage across R for the circuit below:

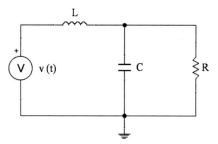

The source voltage $v(t)$ is a step function of 3 volts at t equal to zero. Also, L is 10 millihenry, C is 0.1 microfarad, R is 1000 ohms, and the capacitor initial voltage and inductor initial current are zero. Also plot the time domain solution.

4-32. Determine the voltage across R for the circuit below:

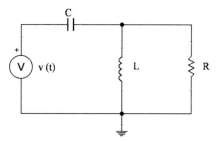

The source voltage $v(t)$ is a pulse at t equal to zero with an amplitude of 2 volts and a width of 1 millisecond. Also, L is 10 millihenry, C is 0.1 microfarad, R is 1000 ohms, and the capacitor initial voltage and inductor initial current are zero. Also plot the time domain solution.

4-33. Determine the voltage across R for the circuit of Problem 4-31 with the source voltage $v(t)$ a pulse at t equal to zero with an amplitude of 3 volts and a width of 200 microseconds. Also, L is 100 millihenry, C is 0.01 microfarad, R is 1000 ohms, and the capacitor initial voltage and inductor initial current are zero. Also plot the time domain solution.

CHAPTER 5
SYSTEM ANALYSIS

5-1. Introduction

A system consists of any device that has an output due to some applied input. The device could be a circuit, a mechanical system, a thermal system or anything that can be described with an integral-differential equation. Figure 5-1 shows a typical system.

Figure 5-1 A Typical System

This book will only be concerned with systems that are circuits. For circuits, the inputs and outputs are voltages or currents.

The analysis of a system involves the determination of the effect the system has on the applied input by observing the output. The transfer function for the system describes how the system input and output are related. The transfer function is the ratio of the output to the input and is usually defined in the Laplace domain. Equation (5-1) describes the transfer function of a circuit.

$$G(s) = \frac{V_{out}(s)}{V_{in}(s)} \tag{5-1}$$

5-2. System Stability

Another way of interpreting the transfer function is that the circuit output equals the circuit input times the transfer function. If the circuit input is a unit impulse function (the Laplace transform equals 1), then the circuit output in the time domain is the inverse Laplace transform of the transfer function and is sometimes referred to as the system impulse response.

One area of concern with circuits and systems in general, especially when feedback is employed, is the stability of the system. If the system is disturbed with an impulse and the response of the system to that impulse over time decays to zero the system is said to be stable. If the response to the impulse causes the system to increase without bound, the system is said to be unstable. If the response to the impulse causes the system to sustain a finite value or oscillate between finite values, the system is said to be marginally stable.

The transfer function $G(s)$ can be represented as a numerator polynomial of s and a denominator polynomial of s. Since the roots of the numerator polynomial (which we refer to as zeros) can only cause $G(s)$ to go to zero, the numerator can never cause the system to be unstable. The roots of the denominator polynomial (which we refer to as poles) can cause $G(s)$ to go to infinity which would cause instability. Thus, from a stability issue only the denominator is of interest.

With the denominator in factored form the real roots yield factors of the form $(s+a)^n$ (where the roots are $s = -a$) and the complex roots yield factors of the form $\left((s+\alpha)^2 + \omega^2\right)^n$ (where the roots are $s = -\alpha \pm j \cdot \omega$). Complex roots are always in conjugate pairs. Partial fraction expansion can be used to expand the factored $G(s)$ into the sum of simpler functions with each one being one of the $G(s)$ denominator factors. Appendix D discusses the method of partial fraction expansion. The expanded $G(s)$ will have a function for each of the denominator pole factors of the form in Equation (5-2)

$$\frac{A}{(s+a)^n} \quad \text{and} \quad \frac{A \cdot s + B}{\left((s+\alpha)^2 + \omega^2\right)^n} \tag{5-2}$$

From Appendix D, the numerators for the real pole functions will be constants and the numerators for the complex pole functions will be $(A \cdot s + B)$ where A and B are constants. Since these are Laplace expressions, the Laplace linearity property applies and the inverse Laplace transform of $G(s)$ is the sum of the inverse Laplace transforms of each of the expanded functions. By examining each type of pole function of Equation (5-2) separately, criteria for stability can be determined.

Any of the pole functions that cause $g(t)$ (the inverse Laplace transform of $G(s)$) to increase without bound will cause $g(t)$ to be unstable. If none of the pole functions increase without bound but any one of the pole functions sustaining a finite value or oscillating between finite values will cause $g(t)$ to be marginally stable. If none of the pole functions increase without bound or sustain a finite value or oscillate between finite values then $g(t)$ is stable.

A diagram used in stability analysis is the pole zero plot. The diagram consists of a set of Cartesian coordinates with the real values of s plotted on the X - axis and the imaginary values of s plotted on the Y - axis. The poles and zeros of $G(s)$ are plotted separately with the zeros represented with a 0 and the poles represented with an x. It is customary to plot both poles and zeros even though only the poles impact the stability. It is also customary to add any constant numerator multipliers somewhere on the plot to insure that the transfer function can be recreated from the plot. This is usually referred to as the gain term.

First, the real pole functions will be examined. The inverse Laplace transform of the real pole functions are of the form in Equation (5-3).

$$g(t) = \mathcal{L}^{-1}\left(\frac{A}{(s+a)^n}\right) = K \cdot t^{(n-1)} \cdot e^{-a \cdot t} \cdot u(t) \tag{5-3}$$

The K in Equation (5-3) is a constant. The inverse Laplace transform of Equation (5-3) has 2 terms that can increase without bound - t^{n-1} and $e^{-a \cdot t}$. For negative values of a, $e^{-a \cdot t}$ increases without bound regardless of n and $g(t)$ is unstable. This corresponds to a real pole (real root of the denominator polynomial) in the right half plane of the pole zero plot.

For a equal to zero, $e^{-a \cdot t}$ is 1 and the t^{n-1} term determines the stability. For n greater than 1, t^{n-1} increases without bound and $g(t)$ is unstable. This corresponds to multiple order real poles at the origin of the pole zero plot. For n equal to 1, t^{n-1} is 1 and $g(t)$ is marginally stable if there are no other poles making it unstable. This corresponds to a first order real pole at the origin of the pole zero plot.

For positive values of a, $e^{-a \cdot t}$ is a decaying exponential that will go to zero with increasing time. For n greater than 1, t^{n-1} will increase without bound. If the inverse Laplace transform of Equation (5-3) is rearranged to Equation (5-4), l'Hospital's rule can be applied to determine stability.

$$K \cdot t^{(n-1)} \cdot e^{-a \cdot t} \cdot u(t) = \frac{K \cdot t^{(n-1)}}{e^{a \cdot t}} \cdot u(t) \tag{5-4}$$

Equation (5-5) is l'Hospital's rule applied to this situation.

$$\lim_{t \to \infty}\left(\frac{K \cdot t^{(n-1)}}{e^{a \cdot t}}\right) = \lim_{t \to \infty}\left(\frac{\dfrac{d(K \cdot t^{(n-1)})}{dt}}{\dfrac{d(e^{a \cdot t})}{dt}}\right) \tag{5-5}$$

Applying l'Hospital's rule multiple times results in multiple derivatives of $K \cdot t^{(n-1)}$ equaling a constant and multiple derivatives of $e^{a \cdot t}$ equaling a constant times $e^{a \cdot t}$. Then as t approaches infinity the limit approaches zero (because $e^{a \cdot t}$ approaches infinity and a constant divided by infinity is zero). Thus, the decaying exponential term overrides the t^{n-1} term. Therefore, for a positive value of a, $g(t)$ will be stable if there are no other poles making it unstable or marginally stable. This corresponds to a real pole in the left half plane of the pole zero plot.

Next, examine the complex pole functions. The inverse Laplace transform of the complex pole functions are of the form in Equation (5-6).

$$g(t) = \mathcal{L}^{-1}\left(\frac{A \cdot s + B}{\left((s+\alpha)^2 + \omega^2 \right)^n} \right)$$

$$= K \cdot t^{(n-1)} \cdot e^{-\alpha \cdot t} \cdot \sin(\omega \cdot t + \theta) \cdot u(t) \tag{5-6}$$

Note that the difference between the real pole function and complex pole function inverse Laplace transforms is that the complex pole function has an additional sinusoidal term. Since the sinusoidal term oscillates between +1 and -1, it can never increase without bound and cause $g(t)$ to be unstable.

The real part of the complex root ($-\alpha$) has the same effect as the real root (-a) in terms of stability. For negative values of α, $e^{-\alpha \cdot t}$ increases without bound regardless of n. This causes the $g(t)$ function to be a sinusoid with an amplitude that increases without bound and $g(t)$ is unstable. This corresponds to a complex pole (complex root of the denominator polynomial) in the right half plane of the pole zero plot.

For α equal to zero, $e^{-\alpha \cdot t}$ is 1 and the t^{n-1} term determines the stability. For n greater than 1, t^{n-1} increases without bound. This causes the $g(t)$ function to be a sinusoid with an amplitude that increases without bound and $g(t)$ is unstable. This corresponds to multiple order complex pole pairs on the imaginary axis of the pole zero plot.

For n equal to 1, t^{n-1} is 1. This causes the $g(t)$ function to be a sinusoid of finite amplitude and $g(t)$ is marginally stable if there are no other poles making it unstable. This corresponds to a first order complex pole pair on the imaginary axis of the pole zero plot.

For positive values of α, $e^{-\alpha \cdot t}$ is a decaying exponential that overrides the t^{n-1} term as was discussed above. This causes $g(t)$ to be a decaying sinusoid that eventually goes to zero and $g(t)$ is stable if there are no other poles making it unstable or marginally stable. This corresponds to a complex pole in the left half plane of the pole zero plot.

Summarizing stability from a pole zero plot interpretation, any poles in the right half plane cause the function to be unstable. Also any multiple order poles on the imaginary axis cause the function to be unstable. If there ore no poles causing the function to be unstable, and there are any first order poles on the imaginary axis, then the function is marginally stable. And finally, if there are no poles causing the function to be unstable and no poles causing the function to be marginally stable, then the function is stable (all poles are in the left half plane).

The advantage of using the pole zero plot to determine stability is that stability can be determined in the Laplace domain and the inverse Laplace transform is not necessary. Why is the stability of a system needed? An unstable system cannot be analyzed for the same reasons that the integral definition of the Laplace transform is only valid if the integral exists.

Example 5-1

Construct a pole zero plot for the transfer function of Equation (5-7) and discuss the stability.

$$G(s) = \frac{3 \cdot (s+2)}{s \cdot (s+3) \cdot \left((s+1)^2 + 36\right)}$$ (5-7)

Figure 5-2 is the pole zero plot for Equation (5-7).

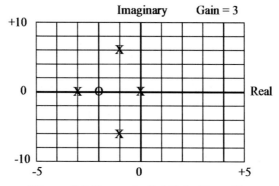

Figure 5-2 Example 5-1 Pole Zero Plot

The gain term is 3. There is a first order real zero at $s = -2$. There are 4 poles, one a first order real pole at the origin, one a first order real pole at $s = -3$, and a pair of first order complex poles at $s = -1 \pm j \cdot 6$. There are no unstable poles and one marginally stable pole, therefore, the system is marginally stable because of the first order real pole at the origin.

Example 5-2

Construct a pole zero plot for the transfer function of Equation (5-8) and discuss the stability.

$$G(s) = \frac{-5 \cdot \left((s-2)^2 + 64\right)}{(s+1) \cdot (s-3) \cdot (s+4)^2} \tag{5-8}$$

Figure 5-3 is the pole zero plot for Equation (5-8).

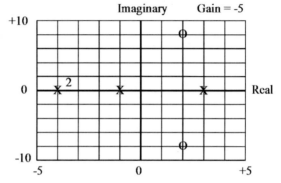

Figure 5-3 Example 5-2 Pole Zero Plot

The gain term is -5. There is a pair of first order complex zeros at $+2 \pm j \cdot 8$. There are 4 poles, one a first order real pole at $s = -1$, one a first order real pole at $s = +3$, and a second order real pole at $s = -4$ (note the 2 indicating 2 poles on top of each other). The system is unstable because of the pole at $s = +3$.

5-3. Frequency Domain Analysis

One of the primary methods for analyzing systems is frequency domain analysis. As was discussed earlier in this chapter, a system is typically represented mathematically by a Laplace transfer function. Since a system has no sources, the system input is the only source and can be any of the sources discussed thus far.

Any source can be considered to be made up of a series of sinusoids and the system can only modify the magnitude and phase of a sinusoid. For frequency domain analysis the input is assumed to be a sinusoid and the magnitude and phase modification caused by the system as the sinusoid frequency is varied is the result of the frequency domain analysis.

The system transfer function is in the Laplace domain and needs to be converted to the frequency domain to do a frequency domain analysis. Chapter 4 Equation (4-5) is the definition of the Fourier transform and is shown here in Equation (5-9).

$$F(\omega) = \int_{-\infty}^{+\infty} \left(f(t) \cdot e^{-j\omega \cdot t} \right) dt \qquad (5\text{-}9)$$

Chapter 4 Equation (4-7) is the definition of the Laplace transform and is shown here in Equation (5-10).

$$F(s) = \int_{0}^{+\infty} \left(f(t) \cdot e^{-s \cdot t} \right) dt \qquad (5\text{-}10)$$

There are two differences between these two equations. The first is that the limits of integration are from $-\infty$ to $+\infty$ for the Fourier transform and from 0 to $+\infty$ for the Laplace transform. Since the Laplace transform defines everything from $-\infty$ to 0 as 0, the integral from $-\infty$ to 0 is also 0 and this is compatible between the two transforms.

The second difference between the two transforms is the $e^{-j\omega \cdot t}$ term for the Fourier transform and the $e^{-s \cdot t}$ term for the Laplace transform. Therefore, in order to convert the Laplace transform to the frequency domain, simply evaluate the transfer function with $s = j \cdot \omega$. This is sometimes referred to as the single sided frequency transform since the integral from $-\infty$ to 0 has been forced to zero. This process has converted the Laplace transform to a steady state response and the transient response has been lost.

The converted Laplace transfer function now becomes the magnitude and phase modification caused by the system as a function of frequency. In the 1930's Hendrik Bode, while working as an engineer for Bell Laboratories, devised a method of plotting the transfer function magnitude and phase as a function of frequency which has been named the bode plot in his honor.

This method of plotting magnitude and phase versus frequency is to use a semi-log graph. This semi-log graph has a linear Y axis and a logarithmic X axis. Frequency is plotted on the X axis and the magnitude Y axis is plotted in decibels or $20 \cdot \log_{10} (\text{magnitude})$. This yields a log-log plot for magnitude. The phase is plotted on the Y axis in degrees and yields a linear-log plot for phase. Typically a computer is used to make these plots but they can be sketched very closely using the bode plot process of straight line segments.

As was stated above, the system can only modify the magnitude and phase of a sinusoid. Therefore, a sinusoid input to the system results in sinusoid output from the system with the magnitude and phase modified. It should be noted at this point that this is only true for a stable system. If the system is unstable, its output increases without bound when an input is applied and the output will not resemble the input. A marginally stable system has a finite output or the output oscillates between finite values and also may not resemble the input. Thus, before proceeding with a frequency domain analysis, make sure the system is stable using the pole zero plot method of Section 5-2.

Section 5-2 also showed that the transfer function numerator and denominator can be factored into real and complex factors of the form $(s+a)^n$ and $((s+\alpha)^2 + \omega^2)^n$. The numerator factors are called zeros and the denominator factors are called poles. Real poles will be considered first.

Equation (5-11) is the transfer function of a normalized $(a=1)$ first order real pole.

$$G(s) = \frac{1}{s+1} \tag{5-11}$$

Equation (5-11) can be converted to the frequency domain by setting $s = j \cdot \omega$ as shown in Equation (5-12).

$$G(j \cdot \omega) = \frac{1}{j \cdot \omega + 1} \tag{5-12}$$

A bode magnitude plot of Equation (5-12) is shown in Figure 5-4.

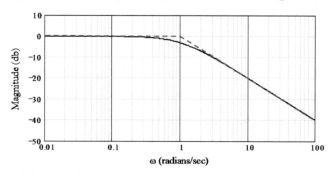

Figure 5-4 Bode Magnitude Plot of Equation (5-12)

An approximation of the magnitude plot of Figure 5-4 can be accomplished using straight line segments. The low frequency magnitudes can be approximated by a straight horizontal line at 0 db and the high frequency magnitudes can be approximated by a sloped straight line with a slope of -20 db/decade. The two straight line approximations intersect at $\omega = 1$ (the normalized value of a) as shown by the red dashed lines in Figure 5-4.

A bode phase plot of Equation (5-12) is shown in Figure 5-5.

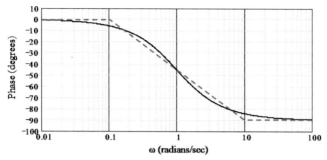

Figure 5-5 Bode Phase Plot of Equation (5-12)

An approximation of the phase plot of Figure 5-5 can also be accomplished using straight line segments. The low frequency phases can be approximated by a straight horizontal line at 0 degrees and the high frequency phases can be approximated by a straight horizontal line at -90 degrees. The middle frequency phases can be approximated by a sloped straight line with a slope of -45 degrees/decade that passes through -45 degrees at $\omega = 1$. The sloped line approximation intersects the two horizontal line approximations at $\omega = 0.1$ and $\omega = 10$ as shown by the red dashed lines in Figure 5-5.

The plots of Figures 5-4 and 5-5 were for normalized $(a = 1)$ real poles. Equation (5-11) can be modified to represent any first order real pole by dividing out the a term and renaming it to ω_b as shown in Equation (5-13).

$$G(s) = \frac{K}{\dfrac{s}{\omega_b} + 1} \qquad (5\text{-}13)$$

Equation (5-13) now represents the bode plot format for any first order real pole. Note that the constant K term in the numerator allows for any gain (K must be converted to db) and the ω_b term in the denominator allows for any break frequency (the point where the straight line segments intersect).

What about higher order real poles? Equation (5-13) can be modified to also include higher order real poles as shown in Equation (5-14).

$$G(s) = \frac{K}{\left(\dfrac{s}{\omega_b} + 1\right)^n} \qquad (5\text{-}14)$$

The higher order has the effect of increasing the slope of the sloped line approximations and the maximum phase angle as shown in Figure 5-6.

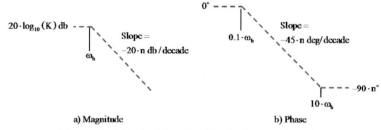

a) Magnitude   b) Phase

Figure 5-6 Bode Plot Real Pole Approximation

Next, consider real zeros. Equation (5-15) is the transfer function of a normalized $(a = 1)$ left half plane first order real zero.

$$G(s) = s + 1 \qquad (5\text{-}15)$$

Equation (5-15) can be converted to the frequency domain by setting $s = j \cdot \omega$ as shown in Equation (5-16).

$$G(j \cdot \omega) = j \cdot \omega + 1 \qquad (5\text{-}16)$$

A bode magnitude plot of Equation (5-16) is shown in Figure 5-7.

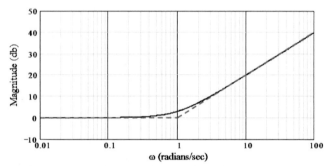

Figure 5-7 Bode Magnitude Plot of Equation (5-16)

An approximation of the magnitude plot of Figure 5-7 can be accomplished using straight line segments. The low frequency magnitudes can be approximated by a straight horizontal line at 0 db and the high frequency magnitudes can be approximated by a sloped straight line with a slope of +20 db/decade. The two straight line approximations intersect at $\omega = 1$ (the normalized value of a) as shown by the red dashed lines in Figure 5-7.

A bode phase plot of Equation (5-16) is shown in Figure 5-8. An approximation of the phase plot of Figure 5-8 can also be accomplished using straight line segments. The low frequency phases can be approximated by a straight horizontal line at 0 degrees and the high frequency phases can be approximated by a straight horizontal line at +90 degrees. The middle frequency phases can be approximated by a sloped straight line with a slope of +45 degrees/decade that passes through +45 degrees at $\omega = 1$. The sloped line approximation intersects the two horizontal line approximations at $\omega = 0.1$ and $\omega = 10$ as shown by the red dashed lines in Figure 5-8.

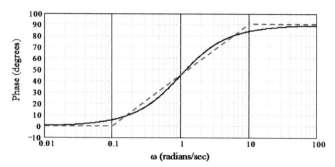

Figure 5-8 Bode Phase Plot of Equation (5-16)

The plots of Figures 5-7 and 5-8 were for normalized $(a = 1)$ real left half plane zeros. Equation (5-15) can be modified to represent any left half plane first order real zero by dividing out the a term and renaming it to ω_b as shown in Equation (5-17).

$$G(s) = K \cdot \left(\frac{s}{\omega_b} + 1 \right) \tag{5-17}$$

Equation (5-17) now represents the bode plot format for any left half plane first order real zero. Note that the constant K term allows for any gain (K must be converted to db) and the ω_b term allows for any break frequency (the point where the straight line segments intersect).

What about higher order left half plane real zeros? Equation (5-17) can be modified to also include higher order left half plane real zeros as shown in Equation (5-18).

$$G(s) = K \cdot \left(\frac{s}{\omega_b} + 1 \right)^n \tag{5-18}$$

The higher order has the effect of increasing the slope of the sloped line approximations and the maximum phase angle as shown in Figure 5-9.

In order for the transfer function to be stable only left half plane poles are permitted. With zeros both left and right half plane zeros as well as zeros at the origin are permitted. How is Equation (5-18) different for right half plane zeros? Equation (5-19) is used for right half plane zeros.

$$G(s) = K \cdot \left(\frac{s}{\omega_b} - 1 \right)^n \qquad (5\text{-}19)$$

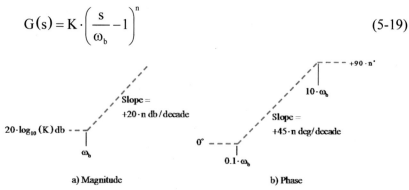

a) Magnitude   b) Phase

Figure 5-9 Bode Plot Left Half Plane Zero Approximation

Figure 5-10 shows the bode plot approximation for right half plane real zeros. Note that only the phase plot has changed from Figure 5-9.

a) Magnitude   b) Phase

Figure 5-10 Bode Plot Right Half Plane Zero Approximation

The starting phase angle is 180 degrees for n odd and 0 degrees for n even. The ending phase angle is $n \cdot 90$ degrees less than the starting phase angle. Real zeros at the origin are shown in Equation (5-20).

$$G(s) = K \cdot \left(\frac{s}{\omega_b} \right)^n \qquad (5\text{-}20)$$

Figure 5-11 shows the bode plot approximation for real zeros at the origin.

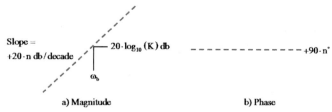

a) Magnitude   b) Phase

Figure 5-11 Bode Plot Zero at Origin Approximation

Note that the magnitude plot is just a sloped line and the phase plot is just a horizontal line. The value of ω_b is the first zero or pole break frequency.

Next, consider complex second order poles. Equation (5-21) is the general transfer function for a second order complex pole.

$$G(s) = \frac{K}{(s+\alpha)^2 + \omega^2} \tag{5-21}$$

This format is ideal for pole zero plots but is not the best form for bode plots. Equation (5-22) is the preferred form for bode plots.

$$G(s) = \frac{K}{\left(\dfrac{s}{\omega_b}\right)^2 + \dfrac{1}{Q}\cdot\left(\dfrac{s}{\omega_b}\right) + 1} \tag{5-22}$$

where :

$$\omega_b = \sqrt{\alpha^2 + \omega^2}$$

$$Q = \frac{\sqrt{\alpha^2 + \omega^2}}{2\cdot\alpha}$$

The ω_b term is the break frequency term as in the previous discussions and the Q term is the quality factor which will be explained later. In order for the roots to be complex, the value of Q must be greater than 0.5. Equation (5-23) is a normalized (K and ω_b equal to 1) second order complex pole evaluated at $s = j\cdot\omega$.

$$G(j\cdot\omega) = \frac{1}{(j\cdot\omega)^2 + \dfrac{j\cdot\omega}{Q} + 1} \tag{5-23}$$

Figure 5-12 is a bode magnitude plot of Equation (5-23) with Q equal to 5, 2, 1, and 0.7.

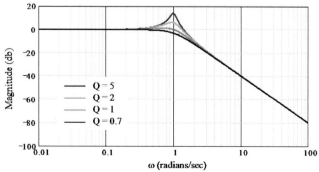

Figure 5-12 Bode Magnitude plot of Equation (5-23)

The higher the value of Q the more peaking that occurs at ω_b. If ω equal to 1 is substituted into Equation (5-23), the result is $G(j\cdot\omega) = -j\cdot Q$. The magnitude of the peaking is then $20\cdot\log_{10}(Q)$. The quality factor Q is then a measure of the amount of peaking seen on the magnitude plot. Note that if Q is between 0.5 and 1, no peaking occurs at ω_b.

A bode phase plot of Equation (5-23) is shown in Figure 5-13.

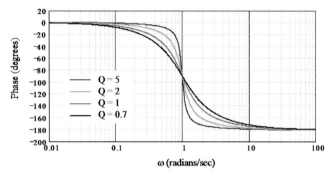

Figure 5-13 Bode Phase plot of Equation (5-23)

The phase goes from 0 degrees to -180 degrees and the value of Q determines how fast the transition occurs. The higher the value of Q, the faster the transition.

The transfer function for higher order complex second order poles is shown in Equation (5-24).

$$G(s) = \frac{K}{\left(\left(\dfrac{s}{\omega_b}\right)^2 + \dfrac{1}{Q}\cdot\left(\dfrac{s}{\omega_b}\right) + 1\right)^n} \qquad (5\text{-}24)$$

The bode straight line approximation for the complex second order poles is similar to the bode straight line approximation for the real poles and can be seen in Figure 5-14.

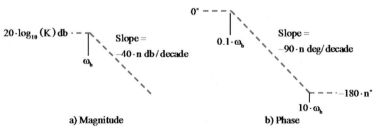

Figure 5-14 Bode Plot Complex Pole Approximation

The magnitude plot of Figure 5-14 has the same shape as the magnitude plot of Figure 5-6, but the slope of the sloped line approximation is −40 db/decade per pole as compared to -20 db/decade per pole for real poles. Also for higher values of Q, the peaking could be hand sketched in to make the plot more realistic.

The phase plot of Figure 5-14 is also the same shape as the phase plot of Figure 5-6, but the phase ranges from 0 to -180 degrees per pole as compared to 0 to -90 degrees for real poles. Also the slope of the sloped line approximation is -90 degrees/decade per pole for complex poles as compared to -45 degrees/decade per pole for real plots. For higher values of Q the slope might need to be increased to make the plot more realistic.

Next, consider complex second order left half plane zeros. Equation (5-25) is the break frequency form for a general left half plane complex second order zero.

$$G(s) = K \cdot \left(\left(\frac{s}{\omega_b} \right)^2 + \frac{1}{Q} \cdot \left(\frac{s}{\omega_b} \right) + 1 \right) \tag{5-25}$$

Equation (5-26) is the normalized left half plane complex second order zero evaluated at $s = j \cdot \omega$.

$$G(j \cdot \omega) = (j \cdot \omega)^2 + \frac{j \cdot \omega}{Q} + 1 \tag{5-26}$$

Figure 5-15 is the bode magnitude plot of Equation (5-26).

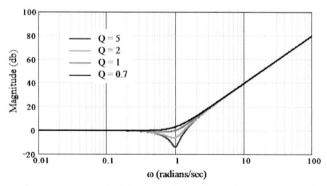

Figure 5-15 Bode Magnitude Plot of Equation (5-26)

The bode magnitude plot for a left half plane complex second order zero is similar to the bode magnitude plot for the complex second order pole of Figure 5-12 – just flipped upside down. The peaking due to Q now becomes a dip rather than a peak. At $\omega = 1$, Equation (5-26) becomes j/Q and the magnitude is $20 \cdot \log_{10}(1/Q) = -20 \cdot \log_{10}(Q)$ or just the negative of a complex second order pole.

Figure 5-16 is the bode phase plot of Equation (5-26).

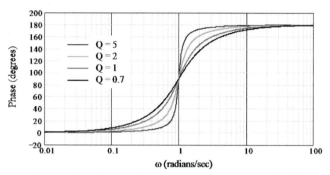

Figure 5-16 Bode Phase Plot of Equation (5-26)

The bode phase plot for a left half plane complex second order zero is also similar to the bode phase plot for the complex second order pole of Figure 5-13 – again just flipped upside down. The phase angle goes from 0 to +180 degrees.

The transfer function for higher order left half plane complex second order zeros is shown in Equation (5-27).

$$G(s) = K \cdot \left(\left(\frac{s}{\omega_b} \right)^2 + \frac{1}{Q} \cdot \left(\frac{s}{\omega_b} \right) + 1 \right)^n \tag{5-27}$$

The bode straight line approximation for the left half plane complex second order zeros is similar to the bode straight line approximation for the real zeros and can be seen in Figure 5-17.

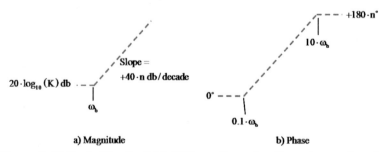

a) Magnitude   b) Phase

Figure 5-17 Bode Plot Left Half Plane Complex Zero Approximation

The magnitude plot of Figure 5-17 has the same shape as the magnitude plot of Figure 5-9, but the slope of the sloped line approximation is +40 db/decade per zero as compared to +20 db/decade per zero for real zeros. Also for higher values of Q, the dipping could be hand sketched in to make the plot more realistic.

The phase plot of Figure 5-17 is also the same shape as the phase plot of Figure 5-9, but the phase ranges from 0 to +180 degrees per zero as compared to 0 to +90 degrees for real zeros. Also the slope of the sloped line approximation is +90 degrees/decade per zero for complex poles as compared to +45 degrees/decade per zero for real zeros. For higher values of Q the slope might need to be increased to make the plot more realistic.

Just as with real zeros, right half plane complex second order zeros and complex second order zeros on the imaginary axis are permitted. Figure 5-18 shows the bode plot approximation for right half plane complex second order zeros. Note that only the phase plot has changed from Figure 5-17.

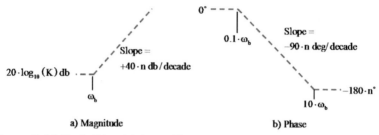

a) Magnitude   b) Phase

Figure 5-18 Bode Plot Right Half Plane Complex Zero Approximation

The phase plot is the same as the phase plot for complex second order poles of Figure 5-14.

Complex second order zeros on the imaginary axis are shown in Equation (5-28).

$$G(s) = K \cdot \left(\left(\frac{s}{\omega_b} \right)^2 + 1 \right)^n \tag{5-28}$$

Equation (5-28) is basically Equation (5-27) with Q equal to infinity.

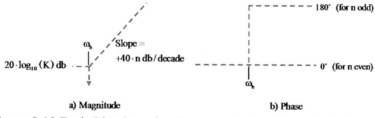

a) Magnitude   b) Phase

Figure 5-19 Bode Plot Complex Zeros on the Imaginary Axis Approx.

Figure 5-19 is the bode straight line approximation for complex second order zeros on the imaginary axis. The magnitude plot is similar to the right half plane complex zero approximation of Figure 5-18. The addition of the downward arrow at ω_b is the difference. The downward arrow indicates that the plot goes to zero at that point. This is because Q is infinite and the dipping goes to zero.

Care must be exercised when computer plotting zeros on the imaginary axis because the log of zero is undefined. Choose a value for ω_b that will not equal any of the ω points that will be plotted. This will avoid points going to zero and yielding an error.

The phase plot of Figure 5-19 has two possible shapes. When n is even, the phase is a constant 0 degrees. When n is odd, the phase changes instantaneously from 0 degrees to 180 degrees at ω_b.

So far only individual poles and zeros have been considered. How are multiple poles and zeros plotted? Another advantage of the bode plot is that cascaded transfer functions can be plotted separately and graphically added.

Consider two cascaded transfer functions $G_1(j \cdot \omega) \cdot G_2(j \cdot \omega)$. For a given ω, the functions can be represented in polar form as $m_1 \angle \theta_1 \cdot m_2 \angle \theta_2 = m_1 \cdot m_2 \angle (\theta_1 + \theta_2)$. Since the magnitude plot is in db $\left(20 \cdot \log_{10}(m) \right)$, the product is the sum of the logarithms of the m's which can easily be accomplished graphically. The phase is plotted on a linear scale but the phases are added which can also easily be accomplished graphically.

The process begins by converting all the poles and zeros to break frequency format with a single constant term in the transfer function numerator. The poles and zeros are arranged in order of break frequencies and each plotted separately on the same plot. The individual plots are then added algebraically. Note that sloped lines are combined by algebraically adding the slopes. The numerator constant is converted to db and the magnitude plot is shifted vertically by that amount. The numerator constant does not change the phase plot.

Example 5-3

Construct a bode straight line approximation plot for the transfer function of Equation (5-29) and compare it to a computer generated bode plot.

$$G(s) = \frac{0.1 \cdot (s+10)}{s+1} = \left(\frac{1}{s+1} \right) \cdot \left(\frac{\frac{s}{10}+1}{1} \right) \tag{5-29}$$

All of the poles are in the left half plane so the system is stable. Equation (5-29) is color coded the same as the bode straight line approximation plot to differentiate the pole and zero plots. Figure 5-20 is the bode straight line magnitude plot for the pole and zero of Equation (5-29).

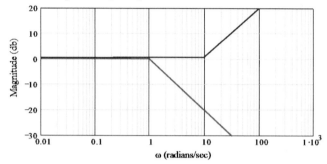

Figure 5-20 Bode Magnitude Plot Approx. for Equation (5-29)

The pole is plotted in red in Figure 5-20 using Figure 5-6a. The zero is plotted in blue in Figure 5-20 using Figure 5-9a. Figure 5-21 algebraically adds the pole and zero plots of Figure 5-20 (the red dashed line) and superimposes it over a computer generated bode plot (the black line).

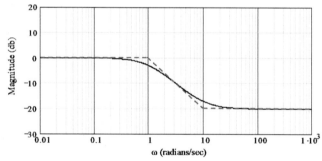

Figure 5-21 Bode Magnitude Plot of Equation (5-29)

The approximation plots start at $\omega = 0.01$ and are both 0 db out to $\omega = 1$. At that point the red plot slopes down at -20 db/decade while the blue plot remains flat at 0 db so the combined plot follows the red plot. At $\omega = 10$ the blue plot slopes up at +20 db/decade cancelling out the negative slope of the red plot for the rest of the plot creating a flat line at -20 db. The magnitude approximation compares quite well with the computer generated plot of Figure 5-21. Figure 5-22 is the bode straight line phase plot for the pole and zero of Equation (5-29).

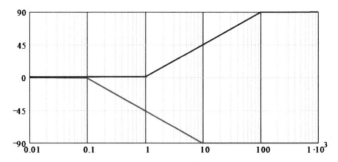

Figure 5-22 Bode Phase Plot Approx. for Equation (5-29)

The pole is plotted in red in Figure 5-22 using Figure 5-6b. The zero is plotted in blue in Figure 5-22 using Figure 5-9b. Figure 5-23 algebraically adds the pole and zero plots of Figure 5-22 (the red dashed line) and superimposes it over a computer generated bode plot (the black line).

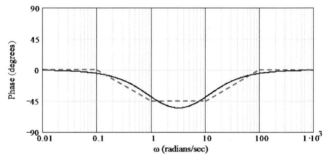

Figure 5-23 Bode Phase Plot of Equation (5-29)

The approximation plots start at $\omega = 0.01$ and are both 0 degrees out to $\omega = 0.1$. At that point the red plot slopes down at -45 deg/decade while the blue plot remains flat at 0 degrees so the combined plot follows the red plot. At $\omega = 1$ the blue plot slopes up at +45 deg/decade cancelling out the negative slope of the red plot out to $\omega = 10$ creating a flat line at -45 degrees. At $\omega = 10$, the red plot flattens out at -90 degrees while the blue plot continues to rise so the plot rises at +45 deg/decade out to $\omega = 100$. At $\omega = 100$, the blue plot flattens out at +90 degrees canceling out the red plot at -90 degrees yielding a flat plot at 0 degrees for the rest of the plot. The phase approximation compares quite well with the computer generated plot.

Example 5-4

Construct a bode straight line approximation plot for the transfer function of Equation (5-30) and compare it to a computer generated bode plot.

$$G(s) = \frac{1000 \cdot s \cdot (20 \cdot s + 200)}{(2 \cdot s + 2) \cdot ((s + 50)^2 + 7500)} \tag{5-30}$$

All of the poles are in the left half plane so the system is stable. Equation (5-31) converts Equation (5-30) to break frequency form and color codes the poles and zeros.

$$G(s) = \frac{10 \cdot \dfrac{s}{1} \cdot \left(\dfrac{s}{10} + 1\right)}{\left(\dfrac{s}{1} + 1\right) \cdot \left(\left(\dfrac{s}{100}\right)^2 + \dfrac{1}{1} \cdot \left(\dfrac{s}{100}\right) + 1\right)}$$

$$= 10 \cdot \left(\frac{s}{1}\right) \cdot \left(\frac{1}{\dfrac{s}{1} + 1}\right) \cdot \left(\frac{\dfrac{s}{10} + 1}{1}\right) \cdot \left(\frac{1}{\left(\dfrac{s}{100}\right)^2 + \dfrac{1}{1} \cdot \left(\dfrac{s}{100}\right) + 1}\right) \tag{5-31}$$

The break frequency for the zero at the origin (the red zero) will be the same break frequency as the pole at $\omega = 1$. Figure 5-24 is the bode straight line magnitude plot for the poles and zeros of Equation (5-31).

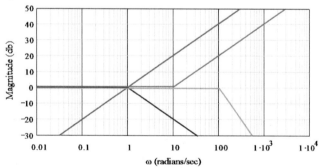

Figure 5-24 Bode Magnitude Plot Approx. for Equation (5-31)

The gain factor of 10 will be dealt with later. The zero at the origin (red line) is plotted using Equation (5-11a). The real pole at $\omega = 1$ (the blue lines) is plotted using Equation (5-6a). The real zero at $\omega = 10$ (magenta lines) is plotted using Equation

(5-9a). The complex pole at $\omega = 100$ (the orange lines) is plotted using Equation (5-14a). Figure 5-25 algebraically combines the pole and zero plots of Figure 5-24.

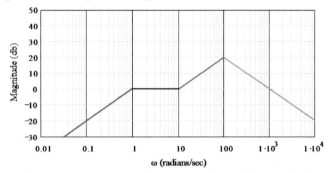

Figure 5-25 Combined Approx. Plot of Figure 5-24

Starting at $\omega = 0.01$, all of the poles and zeros are 0 db but the zero at the origin (red line) which is a sloped line of +20 db/decade . The combined plot follows this line out to $\omega = 1$. At $\omega = 1$, the pole at $\omega = 1$ (blue line) slopes down at -20 db/decade canceling out the +20 db/decade slope creating a flat line out to $\omega = 10$. At $\omega = 10$, the zero at $\omega = 10$ (magenta line) slopes up at +20 db/decade out to $\omega = 100$. At $\omega = 100$, the pole at $\omega = 100$ (the orange line) slopes down at -40 db/decade combining with the +20 db/decade slope to yield a -20 db/decade slope for the rest of the plot. At this point the gain factor of 10 is taken into account. Converting this to db yields +20 db which causes the entire Figure 5-25 to be shifted vertically by +20 db. Figure 5-26 shows the final approximation superimposed over the computer generated bode plot of Equation (5-30). The approximation compares quite well with the computer generated plot.

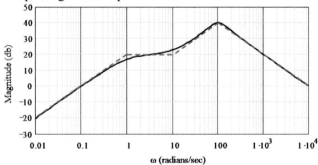

Figure 5-26 Bode Magnitude Plot of Equation (5-30)

Figure 5-27 is the bode straight line phase plot for the poles and zeros of Equation (5-31).

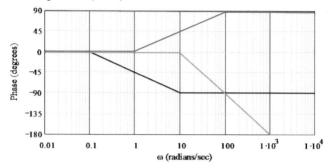

Figure 5-27 Bode Phase Plot Approx. for Equation (5-31)

The zero at the origin (red line) is plotted using Equation (5-11b). The real pole at $\omega = 1$ (the blue lines) is plotted using Equation (5-6b). The real zero at $\omega = 10$ (magenta lines) is plotted using Equation (5-9b). The complex pole at $\omega = 100$ (the orange lines) is plotted using Equation (5-14b). Figure 5-28 algebraically combines the pole and zero plots of Figure 5-27.

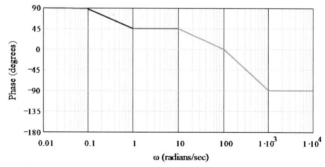

Figure 5-28 Combined Approx. Plot of Figure 5-27

Starting at $\omega = 0.01$, all of the poles and zeros are 0 degrees but the zero at the origin (red line) which is at 90 degrees. The combined plot follows this line out to $\omega = 0.1$. At $\omega = 0.1$, the pole at $\omega = 1$ (blue line) slopes down at -45 deg/decade out to $\omega = 1$. At $\omega = 1$, the zero at $\omega = 10$ (magenta line) slopes up at +45 deg/decade canceling out the -45 deg/decade slope creating a flat line out to $\omega = 10$. At $\omega = 10$, the pole at $\omega = 1$ flattens out at -90 degrees while the zero at $\omega = 10$ continues to slope up at +45 deg/decade and the pole at $\omega = 100$ (the orange line) slopes down at -90 deg/decade combining with the +45 deg/decade slope to yield a -45 deg/decade slope out to $\omega = 100$. At $\omega = 100$, the zero at $\omega = 10$ flattens out to +90 degrees and the pole at $\omega = 100$ continues to slope down at -90 deg/decade out to $\omega = 1000$ where it flattens out to -90 degrees for the rest of the plot.

Figure 5-29 shows the final approximation superimposed over the computer generated bode plot of Equation (5-30). The approximation compares quit well with the computer generated plot with the exception of the area near $\omega = 100$. This is due to the Q of 1 for the pole at $\omega = 100$. Had a steeper slope been used for the approximation, it would have compared more favorably.

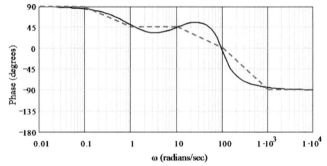

Figure 5-29 Bode Phase Plot of Equation (5-30)

The last two examples have shown that the Bode straight line approximation method yields a close approximation to a computer generated plot. But why use an approximation method when the computer generated method is much more accurate and easier to use?

The obvious reason is that computer software that does bode plots may not always be available. There is another more important reason though. There may be a portion of either the magnitude or phase plot that is undesirable and needs to be corrected by modifying a pole or zero value or by adding a pole or zero to the system to cancel out the undesirable effect of an existing pole or zero. The computer generated bode plot does not yield enough information to identify the source of the undesirable effect.

As an example, consider the bode magnitude plot of Figure 5-26 in Example 5-4. Assume that the 20 db peaking at $\omega = 100$ is an undesirable effect. What is desired is a flat response between $\omega = 1$ and $\omega = 100$. The computer generated bode magnitude plot would indicate that the pole at

$\omega = 100$ would be causing the problem. This is incorrect. Examination of Figure 5-25 reveals that the zero at $\omega = 10$ is the actual cause.

To correct the problem, add a pole at $\omega = 10$ to the system to cancel out the zero at $\omega = 10$. Figure 5-30 is a computer generated bode magnitude plot of the corrected system transfer function. Note that the 20 db peaking at $\omega = 100$ has been removed.

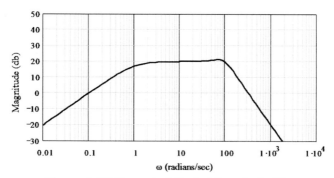

Figure 5-30 Corrected Bode Magnitude Plot

PROBLEMS

5-1. Construct the pole zero plot for the system transfer function below and discuss the system stability.

$$G(s) = \frac{s+1}{s \cdot (s+2) \cdot \left((s+3)^2 + 16\right)}$$

5-2. Construct the pole zero plot for the system transfer function below and discuss the system stability.

$$G(s) = \frac{s^2 \cdot (s-4)}{(s+3) \cdot (s+2)^2}$$

5-3. Construct the pole zero plot for the system transfer function below and discuss the system stability.

$$G(s) = \frac{(s-3)^2 + 4}{(s+3) \cdot (s-4) \cdot (s+1)^2}$$

5-4. Construct the pole zero plot for the system transfer function below and discuss the system stability.

$$G(s) = \frac{(s+2)\cdot(s-4)}{(s+1)\cdot(s^2+9)}$$

5-5. Construct the pole zero plot for the system transfer function below and discuss the system stability.

$$G(s) = \frac{(s^2+4)\cdot((s-1)^2+9)}{s\cdot(s+5)\cdot(s+4)}$$

5-6. Construct the bode straight line approximation plot for the system transfer function below and compare it with a computer generated bode plot.

$$G(s) = \frac{10}{s+10}$$

5-7. Construct the bode straight line approximation plot for the system transfer function below and compare it with a computer generated bode plot.

$$G(s) = \frac{10\cdot(s+10)}{s+100}$$

5-8. Construct the bode straight line approximation plot for the system transfer function below and compare it with a computer generated bode plot.

$$G(s) = \frac{10,000}{(s+50)^2+7500}$$

5-9. Construct the bode straight line approximation plot for the system transfer function below and compare it with a computer generated bode plot.

$$G(s) = \frac{1000\cdot s}{(s+10)\cdot(s+100)}$$

5-10. Construct the bode straight line approximation plot for the system transfer function below and compare it with a computer generated bode plot.

$$G(s) = \frac{10\cdot((s+7)^2+51)}{(s+100)\cdot(s+1)}$$

CHAPTER 6
INTRODUCTION TO FILTERS

6-1. Why Filters

One of the most common applications in circuit design is filtering. The process of filtering involves modification of the frequency components of a signal. Some frequency components are passed through the circuit while others are attenuated or blocked.

Before a filter can be designed a process must be established to measure the actual performance of the filter against the desired performance. This process is generally referred to as the filter specification. Since a filter modifies the frequency spectrum of a given signal, this specification is best visualized on a sketch of the frequency spectrum of the filter response similar to a bode plot.

This chapter will examine the different types of filters, how to develop the filter specification, and what the filter transfer function needs to be to meet the filter specification.

6-2. Filter Types

There are four basic types of filters that are characterized by the shape of their magnitude-frequency spectrum. The magnitude-frequency spectrum of a filter can be thought of as a multiplier of any applied signal magnitude-frequency spectrum. That is, the magnitude-frequency spectrum of the output signal is the product of the input signal magnitude-frequency spectrum and the magnitude-frequency spectrum of the filter.

The filter magnitude-frequency spectrum is the same concept as the transfer function and the filter design is actually a Laplace transfer function. The four basic filter types based on the shape of the frequency spectrum are:

> *Low Pass* - A low pass filter passes frequencies below a pass band frequency and attenuates or blocks frequencies above a stop band frequency.

High Pass - A high pass filter passes frequencies above a pass band frequency and attenuates or blocks frequencies below a stop band frequency.

Band Pass - A band pass filter passes a range or band of frequencies between two pass band frequencies and attenuates or blocks frequencies above an upper stop band frequency and below a lower stop band frequency.

Band Stop, Band Reject, or Notch - A band stop, band reject, or notch filter (hereafter referred to as a notch filter) is the opposite of a band pass filter. It attenuates or blocks a range or band of frequencies between two stop band frequencies and passes frequencies above an upper pass band frequency and below a lower pass band frequency.

Ideally the pass and stop band frequencies of the filter type definitions would be identical. That is there would be no frequencies that fell in between the pass and stop bands (referred to as the transition band). However, this would require an infinite amount of hardware which is not practical. The challenge in designing filters then becomes how close do the pass and stop band frequencies have to be to meet the filter specification, and what specific filter design will achieve this with the least amount of hardware.

6-3. Filter Specification

Based on the definitions of the basic filter types, it will be necessary to define the location of the pass and stop band frequencies for a filter specification as well as the magnitude of the frequency response at other frequencies.

This specification can be visualized on a bode magnitude-frequency plot by indicating the region of frequencies and magnitudes that must be excluded by the filter transfer function. Filter design usually only involves the magnitude response and not the phase response. Figure 6-1 illustrates a filter specification graph for a low pass filter. The transfer function must lie in the green area of the graph and not cross into any of the red areas. Figures 6-2 through 6-4 illustrate filter specification graphs for a high pass filter, a band pass filter, and a notch filter respectively.

In order to design a filter, it will be necessary to specify the type of filter and to sketch a filter specification graph for that type of filter identifying all of the frequencies and magnitudes indicated for the appropriate filter specification graph as illustrated in Figures 6-1 through 6-4.

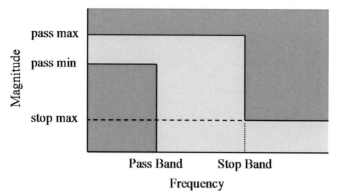

Figure 6-1 Typical Low Pass Filter Specification

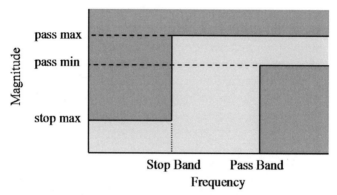

Figure 6-2 Typical High Pass Filter Specification

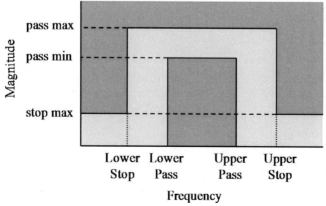

Figure 6-3 Typical Band Pass Filter Specification

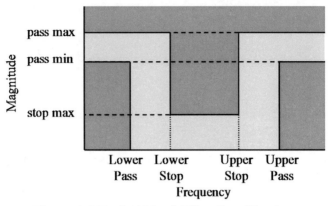

Figure 6-4 Typical Notch Filter Specification

The next step is to determine the filter transfer function that will meet but not exceed the filter specification. A filter design that exceeds the specification will require more components than one that meets the specification and be more expensive and larger in size.

The low pass filter type will be considered first and then methods to convert the low pass design to high pass, band pass, and notch will be developed. The real and complex poles of Chapter 5 (see Figures 5-6a and 5-14a) will fit in the green area of the low pass filter specification of Figure 6-1.

The optimum filter design will consist of a combination of real and complex poles that will just meet the filter specification. The horizontal line portion of the bode straight line approximation will be just below the pass max value and the sloped line portion will just miss the (pass min)/(pass band) corner and the (stop max)/(stop band) corner as shown by the blue line in Figure 6-5.

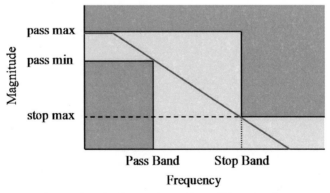

Figure 6-5 Optimum Low Pass Filter Design

The design goal is to find the transfer function with the minimum number of poles that just meets the filter specification. The larger the number of poles of the transfer function, the more components will be required to implement the design.

The task of determining the transfer function that meets and not exceeds the filter specification is a difficult task. Fortunately several people have done extensive research on how to design the optimum filter. This book will focus on the two most common filter approximations to the ideal filter (zero transition band width) – the Butterworth and the Chebyshev approximations.

6-4. The Butterworth Approximation

In 1930 Stephen Butterworth, a British engineer and physicist, developed a low pass filter approximation that has as flat a frequency response as possible in the pass band. The approximation is also referred to as the maximally flat response filter. His approach started with a normalized transfer function (max amplitude equal 1 and pass band frequency equal to 1 radian/second). The pass band frequency is the same as the break frequency for bode plots. When the transfer function is converted to the frequency domain, the frequency domain magnitude approximation is shown in Equation (6-1).

$$G(\omega) = \frac{1}{\sqrt{1 + \omega^{2 \cdot n}}} \quad \text{where n is the number of poles} \qquad (6\text{-}1)$$

The pole locations that meet Equation (6-1) lie equally spaced on a unit circle on the pole zero plot. Since the transfer function must be stable, all the poles must be in the left half plane or in quadrants 2 and 3. The poles in quadrant 3 are the complex conjugates of the poles in quadrant 2, so only quadrant 2 must be considered.

The positive imaginary axis of the pole zero plot will be the angular reference to determine the pole locations. The angle θ will be measured counterclockwise from the positive imaginary axis of the pole zero plot. The value of θ is determined By Equation (6-2).

$$\theta = 90 \cdot \frac{(2 \cdot k - 1)}{n} \quad \text{where n is the number of poles} \qquad (6\text{-}2)$$

The parameter k varies from 1 to n/2 in order to generate the angular locations for quadrant 2. When a pole falls on the negative real axis it is a real pole. The poles not on the negative real axis are complex second order poles. Since the poles lie on a unit circle, the real and imaginary coordinates for the poles can be calculated from Equation (6-3).

$$\alpha_k = \sin(\theta) = \sin\left(90 \cdot \frac{(2 \cdot k - 1)}{n}\right)$$

$$\omega_k = \cos(\theta) = \cos\left(90 \cdot \frac{(2 \cdot k - 1)}{n}\right)$$

(6-3)

Figure 6-6 shows pole locations for n equal to 3 and 4.

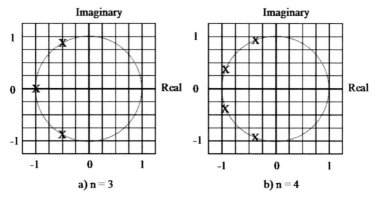

a) n = 3   b) n = 4

Figure 6-6 Butterworth Pole Locations

The pole factors, in pole zero format, will be of the form $(s + \alpha_k)$ for a real pole and $\left((s + \alpha_k)^2 + \omega_k^2\right)$ for complex second order poles. The filter transfer function has a numerator of 1 and a denominator of the product of the Butterworth pole factors in break frequency format. Table 6-1 gives the pole expressions in both pole zero format and break frequency format for several values of n.

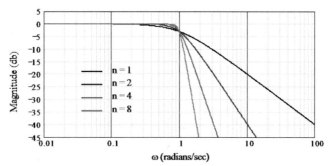

Figure 6-7 Butterworth Transfer Function Bode Magnitude Plot

Four of the Butterworth transfer functions are shown in the bode magnitude plot of Figure 6-7. Note that all of the Butterworth transfer functions

86

pass through a common point at 1 radian per second. Substituting $\omega = 1$ in Equation (6-1) and converting to db yields -3.0103 db $\left(20 \cdot \log_{10}\left(\sqrt{1/2}\right)\right)$. This point is defined as the cutoff frequency (ω_c) for the Butterworth filter.

Table 6-1 Butterworth Pole Values

n	Butterworth Pole Values
1	$(s+1)$
2	$\left((s+0.7071)^2 + 0.7071^2\right) = \left(s^2 + \dfrac{s}{0.7071} + 1\right)$
3	$(s+1)\cdot\left((s+0.5)^2 + 0.866^2\right) = (s+1)\cdot\left(s^2 + \dfrac{s}{1} + 1\right)$
4	$\left((s+0.3827)^2 + 0.9239^2\right)\cdot\left((s+0.9239)^2 + 0.3827^2\right) =$ $\left(s^2 + \dfrac{s}{1.3066} + 1\right)\cdot\left(s^2 + \dfrac{s}{0.5412} + 1\right)$
5	$(s+1)\cdot\left((s+0.309)^2 + 0.9511^2\right)\cdot\left((s+0.809)^2 + 0.5878^2\right) =$ $(s+1)\cdot\left(s^2 + \dfrac{s}{1.618} + 1\right)\cdot\left(s^2 + \dfrac{s}{0.618} + 1\right)$
6	$\left((s+0.2588)^2 + 0.9659^2\right)\cdot\left((s+0.7071)^2 + 0.7071^2\right)\cdot$ $\left((s+0.9659)^2 + 0.2588^2\right) =$ $\left(s^2 + \dfrac{s}{1.9319} + 1\right)\cdot\left(s^2 + \dfrac{s}{0.7071} + 1\right)\cdot\left(s^2 + \dfrac{s}{0.5176} + 1\right)$
7	$(s+1)\cdot\left((s+0.2225)^2 + 0.9749^2\right)\cdot\left((s+0.6235)^2 + 0.7818^2\right)\cdot$ $\left((s+0.901)^2 + 0.4339^2\right) =$ $(s+1)\left(s^2 + \dfrac{s}{2.247} + 1\right)\cdot\left(s^2 + \dfrac{s}{0.8019} + 1\right)\cdot\left(s^2 + \dfrac{s}{0.555} + 1\right)$
8	$\left((s+0.1951)^2 + 0.9808^2\right)\cdot\left((s+0.5556)^2 + 0.8315^2\right)\cdot$ $\left((s+0.8315)^2 + 0.5556^2\right)\cdot\left((s+0.9808)^2 + 0.1951^2\right) =$ $\left(s^2 + \dfrac{s}{2.5629} + 1\right)\cdot\left(s^2 + \dfrac{s}{0.9} + 1\right)\cdot\left(s^2 + \dfrac{s}{0.6013} + 1\right)\cdot\left(s^2 + \dfrac{s}{0.5098} + 1\right)$

There are two issues with the bode magnitude plot of Figure 6-7. One is that the transfer functions have a cutoff frequency of 1 and the other is that the pass min value is fixed at -3.0103 db. In order to make the Butterworth transfer functions more useful a method of translating the cutoff frequency to any value is needed. The cutoff frequency can be translated to any value by making the substitution in Equation (6-4).

$$G(s) \Rightarrow G(s)\big|_{s=(s/\omega_c)} \tag{6-4}$$

This translation has the effect of modifying Equation (6-1) to Equation (6-5).

$$G(\omega) = \frac{1}{\sqrt{1 + \left(\dfrac{\omega}{\omega_c}\right)^{2 \cdot n}}} \quad \text{where n is the number of poles} \tag{6-5}$$

In order to meet the pass min specification value, the cutoff frequency can't simply be the pass band frequency unless the pass min value is -3.0103 db. The design task then becomes the selection of the cutoff frequency and number of poles that will allow all of the filter specification values to be met.

The magnitude values from the filter specification are in db and must be converted to the actual magnitude value. Since the Butterworth approximation has a gain normalization of 1, the pass max specification is always 0 db and the pass min and stop max values will always be negative. Filters with a gain other than 1 can be created by cascading the filter with an amplifier or an attenuator. If the bode magnitude value to be converted is $-A$, Equation (6-6) is the conversion from db to the actual magnitude value.

$$-A \text{ db} \Rightarrow 10^{(-A/20)} = \frac{1}{10^{(A/20)}} \tag{6-6}$$

Equation (6-5) can be used to establish two equations, one for the pass min specification and one for the stop max specification.

$$\frac{1}{10^{(A_p/20)}} = \frac{1}{\sqrt{1 + \left(\dfrac{\omega_p}{\omega_c}\right)^{2 \cdot n}}} \tag{6-7}$$

where: A_p is the pass min value

ω_p is the pass band frequency

$$\frac{1}{10^{(A_s/20)}} = \frac{1}{\sqrt{1+\left(\dfrac{\omega_s}{\omega_c}\right)^{2\cdot n}}} \tag{6-8}$$

where: A_s is the stop max value

ω_s is the stop band frequency

Squaring both sides of Equations (6-7) and (6-8) yields Equations (6-9) and (6-10).

$$\frac{1}{10^{(0.1\cdot A_p)}} = \frac{1}{1+\left(\dfrac{\omega_p}{\omega_c}\right)^{2\cdot n}} \tag{6-9}$$

$$\frac{1}{10^{(0.1\cdot A_s)}} = \frac{1}{1+\left(\dfrac{\omega_s}{\omega_c}\right)^{2\cdot n}} \tag{6-10}$$

Cross multiplying Equations (6-9) and (6-10) yields Equations (6-11) and (6-12).

$$1+\left(\frac{\omega_p}{\omega_c}\right)^{2\cdot n} = 10^{(0.1\cdot A_p)} \Rightarrow \left(\frac{\omega_p}{\omega_c}\right)^{2\cdot n} = 10^{(0.1\cdot A_p)} - 1 \tag{6-11}$$

$$1+\left(\frac{\omega_s}{\omega_c}\right)^{2\cdot n} = 10^{(0.1\cdot A_s)} \Rightarrow \left(\frac{\omega_s}{\omega_c}\right)^{2\cdot n} = 10^{(0.1\cdot A_s)} - 1 \tag{6-12}$$

This creates two equations and two unknowns (n and ω_c). To solve for n, divide Equation (6-12) by Equation (6-11) in order to eliminate ω_c and yield Equation (6-13).

$$\left(\frac{\omega_s}{\omega_p}\right)^{2\cdot n} = \frac{10^{(0.1\cdot A_s)} - 1}{10^{(0.1\cdot A_p)} - 1} \tag{6-13}$$

Taking the log of both sides of Equation (6-13) yields Equation (6-14).

$$2\cdot n \cdot \log_{10}\left(\frac{\omega_s}{\omega_p}\right) = \log_{10}\left(\frac{10^{(0.1\cdot A_s)} - 1}{10^{(0.1\cdot A_p)} - 1}\right)$$

$$\Rightarrow n = \frac{\log_{10}\left(\dfrac{10^{(0.1 \cdot A_s)} - 1}{10^{(0.1 \cdot A_p)} - 1}\right)}{2 \cdot \log_{10}\left(\dfrac{\omega_s}{\omega_p}\right)} \tag{6-14}$$

Since n is the number of poles it must be an integer and must be rounded up to the next integer value. Rounding up the value of n creates multiple solutions for ω_c. The common solution is to match the filter transfer function to the pass min value. To accomplish this solve Equation (6-11) for ω_c as shown in Equation (6-15).

$$\left(\frac{\omega_p}{\omega_c}\right)^{2 \cdot n} = 10^{(0.1 \cdot A_p)} - 1$$

$$\left(\left(\frac{\omega_p}{\omega_c}\right)^{2 \cdot n}\right)^{1/2 \cdot n} = \left(10^{(0.1 \cdot A_p)} - 1\right)^{1/2 \cdot n}$$

$$\left(\frac{\omega_p}{\omega_c}\right) = \left(10^{(0.1 \cdot A_p)} - 1\right)^{1/2 \cdot n}$$

$$\omega_c = \frac{\omega_p}{\left(10^{(0.1 \cdot A_p)} - 1\right)^{1/2 \cdot n}} \tag{6-15}$$

The procedure to determine the Butterworth low pass filter transfer function is to first create a filter specification graph, then use Equation (6-14) to determine the number of poles, then use Equation (6-15) to determine the filter cutoff frequency ω_c, and finally select the normalized poles from Table 6-1 and make the substitution $s = s/\omega_c$.

Example 6-1

Determine the transfer function of a Butterworth low pass filter with a maximum pass band magnitude of 0 db, a minimum pass band magnitude of -2 db out to 1000 hertz, and a maximum stop band magnitude of -40 db from 4000 hertz on. Create a computer generated bode plot of the transfer function and compare it with the filter specification.

First create the filter specification graph shown in Figure 6-8.

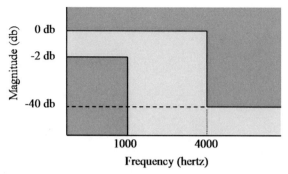

Figure 6-8 Filter Specification Graph for Example 6-1

Note that the frequencies are in hertz and not radians/sec. This is typical in filter designs and the frequencies must be converted from hertz to radians/sec to determine the filter transfer function:

$$\text{pass band frequency} = \omega_p = 2 \cdot \pi \cdot 1000 = 6{,}283$$

$$\text{stop band frequency} = \omega_s = 2 \cdot \pi \cdot 4000 = 25{,}130$$

The magnitude axis is already normalized to pass max equal to 0 db. The pass min value (A_p) is 2 db and the stop max value (A_s) is 40 db. Use Equation (6-14) to determine the number of poles:

$$n = \frac{\log_{10}\left(\dfrac{10^{(0.1 \cdot A_s)} - 1}{10^{(0.1 \cdot A_p)} - 1}\right)}{2 \cdot \log_{10}\left(\dfrac{\omega_s}{\omega_p}\right)} = \frac{\log_{10}\left(\dfrac{10^4 - 1}{10^{0.2} - 1}\right)}{2 \cdot \log_{10}\left(\dfrac{25{,}130}{6{,}283}\right)} = 3.515$$

Rounding n up to the next highest integer means that 4 poles will be required. Next use Equation (6-15) to determine the filter cutoff frequency ω_c:

$$\omega_c = \frac{\omega_p}{\left(10^{(0.1 \cdot A_p)} - 1\right)^{1/2 \cdot n}} = \frac{6{,}283}{\left(10^{0.2} - 1\right)^{1/8}} = 6{,}719$$

Using the Table 6-1 for the 4 pole entry and ω_c from above determine the filter transfer function:

$$G(s) = \cfrac{\cfrac{1}{\left(\left(\cfrac{s}{6,719}\right)^2 + \cfrac{1}{1.3066}\cdot\left(\cfrac{s}{6,719}\right)+1\right)}}{\left(\left(\cfrac{s}{6,719}\right)^2 + \cfrac{1}{0.5412}\cdot\left(\cfrac{s}{6,719}\right)+1\right)}.$$

Figure 6-9 is a computer generated bode magnitude plot of the filter transfer function with the specification superimposed in red lines.

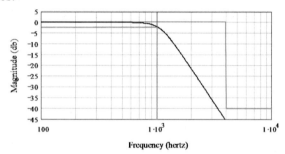

Figure 6-9 Bode Plot of Filter Transfer Function

Note that the filter design meets the pass band specification and exceeds the stop band specification. Also, the bode plot was done in hertz rather than radians/sec so the specification could be easily compared. This is typical with filter design.

6-5. The Chebyshev Approximation

The Chebyshev filter approximation which is based on the Chebyshev polynomials and named for Pafnuty Chebyshev, a Russian mathematician, was developed in the 1950's. The Chebyshev approximation has the advantage of a faster roll off at the end of the pass band as compared to the Butterworth approximation at the expense of ripple in the pass band. The Chebyshev approximation is also referred to as the equal ripple approximation. There are actually two forms of Chebyshev approximations – a type 1 and a type 2. The type 1 has ripple in the pass band and the type 2 has ripple in the stop band. This book will only focus on the type 1 Chebyshev approximation.

When the Chebyshev approximation transfer function is converted to the frequency domain, the frequency domain magnitude approximation is shown in Equation (6-16).

$$G(\omega) = \frac{1}{\sqrt{1 + \varepsilon^2 \cdot C_n^2\left(\dfrac{\omega}{\omega_p}\right)}} \tag{6-16}$$

In Equation (6-16) ω_p is the pass band frequency from the filter specification, ε is a pass band ripple factor, and $C_n()$ is a Chebyshev polynomial. The Chebyshev polynomial can be expressed as shown in Equation (6-17).

$$C_n\left(\frac{\omega}{\omega_p}\right) = \cos\left(n \cdot \cos^{-1}\left(\frac{\omega}{\omega_p}\right)\right) \quad \text{for } \omega \leq 1$$

$$C_n\left(\frac{\omega}{\omega_p}\right) = \cosh\left(n \cdot \cosh^{-1}\left(\frac{\omega}{\omega_p}\right)\right) \quad \text{for } \omega > 1 \tag{6-17}$$

The Chebyshev approximation will be normalized the same as the Butterworth approximation with the pass max value equal to 1 and the pass band frequency ω_p equal to 1. The pass band ripple factor can be calculated from Equation (6-16) by setting it equal to the pass min value and solving for ε. From Equation (6-6), the pass min value is $1/10^{(A_p/20)}$. Also, From Equation (6-17), $C_n^2(1) = 1$. Equation (6-16) then becomes Equation (6-18).

$$\frac{1}{10^{(A_p/20)}} = \frac{1}{\sqrt{1 + \varepsilon^2}} \tag{6-18}$$

Solving for ε yields Equation (6-19).

$$\sqrt{1 + \varepsilon^2} = 10^{(A_p/20)}$$

$$1 + \varepsilon^2 = 10^{(0.1 \cdot A_p)}$$

$$\varepsilon = \sqrt{10^{(0.1 \cdot A_p)} - 1} \tag{6-19}$$

The pole locations that meet Equation (6-16) lie on an ellipse as compared with the Butterworth approximation where the poles lie on a circle. The Chebyshev poles are the Butterworth poles transformed to an ellipse. The ellipse can be thought of as the circle being distorted by compressing or stretching the circle along the real and imaginary axes.

Figure 6-10 shows the normalized 5 Butterworth pole locations of Table 6-1 superimposed on a unit circle along with the normalized 5 Chebyshev 2db pass band ripple pole locations superimposed on an ellipse.

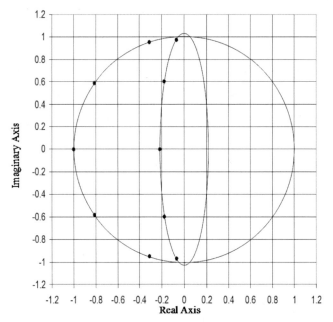

Figure 6-10   5 Butterworth and 5 Chebyshev Pole Locations

The normalized Chebyshev ellipse is determined by n (the number of poles) and ε (the pass band ripple factor). The equation for the normalized Chebyshev ellipse is shown in Equation (6-20).

$$\left(\frac{\alpha}{\sinh\left(\frac{1}{n}\cdot\sinh^{-1}\left(\frac{1}{\varepsilon}\right)\right)}\right)^2 + \left(\frac{\omega}{\cosh\left(\frac{1}{n}\cdot\sinh^{-1}\left(\frac{1}{\varepsilon}\right)\right)}\right)^2 = 1 \qquad (6\text{-}20)$$

The α term is the real axis value and the ω term is the imaginary axis value. The α denominator is the ellipse semi-minor axis value and the ω denominator is the semi-major axis value. The poles of the Butterworth approximation can be transformed to Chebyshev poles by multiplying the real component by the semi-minor axis value and multiplying the imaginary component by the semi-major axis value. Equation (6-21) shows the Chebyshev transformation equations for the Butterworth Equation (6-3).

$$\alpha_k = \sin\left(90\cdot\frac{(2\cdot k - 1)}{n}\right)\cdot\sinh\left(\frac{1}{n}\cdot\sinh^{-1}\left(\frac{1}{\varepsilon}\right)\right)$$

$$\omega_k = \cos\left(90\cdot\frac{(2\cdot k - 1)}{n}\right)\cdot\cosh\left(\frac{1}{n}\cdot\sinh^{-1}\left(\frac{1}{\varepsilon}\right)\right)$$

$$(6\text{-}21)$$

What effect do ε and n have on α and ω? The larger the value of ε, the larger the pass band ripple, and the smaller the values of α and ω. Also, the larger the value of n, the smaller the values of α and ω.

The pole factors of the Chebyshev transfer function, in pole zero format, will be of the form $(s + \alpha_k)$ for a real pole and $\left((s + \alpha_k)^2 + \omega_k^2\right)$ for complex second order poles.

The Chebyshev transfer function has a denominator of the product of the Chebyshev pole factors in break frequency format. The numerator of the Chebyshev transfer function depends on whether n is odd or even. For n odd, the Chebyshev bode plot starts at the pass max value and varies between the pass max and pass min values. For n even, the Chebyshev bode plot starts at the pass min value and varies between the pass max and pass min values. Figure 6-11 shows the Chebyshev bode plot for an even and odd value of n.

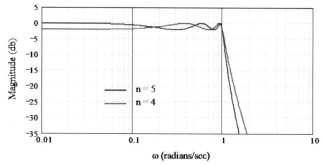

Figure 6-11 Bode Plot of 4 and 5 Pole Chebyshev Transfer Functions

With the transfer function denominator in break frequency format, the low frequency magnitude value (ω equal to 0) is 1. In order for the odd n bode plot to start at 1 (the pass max value), the transfer function numerator must be 1. In order for the even n bode plot to start at the pass min value, the transfer function numerator must be the pass min value. Since the pass min value is in db it must be converted to a magnitude value using Equation (6-6) which yields a numerator factor of $10^{(-A_p/20)}$.

The normalized Butterworth poles all have the same break frequency of 1. The normalized Chebyshev poles all have a different break frequency in order to create the ripple in the pass band. Therefore, each s in the transfer function break frequency format denominator will be s/ω_b where the ω_b is different for each pole.

This can be visualized by examining Figure 6-10 which shows the pole locations on a pole zero plot for a normalized 5 pole Butterworth and Chebyshev transfer function. The pole break frequency is the distance from the pole zero plot origin to the pole $\left(\sqrt{\alpha^2+\omega^2}\right)$. For the Butterworth transfer function, the poles lie on a circle with a radius of 1 for the normalized case. The Chebyshev transfer function poles lie on an ellipse. The ellipse can be considered a circle with a varying radius so the distance to the poles is a variable and thus, the break frequency is a variable. For Figure 6-10 the Chebyshev break frequencies vary from 0.2183 to 0.9758.

The Chebyshev approximation meets the pass band specifications exactly at the pass band frequency of ω_p. The normalized Chebyshev transfer function can be translated to the pass band frequency by setting $s = s/\omega_p$. This makes each s in the transfer function break frequency format denominator $s/(\omega_b \cdot \omega_p)$.

The only remaining item is to determine the value of n that will meet the stop band specification. This can be accomplished by evaluating Equation (6-16) at the stop band specification as shown in Equation (6-22).

$$\frac{1}{10^{(A_s/20)}} = \frac{1}{\sqrt{1+\varepsilon^2 \cdot C_n^2\left(\frac{\omega_s}{\omega_p}\right)}} \tag{6-22}$$

where: A_s is the stop max value

ω_s is the stop band frequency

Squaring both sides of Equation (6-22) yields Equation (6-23).

$$\frac{1}{10^{(0.1A_s)}} = \frac{1}{1+\varepsilon^2 \cdot C_n^2\left(\frac{\omega_s}{\omega_p}\right)} \tag{6-23}$$

Cross multiplying Equation (6-23) yields Equation (6-24).

$$1+\varepsilon^2 \cdot C_n^2\left(\frac{\omega_s}{\omega_p}\right) = 10^{(0.1A_s)} \Rightarrow C_n^2\left(\frac{\omega_s}{\omega_p}\right) = \frac{10^{(0.1A_s)}-1}{\varepsilon^2} \tag{6-24}$$

Taking the square root of Equation (6-24) yields Equation (6-25).

$$C_n\left(\frac{\omega_s}{\omega_p}\right) = \frac{\sqrt{10^{(0.1A_s)}-1}}{\varepsilon} \tag{6-25}$$

Substituting the expression for ε from Equation (6-19) into Equation (6-25) yields Equation 6-26.

$$C_n\left(\frac{\omega_s}{\omega_p}\right)=\frac{\sqrt{10^{(0.1A_s)}-1}}{\sqrt{10^{(0.1A_p)}-1}}=\sqrt{\frac{10^{(0.1A_s)}-1}{10^{(0.1A_p)}-1}} \qquad (6\text{-}26)$$

Substituting the hyperbolic form of the Chebyshev polynomial $\left(\omega_s>\omega_p\right)$ from Equation (6-17) into Equation (6-26) yields Equation (6-27).

$$\cosh\left(n\cdot\cosh^{-1}\left(\frac{\omega_s}{\omega_p}\right)\right)=\sqrt{\frac{10^{(0.1A_s)}-1}{10^{(0.1A_p)}-1}} \qquad (6\text{-}27)$$

Taking the $\cosh^{-1}$ of both sides of Equation (6-27) yields Equation (6-28).

$$n\cdot\cosh^{-1}\left(\frac{\omega_s}{\omega_p}\right)=\cosh^{-1}\left(\sqrt{\frac{10^{(0.1A_s)}-1}{10^{(0.1A_p)}-1}}\right)$$

$$\Rightarrow n=\frac{\cosh^{-1}\left(\sqrt{\frac{10^{(0.1A_s)}-1}{10^{(0.1A_p)}-1}}\right)}{\cosh^{-1}\left(\frac{\omega_s}{\omega_p}\right)} \qquad (6\text{-}28)$$

The value of n must be an integer so it must be rounded up to the next highest integer value.

The procedure to determine the Chebyshev low pass filter transfer function is to first construct a filter specification graph, then determine the number of poles from Equation (6-28), then either use Equation (6-21) or select the pole values from Table 6-1 and multiply them by the ellipse semi-major and semi-minor axis values, then convert the transfer function poles to break frequency format and calculate the transfer function numerator if n is even, then translate the transfer function to the pass band frequency by setting $s=s/\omega_p$.

Example 6-2

Determine the transfer function of a Chebyshev low pass filter with a maximum pass band magnitude of 0 db, a minimum pass band magnitude of -2 db out to 1000 hertz, and a maximum stop band magnitude of -40 db from 3000 hertz on. Create a computer generated bode plot of the transfer function and compare it with the filter specification.

First create the filter specification graph shown in Figure 6-12.

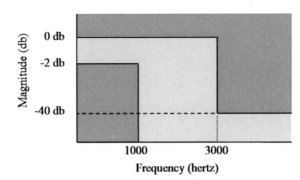

Figure 6-12 Filter Specification Graph for Example 6-2

Note that the frequencies are in hertz and not radians/sec. This is typical in filter designs and the frequencies must be converted from hertz to radians/sec to determine the filter transfer function:

$$\text{pass band frequency} = \omega_p = 2 \cdot \pi \cdot 1000 = 6,283$$
$$\text{stop band frequency} = \omega_s = 2 \cdot \pi \cdot 3000 = 18,850$$

The magnitude axis is already normalized to pass max equal to 0 db. The pass min value (A_p) is 2 db and the stop max value (A_s) is 40 db. Use Equation (6-28) to determine the number of poles:

$$n = \frac{\cosh^{-1}\left(\sqrt{\dfrac{10^{(0.1 \cdot A_s)} - 1}{10^{(0.1 \cdot A_p)} - 1}}\right)}{\cosh^{-1}\left(\dfrac{\omega_s}{\omega_p}\right)} = \frac{\cosh^{-1}\left(\sqrt{\dfrac{10^4 - 1}{10^{0.2} - 1}}\right)}{\cosh^{-1}\left(\dfrac{18,850}{6,238}\right)} = 3.158$$

Rounding n up to the next highest integer means that 4 poles will be required. Using Equation (6-19), determine ε:

$$\varepsilon = \sqrt{10^{(0.1 \cdot A_p)} - 1} = \sqrt{10^{0.2} - 1} = 0.7648$$

Using the Table 6-1 for the 4 pole entry, determine the Butterworth pole values:

$$\alpha_1 = 0.3827 \qquad \omega_1 = 0.9239$$
$$\alpha_2 = 0.9239 \qquad \omega_2 = 0.3827$$

The Chebyshev transformation factors are:

$$K_\alpha = \sinh\left(\frac{1}{n} \cdot \sinh^{-1}\left(\frac{1}{\varepsilon}\right)\right)$$

$$= \sinh\left(\frac{1}{4} \cdot \sinh^{-1}\left(\frac{1}{0.7648}\right)\right) = 0.2741$$

$$K_\omega = \cosh\left(\frac{1}{n} \cdot \sinh^{-1}\left(\frac{1}{\varepsilon}\right)\right)$$

$$= \cosh\left(\frac{1}{4} \cdot \sinh^{-1}\left(\frac{1}{0.7648}\right)\right) = 1.0369$$

The Chebyshev pole values are then:

$$\alpha_1 = 0.3827 \cdot 0.2741 = 0.1049$$
$$\omega_1 = 0.9239 \cdot 1.0369 = 0.958$$
$$\alpha_2 = 0.9239 \cdot 0.2741 = 0.2532$$
$$\omega_2 = 0.3827 \cdot 1.0369 = 0.3968$$

The normalized Chebyshev poles are then:

$$\left((s+0.1049)^2 + 0.958^2\right) \cdot \left((s+0.2532)^2 + 0.3968^2\right)$$

Converting the poles to break frequency format yields:

$$\left(\left(\frac{s}{0.9637}\right)^2 + \frac{1}{4.5939}\left(\frac{s}{0.9637}\right) + 1\right) \cdot$$

$$\left(\left(\frac{s}{0.4707}\right)^2 + \frac{1}{0.9294}\left(\frac{s}{0.4707}\right) + 1\right)$$

Translate the poles to the pass band frequency by setting $s = s/\omega_p$:

$$\left(\left(\frac{s}{6055}\right)^2 + \frac{1}{4.5939}\left(\frac{s}{6055}\right) + 1\right) \cdot$$

$$\left(\left(\frac{s}{2957}\right)^2 + \frac{1}{0.9294}\left(\frac{s}{2957}\right) + 1\right)$$

Since n is even the numerator of the transfer function must be the pass min value converted using Equation (6-6):

$$\text{numerator} = \frac{1}{10^{(A_p/20)}} = \frac{1}{10^{0.1}} = 0.7943$$

The Chebyshev low pass filter transfer function then is:

$$G(s) = \frac{0.7943}{\left(\left(\dfrac{s}{6055}\right)^2 + \dfrac{1}{4.5939} \cdot \left(\dfrac{s}{6055}\right) + 1\right)} \cdot \frac{1}{\left(\left(\dfrac{s}{2957}\right)^2 + \dfrac{1}{0.9294} \cdot \left(\dfrac{s}{2957}\right) + 1\right)}$$

Figure 6-13 is a computer generated bode magnitude plot of the filter transfer function with the specification superimposed in red lines.

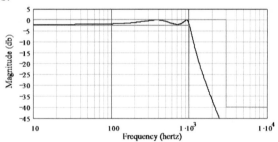

Figure 6-13 Bode Plot of Filter Transfer Function

Note that the filter design meets the pass band specification and exceeds the stop band specification. Also, the bode plot was done in hertz rather than radians/sec so the specification could be easily compared. This is typical with filter design.

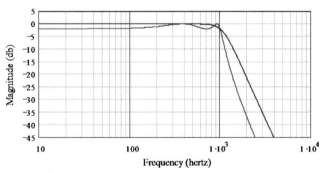

Figure 6-14 Butterworth and Chebyshev 4 Pole Transfer Functions

In order to better compare the Butterworth and Chebyshev 4 pole low pass filter transfer functions, Figure 6-14 shows both transfer function bode plots on the same plot. Note that the Chebyshev approximation reaches the stop max value of -40 db much quicker than the Butterworth approximation. That is why the Chebyshev approximation usually requires a fewer number of poles than the Butterworth approximation.

The disadvantage of the Chebyshev approximation is the ripple in the pass band as compared to the flat response of the Butterworth approximation. One of the typical mistakes exhibited by filter designers is that they choose a Chebyshev design for the lower number of poles but then are dissatisfied with the pass band ripple and change the pass band ripple to a lower value which increases the number of poles. An acceptable pass band ripple for a Butterworth approximation may be larger than the acceptable pass band ripple for a Chebyshev approximation. The lesson here is to make sure the pass band ripple is acceptable before choosing an approximation for the design.

Another disadvantage of the Chebyshev approximation is that the pole Q values are higher than the Butterworth approximation. For Example 6-1 and 6-2, the highest Q value for the Chebyshev transfer function is 4.59 as compared to the highest Q value for the Butterworth transfer function of 1.31. The higher the Q value, the higher precision required for the circuit component values. In fact, high Q values may require multiple components in series or parallel to get close enough to the design value due to available standard component values.

6-6. Low Pass to High Pass Transformation

It is not necessary to go through the same analysis process for the high pass filter that was done for the low pass filter. The low pass transfer function just needs to be transformed into a high pass form. The low pass to high pass transformation is simply replacing s with 1/s. Start with a normalized low pass transfer function, replace s with 1/s, and then translate to the proper pass band frequency.

Consider the Butterworth approximation first. From Table 6-1, the first order real poles are all of the form $1/(s+1)$. Replacing s with 1/s yields Equation (6-29).

$$\frac{1}{s+1} \Rightarrow \frac{1}{\frac{1}{s}+1} = \frac{s}{s+1} \qquad (6\text{-}29)$$

Note that the denominator (the pole) is identical to the low pass real pole but the numerator is now s. The second order complex poles are all of the form $1/(s^2 + s/Q + 1)$ and replacing s with $1/s$ yields Equation (6-30).

$$\frac{1}{s^2 + \dfrac{s}{Q} + 1} \Rightarrow \frac{1}{\left(\dfrac{1}{s}\right)^2 + \dfrac{1}{Q \cdot s} + 1} = \frac{s^2}{s^2 + \dfrac{s}{Q} + 1} \tag{6-30}$$

Again, the denominator (the pole) is identical to the low pass complex pole but the numerator is now s^2. When multiple poles are multiplied together as in Table 6-1, the denominators will be the same as for the low pass transfer function but the numerator will be s^n. Thus, Table 6-1 can be used for either a low pass or high pass filter just by using the proper numerator.

How is the number of poles determined? Equation (6-14) yields the number of poles required for a low pass filter. Since the transformation to a high pass filter involves replacing s with $1/s$, the number of poles equation must replace ω with $1/\omega$ and the high pass number of poles equation becomes Equation (6-31).

$$n = \frac{\log_{10}\left(\dfrac{10^{(0.1 \cdot A_s)} - 1}{10^{(0.1 \cdot A_p)} - 1}\right)}{2 \cdot \log_{10}\left(\dfrac{\omega_p}{\omega_s}\right)} \tag{6-31}$$

Again n must be rounded up to the next highest integer.

Table 6-1 just provides normalized pole values and they must be translated to the proper cutoff frequency. Equation (6-15) shows the calculation of the cutoff frequency ω_c for a low pass filter. Equation (6-32) shows the replacement of ω with $1/\omega$ to yield the high pass cutoff frequency.

$$\frac{1}{\omega_c} = \frac{\dfrac{1}{\omega_p}}{\left(10^{(0.1 \cdot A_p)} - 1\right)^{1/2n}} \Rightarrow \omega_c = \omega_p \cdot \left(10^{(0.1 \cdot A_p)} - 1\right)^{1/2n} \tag{6-32}$$

When the real high pass pole is translated to ω_c, the denominator format needs to be changed to polynomial form as shown in Equation (6-33). Note that for the low pass break frequency format, the constant term is 1 and for the high pass polynomial format the s coefficient is 1.

$$\frac{\dfrac{s}{\omega_c}}{\dfrac{S}{\omega_c}+1}=\frac{s}{s+\omega_c} \tag{6-33}$$

When the complex high pass pole is translated to ω_c, the denominator format also needs to be changed to polynomial form as shown in Equation (6-34). Again, the s^2 term has the coefficient of 1.

$$\frac{\left(\dfrac{s}{\omega_c}\right)^2}{\left(\dfrac{s}{\omega_c}\right)^2+\dfrac{\dfrac{s}{\omega_c}}{Q}+1}=\frac{s^2}{s^2+\dfrac{\omega_c}{Q}\cdot s+\omega_c^2} \tag{6-34}$$

Example 6-3

Determine the transfer function of a Butterworth high pass filter with a maximum pass band magnitude of 0 db, a minimum pass band magnitude of -2 db from 2000 hertz on out, and a maximum stop band magnitude of -40 db below 500 hertz. Create a computer generated bode plot of the transfer function and compare it with the filter specification.

First create the filter specification graph shown in Figure 6-15.

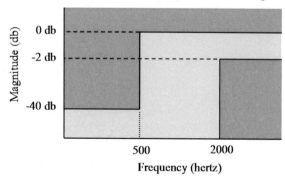

Figure 6-15 Filter Specification Graph for Example 6-3

The frequencies must be converted from hertz to radians/sec to determine the filter transfer function:

$$\text{pass band frequency} = \omega_p = 2\cdot\pi\cdot 2000 = 12{,}566$$
$$\text{stop band frequency} = \omega_s = 2\cdot\pi\cdot 500 = 3{,}142$$

The magnitude axis is already normalized to pass max equal to 0 db. The pass min value (A_p) is 2 db and the stop max value (A_s) is 40 db. Use Equation (6-31) to determine the number of poles:

$$n = \frac{\log_{10}\left(\frac{10^{(0.1 A_s)}-1}{10^{(0.1 A_p)}-1}\right)}{2 \cdot \log_{10}\left(\frac{\omega_p}{\omega_s}\right)} = \frac{\log_{10}\left(\frac{10^4-1}{10^{0.2}-1}\right)}{2 \cdot \log_{10}\left(\frac{12{,}566}{3{,}142}\right)} = 3.515$$

Rounding n up to the next highest integer means that 4 poles will be required. Next use Equation (6-32) to determine the filter cutoff frequency ω_c:

$$\omega_c = \omega_p \cdot \left(10^{(0.1 A_p)}-1\right)^{1/2 \cdot n} = 12{,}566 \cdot \left(10^{(0.2)}-1\right)^{1/8} = 11{,}752$$

Using the Table 6-1 for the 4 pole entry, Equation (6-34), and ω_c from above determine the filter transfer function:

$$G(s) = \frac{s^4}{\left(s^2 + \frac{11{,}752}{1.3066} \cdot s + 11{,}752^2\right) \cdot \left(s^2 + \frac{11{,}752}{0.5412} \cdot s + 11{,}752^2\right)}$$

Figure 6-16 is a computer generated bode magnitude plot of the filter transfer function with the specification superimposed in red lines.

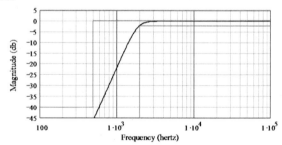

Figure 6-16 Bode Plot of Filter Transfer Function

Note that the filter design meets the pass band specification and exceeds the stop band specification.

The same approach applies to the high pass Chebyshev approximation. Replacing s with 1/s for the Chebyshev low pass real poles yields Equation (6-35).

$$\frac{1}{\dfrac{s}{\omega_b}+1} \Rightarrow \frac{1}{\dfrac{1}{\dfrac{s}{\omega_b}}+1} = \frac{\omega_b \cdot s}{\omega_b \cdot s + 1} = \frac{s}{s+\dfrac{1}{\omega_b}} \qquad (6\text{-}35)$$

Recall, that the break frequency for the Butterworth poles are all 1 but the break frequency for the Chebyshev poles are all different. Note that in Equation (6-35) that the denominator ω_b factor is now under the constant term rather than the s term and the numerator is now s. Translating Equation (6-35) to the pass band frequency ω_p yields Equation (6-36).

$$\frac{s}{s+\dfrac{1}{\omega_b}} \Rightarrow \frac{\dfrac{s}{\omega_p}}{\dfrac{s}{\omega_p}+\dfrac{1}{\omega_b}} = \frac{s}{s+\dfrac{\omega_p}{\omega_b}} \qquad (6\text{-}36)$$

Replacing s with 1/s for the Chebyshev low pass complex poles yields Equation (6-37).

$$\frac{1}{\left(\dfrac{s}{\omega_b}\right)^2 + \dfrac{1}{Q}\cdot\left(\dfrac{s}{\omega_b}\right)+1} \Rightarrow \frac{1}{\left(\dfrac{1}{\dfrac{s}{\omega_b}}\right)^2 + \dfrac{1}{Q}\cdot\left(\dfrac{1}{\dfrac{s}{\omega_b}}\right)+1}$$

$$= \frac{\omega_b^2 \cdot s^2}{\omega_b^2 \cdot s^2 + \dfrac{1}{Q}\cdot \omega_b \cdot s + 1} = \frac{s^2}{s^2 + \dfrac{1}{Q\cdot\omega_b}\cdot s + \left(\dfrac{1}{\omega_b}\right)^2} \qquad (6\text{-}37)$$

Note the change similar to the real pole. Translating Equation (6-37) to the pass band frequency ω_p yields Equation (6-38).

$$\frac{s^2}{s^2 + \dfrac{1}{Q\cdot\omega_b}\cdot s + \left(\dfrac{1}{\omega_b}\right)^2} \Rightarrow \frac{\left(\dfrac{s}{\omega_p}\right)^2}{\left(\dfrac{s}{\omega_p}\right)^2 + \dfrac{1}{Q\cdot\omega_b}\cdot\left(\dfrac{s}{\omega_p}\right) + \left(\dfrac{1}{\omega_b}\right)^2}$$

$$= \frac{s^2}{s^2 + \dfrac{1}{Q}\cdot\left(\dfrac{\omega_p}{\omega_b}\right)\cdot s + \left(\dfrac{\omega_p}{\omega_b}\right)^2} \qquad (6\text{-}38)$$

The number of poles is determined similar to the Butterworth high pass method. Equation (6-28) yields the number of poles required for a low pass

Chebyshev filter. Since the transformation to a high pass filter involves replacing s with 1/s, the number of poles equation must replace ω with $1/\omega$ and the high pass number of poles equation becomes Equation (6-39).

$$n = \frac{\cosh^{-1}\left(\sqrt{\dfrac{10^{(0.1 A_s)} - 1}{10^{(0.1 A_p)} - 1}}\right)}{\cosh^{-1}\left(\dfrac{\omega_p}{\omega_s}\right)} \qquad (6\text{-}39)$$

Again n must be rounded up to the next highest integer. The Chebyshev low pass function starts at the pass max value for n odd and the pass min value for n even. The high pass function has a similar condition but for n odd the function finishes at the pass max value and for n even finishes at the pass min value. The numerator constant factor must be implemented the same as for the low pass function.

Example 6-4

Determine the transfer function of a Chebyshev high pass filter with a maximum pass band magnitude of 0 db, a minimum pass band magnitude of -2 db from 3000 hertz on out, and a maximum stop band magnitude of -40 db below 1000 hertz. Create a computer generated bode plot of the transfer function and compare it with the filter specification.

First create the filter specification graph shown in Figure 6-17.

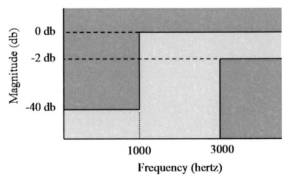

Figure 6-17 Filter Specification Graph for Example 6-3

The frequencies must be converted from hertz to radians/sec to determine the filter transfer function:

$$\text{pass band frequency} = \omega_p = 2 \cdot \pi \cdot 3000 = 18,850$$
$$\text{stop band frequency} = \omega_s = 2 \cdot \pi \cdot 1000 = 6,283$$

The magnitude axis is already normalized to pass max equal to 0 db. The pass min value (A_p) is 2 db and the stop max value (A_s) is 40 db. Use Equation (6-39) to determine the number of poles:

$$n = \frac{\cosh^{-1}\left(\sqrt{\dfrac{10^{(0.1 A_s)} - 1}{10^{(0.1 A_p)} - 1}}\right)}{\cosh^{-1}\left(\dfrac{\omega_p}{\omega_s}\right)} = \frac{\cosh^{-1}\left(\sqrt{\dfrac{10^4 - 1}{10^{0.2} - 1}}\right)}{\cosh^{-1}\left(\dfrac{18,850}{6,283}\right)} = 3.515$$

Rounding n up to the next highest integer means that 4 poles will be required. Using Equation (6-19), determine ε:

$$\varepsilon = \sqrt{10^{(0.1 A_p)} - 1} = \sqrt{10^{0.2} - 1} = 0.7648$$

Using the Table 6-1 for the 4 pole entry, determine the Butterworth pole values:

$$\alpha_1 = 0.3827 \qquad \omega_1 = 0.9239$$
$$\alpha_2 = 0.9239 \qquad \omega_2 = 0.3827$$

The Chebyshev transformation factors are:

$$K_\alpha = \sinh\left(\frac{1}{n} \cdot \sinh^{-1}\left(\frac{1}{\varepsilon}\right)\right)$$

$$= \sinh\left(\frac{1}{4} \cdot \sinh^{-1}\left(\frac{1}{0.7648}\right)\right) = 0.2741$$

$$K_\omega = \cosh\left(\frac{1}{n} \cdot \sinh^{-1}\left(\frac{1}{\varepsilon}\right)\right)$$

$$= \cosh\left(\frac{1}{4} \cdot \sinh^{-1}\left(\frac{1}{0.7648}\right)\right) = 1.0369$$

The Chebyshev pole values are then:

$$\alpha_1 = 0.3827 \cdot 0.2741 = 0.1049$$
$$\omega_1 = 0.9239 \cdot 1.0369 = 0.958$$
$$\alpha_2 = 0.9239 \cdot 0.2741 = 0.2532$$
$$\omega_2 = 0.3827 \cdot 1.0369 = 0.3968$$

The normalized Chebyshev poles are then:

$$\left((s+0.1049)^2 + 0.958^2\right) \cdot \left((s+0.2532)^2 + 0.3968^2\right)$$

Converting the poles to break frequency format yields:

$$\left(\left(\frac{s}{0.9637}\right)^2 + \frac{1}{4.5939} \cdot \left(\frac{s}{0.9637}\right) + 1\right) \cdot$$

$$\left(\left(\frac{s}{0.4707}\right)^2 + \frac{1}{0.9294} \cdot \left(\frac{s}{0.4707}\right) + 1\right)$$

Using Equation (6-38), convert these poles to the high pass form translated to ω_p.

$$s^2 + \frac{1}{4.5939} \cdot \left(\frac{18,850}{0.9637}\right) \cdot s + \left(\frac{18,850}{0.9637}\right)^2 = s^2 + \frac{19,560}{4.5939} \cdot s + 19,560^2$$

$$s^2 + \frac{1}{0.9294} \cdot \left(\frac{18,850}{0.4707}\right) \cdot s + \left(\frac{18,850}{0.4707}\right)^2 = s^2 + \frac{40,047}{0.9294} \cdot s + 40,047^2$$

Since n is even the numerator constant of the transfer function must be the pass min value converted using Equation (6-6):

$$\text{numerator} = \frac{1}{10^{(A_p/20)}} = \frac{1}{10^{0.1}} = 0.7943$$

The Chebyshev high pass transfer function then becomes:

$$G(s) = \frac{0.7943 \cdot s^4}{\left(s^2 + \frac{19,560}{4.5939} \cdot s + 19,560^2\right) \cdot \left(s^2 + \frac{40,047}{0.9294} \cdot s + 40,047^2\right)}$$

Figure 6-18 is a computer generated bode magnitude plot of the filter transfer function with the specification superimposed in red lines.

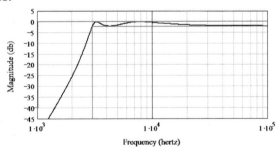

Figure 6-18 Bode Plot of Filter Transfer Function

Note that the filter design meets the pass band specification and exceeds the stop band specification.

6-7. Low Pass to Band Pass Transformation

The band pass filter actually consists of a high pass filter followed by a low pass filter with a higher pass band frequency. The high pass filter blocks low frequencies and passes the high frequencies on to the low pass filter which passes some of the high frequencies and blocks the rest creating a band of passed frequencies.

Wide band band pass filters are actually designed with a high pass design and a low pass design. With narrow band band pass filters the transition bands of the two designs interact and a different approach must be used. This approach uses a transformation of the low pass filter transfer function. However, the process is a little more complicated than the low pass to high pass transformation.

When is a band pass design a wide band design and when is it a narrow band design? The concept of the band pass filter bandwidth must first be introduced. From the band pass filter specification graph of Figure 6-3, it can be determined that the filter has a lower pass frequency and an upper pass frequency. The difference between these two is the filter bandwidth (upper pass – lower pass). The center frequency of a band pass filter is not the arithmetic mean of the upper and lower pass frequencies but rather the geometric mean and is shown in Equation (6-40).

$$\text{bandwidth} = B = \omega_{ph} - \omega_{pl} = 2 \cdot \pi \cdot \left(f_{ph} - f_{pl} \right)$$
$$\text{center frequency} = \omega_0 = \sqrt{\omega_{ph} \cdot \omega_{pl}} = 2 \cdot \pi \cdot f_0 = 2 \cdot \pi \cdot \sqrt{f_{ph} \cdot f_{pl}}$$
(6-40)

The bandwidth of a band pass filter is usually defined as the point where the magnitude is 3.0103 db down from the magnitude at the center frequency. The quality factor, Q, of a band pass filter is the ratio of the center frequency to the bandwidth as shown in Equation (6-41).

$$Q = \frac{\omega_0}{B} = \frac{2 \cdot \pi \cdot f_0}{2 \cdot \pi \cdot \left(f_{ph} - f_{pl} \right)} = \frac{f_0}{f_{ph} - f_{pl}}$$
(6-41)

Wide band band pass filter designs are used when Q is less than or equal to 0.5 and narrow band band pass filter designs are used when Q is greater than 0.5.

The low pass to band pass transformation must provide a low pass and a high pass component. The low pass to high pass transformation involved replacing the s in the low pass transfer function with 1/s. This can be thought of as the high pass component. To provide the low pass component s must be replaced with s. Combining these two substitutions yields the low pass to band pass transformation of Equation (6-42).

$$S_{LP} \Rightarrow \left(S_{BP} + \frac{1}{S_{BP}}\right) = \frac{S_{BP}^2 + 1}{S_{BP}} \qquad (6\text{-}42)$$

The s^2 term in Equation (6-42) results in two band pass frequencies for each low pass frequency which doubles the number of poles in the band pass transfer function.

Just as the low pass to high pass transformation started with a normalized specification, the low pass to band pass transformation also needs to start with a normalized specification. The magnitudes are normalized to 0 db at ω_0. The pass and stop band frequencies must be normalized to the center frequency by dividing all frequencies by the center frequency. It doesn't matter whether the filter specification is in hertz or radians per second, just divide by the center frequency of the same units and the $2 \cdot \pi$ factors will cancel out.

In order to determine the low pass transfer function to start with, the normalized band pass frequencies need to be transformed to the low pass equivalent. Going back to Equation (6-42) and replacing s with $j \cdot \omega$ yields the band pass to low pass frequency transformation of Equation (6-43).

$$\frac{\left(j \cdot \omega_{BP}\right)^2 + 1}{j \cdot \omega_{BP}} \Rightarrow j \cdot \omega_{LP}$$

$$\frac{-\omega_{BP}^2 + 1}{j \cdot \omega_{BP}} = j \cdot \frac{\omega_{BP}^2 - 1}{\omega_{BP}} \Rightarrow j \cdot \omega_{LP}$$

$$\frac{\omega_{BP}^2 - 1}{\omega_{BP}} \Rightarrow \omega_{LP} \qquad (6\text{-}43)$$

First transform the normalized band pass upper pass frequency to the low pass equivalent as shown in Equation (6-44).

$$\frac{\omega_{ph}^2 - 1}{\omega_{ph}} = \omega_{ph} - \frac{1}{\omega_{ph}} \Rightarrow \omega_{PLP}$$

But from Equation (6-40) $\dfrac{1}{\omega_{ph}} = \omega_{pl} \left(\omega_0 = 1 = \sqrt{\omega_{ph} \cdot \omega_{pl}}\right)$

$$\omega_{ph} - \omega_{pl} = B \Rightarrow \omega_{PLP} \qquad (6\text{-}44)$$

Thus, the low pass equivalent pass band frequency is the band pass bandwidth. The geometric symmetry of the band pass response can be extended to all points on the filter spectrum. That is Equation (6-45) must be satisfied for any pair of frequencies that have the same magnitude value.

$$\omega_0 = \sqrt{\omega_1 \cdot \omega_2}$$

And for the normalized band pass function:

$$1 = \sqrt{\omega_1 \cdot \omega_2} = \omega_1 \cdot \omega_2 \quad \text{or} \quad \omega_1 = \frac{1}{\omega_2} \quad \text{and} \quad \omega_2 = \frac{1}{\omega_1} \qquad (6\text{-}45)$$

In transforming the stop frequency back to the low pass equivalent there are two possibilities. Since the stop band specification is determined by a desired performance rather than the geometric symmetry like the pass band specification, either stop frequency could define the highest number of poles. Geometric symmetry can be used to determine which stop frequency to use. If the reciprocal of the lower stop frequency is lower than the upper stop frequency, then use the lower stop frequency for ω_{sl} and the reciprocal of the lower stop frequency for ω_{sh}. If the reciprocal of the lower stop frequency is higher than the upper stop frequency, then use the higher stop frequency for ω_{sh} and the reciprocal of the higher stop frequency for ω_{sl}. Equation (6-46) shows that the low pass equivalent stop band frequency is the band pass geometric stop width.

$$\frac{\omega_{sh}^2 - 1}{\omega_{sh}} = \omega_{sh} - \frac{1}{\omega_{sh}} \Rightarrow \omega_{SLP}$$

But from Equation (6-45) $\dfrac{1}{\omega_{sh}} = \omega_{sl}$

$$\omega_{sh} - \omega_{sl} = \text{geometric stop width} \Rightarrow \omega_{SLP} \qquad (6\text{-}46)$$

This defines the low pass equivalent specification. The Butterworth approximation is usually used for the narrow band design since the math is simpler and the narrow peak is smoother with no ripple. The equation for the number of poles for a Butterworth low pass filter (Equation (6-14)) can be used to determine the number of low pass equivalent poles required as shown in Equation (6-47).

$$n = \frac{\log_{10}\left(\dfrac{10^{(0.1 \cdot A_s)} - 1}{10^{(0.1 \cdot A_p)} - 1}\right)}{2 \cdot \log_{10}\left(\dfrac{\omega_s}{\omega_p}\right)}$$

$$n = \frac{\log_{10}\left(\dfrac{10^{(0.1 \cdot A_s)} - 1}{10^{(0.1 \cdot A_p)} - 1}\right)}{2 \cdot \log_{10}\left(\dfrac{\text{geometric stop width}}{\text{bandwidth}}\right)} \tag{6-47}$$

Once the number of poles is determined the normalized transfer function for the low pass equivalent can be determined either from Table 6-1 or Equation (6-3). Instead of translating the transfer function to ω_c, It will be translated to B. The transfer function for a real pole translated to B and transformed to a band pass transfer function using Equation (6-42) is shown in Equation (6-48).

$$G(s) = \frac{1}{\dfrac{s_{LP}}{B} + 1}\Bigg|_{s_{LP} = \frac{s^2+1}{s}} = \frac{1}{\left(\dfrac{s^2+1}{B \cdot s}\right) + 1} = \frac{B \cdot s}{s^2 + B \cdot s + 1} \tag{6-48}$$

Analyzing the bode plot of the band pass transfer function of Equation (6-48) yields a plot that starts with a rising +20 db per decade slope out to a frequency of 1 radian per second and a magnitude of 0 db, and then falling at a -20 db per decade slope. The Q of the transfer function is 1/B which from Equation (6-41) is Q_{BP}, the Q of the band pass specification.

The transfer function for a low pass equivalent complex pole is of the form of Equation (6-49).

$$G(s_{LP}) = \frac{1}{s_{LP}^2 + \dfrac{1}{Q_{LP}} \cdot s_{LP} + 1} \tag{6-49}$$

Translating this to B and transforming to a band pass transfer function using Equation (6-42) yields Equation (6-50).

$$G(s) = \frac{1}{\left(\dfrac{s_{LP}}{B}\right)^2 + \dfrac{1}{Q_{LP}} \cdot \left(\dfrac{s_{LP}}{B}\right) + 1}\Bigg|_{s_{LP} = \frac{s^2+1}{s}}$$

$$= \frac{1}{\left(\dfrac{s^2+1}{B \cdot s}\right)^2 + \dfrac{1}{Q_{LP}} \cdot \left(\dfrac{s^2+1}{B \cdot s}\right) + 1}$$

$$= \frac{B^2 \cdot s^2}{s^4 + \dfrac{B}{Q_{LP}} \cdot s^3 + (2 + B^2) \cdot s^2 + \dfrac{B}{Q_{LP}} \cdot s + 1} \tag{6-50}$$

Equation (6-50) can be simplified to the product of two second order band pass functions, one on either side of the normalized center frequency of 1 as shown in Equation (6-51).

$$G(s) = \frac{B \cdot s}{\left(s^2 + \frac{\omega_b}{Q_b} \cdot s + \omega_b^2\right)} \cdot \frac{B \cdot s}{\left(s^2 + \frac{1}{\omega_b \cdot Q_b} \cdot s + \left(\frac{1}{\omega_b}\right)^2\right)} \qquad (6\text{-}51)$$

The geometric symmetry of the band pass function creates the ω_b and $1/\omega_b$ relationship between the two second order factors as well as both Q's being equal. Solving for ω_b and Q_b symbolically results in some very complex equations. The simpler solution is to use actual values before factoring the denominator of Equation (6-50). P. R. Geffe published the symbolic solution in "Designers Guide to Active Band Pass Filters" published in EDN April 5, 1974.

Once the normalized transfer function of the band pass filter is determined, it can be translated to any desired center frequency by replacing s with s/ω_0 where ω_0 is the desired center frequency.

Example 6-5

Determine the transfer function of a band pass filter with a maximum pass band magnitude of 0 db at a center frequency of 1000 hertz and a 200 hertz bandwidth at the -3.0103 db points. There shall be a maximum stop band magnitude of -40 db below 500 hertz and above 2500 hertz. Create a computer generated bode plot of the transfer function and compare it with the filter specification.

First determine the filter Q using Equation (6-41):

$$Q = \frac{\omega_0}{B} = \frac{2 \cdot \pi \cdot 1000}{2 \cdot \pi \cdot 200} = 5$$

Since Q is greater than 0.5, use the narrow band approach.

Next, create the filter specification graph. Since the center frequency and bandwidth are given, they must be normalized and converted to the upper and lower pass frequencies. The normalized center frequency and bandwidth are:

$$\omega_0 = \frac{f_0}{1000} = 1 \qquad B = \frac{200}{1000} = 0.2$$

113

The upper and lower pass frequencies can be calculated from Equation (6-40):

$$f_0 = 1000 = \sqrt{f_{ph} \cdot f_{pl}} \Rightarrow f_{pl} = \frac{1000^2}{f_{ph}}$$

$$B_f = 200 = f_{ph} - f_{pl} = f_{ph} - \frac{1000^2}{f_{ph}}$$

$$f_{ph}^2 - 200 \cdot f_{ph} - 1000^2 = 0 \Rightarrow f_{ph} = 1105 \quad \text{and} \quad f_{pl} = 905$$

The specification graph is shown in Figure 6-19:

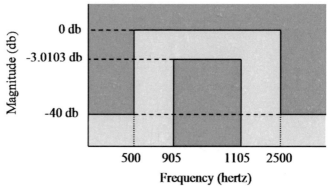

Figure 6-19 Filter Specification Graph for Example 6-5

The next step is to normalize the specifications. The magnitude axis is already normalized since the pass max value is 0 db. To normalize the frequency axis divide all of the specifications by f_0 or 1000. Since this is a frequency divided by a frequency, the normalized axis can be considered either hertz or radians per second. The normalized frequency values are:

$$\omega_{pl} = \frac{905}{1000} = 0.905 \quad \omega_{ph} = \frac{1105}{1000} = 1.105$$

$$\text{lower stop} = \frac{500}{1000} = 0.5 \quad \text{upper stop} = \frac{2500}{1000} = 2.5$$

The next step is to determine the low pass equivalent. To determine the low pass equivalent number of poles, determine the pass bandwidth B and the geometric stop bandwidth:

$$B = \frac{B_f}{f_0} = \frac{200}{1000} = 0.2 \text{ or } B = \omega_{ph} - \omega_{pl} = 1.105 - 0.905 = 0.2$$

$1/(\text{lower stop}) = 1/0.5 = 2$ which is less than the upper stop

Therefore: $\omega_{sl} = 0.5$ and $\omega_{sh} = \frac{1}{0.5} = 2$

geometric stop width $= \omega_{sh} - \omega_{sl} = 2 - 0.5 = 1.5$

Determine the number of low pass equivalent poles using Equation (6-47):

$$n = \frac{\log_{10}\left(\frac{10^{(0.1 \cdot A_s)} - 1}{10^{(0.1 \cdot A_p)} - 1}\right)}{2 \cdot \log_{10}\left(\frac{\text{geometric stop width}}{\text{bandwidth}}\right)}$$

$$= \frac{\log_{10}\left(\frac{10^{(4)} - 1}{10^{(0.30103)} - 1}\right)}{2 \cdot \log_{10}\left(\frac{1.5}{0.2}\right)} = 2.285$$

Rounding up to the next highest integer yields n equal to 3. Using Table 6-1 for the 3 pole entry, the normalized low pass equivalent transfer function is:

$$G(s)_{LP} = \frac{1}{(s+1)} \cdot \frac{1}{\left(s^2 + \frac{s}{1} + 1\right)}$$

Translating this to the pass band frequency of B yields:

$$G(s)_{LP} = \frac{1}{\left(\frac{s}{0.2} + 1\right)} \cdot \frac{1}{\left(\left(\frac{s}{0.2}\right)^2 + \frac{1}{1} \cdot \left(\frac{s}{0.2}\right) + 1\right)}$$

Transforming this to band pass using Equations (6-48) and (6-50) yields:

$$G(s) = \frac{0.2 \cdot s}{(s^2 + 0.2 \cdot s + 1)} \cdot \frac{0.04 \cdot s^2}{(s^4 + 0.2 \cdot s^3 + 2.04 \cdot s^2 + 0.2 \cdot s + 1)}$$

Factoring the 4th order denominator yields:

$$G(s) = \frac{0.2 \cdot s}{(s^2 + 0.2 \cdot s + 1)} \cdot$$

$$\frac{0.2 \cdot s}{(s^2 + 0.1086 \cdot s + 1.1891)} \cdot \frac{0.2 \cdot s}{(s^2 + 0.0914 \cdot s + 0.841)}$$

Translating this out to the specification center frequency of 1000 hertz which is $2000 \cdot \pi$ radians per second yields:

$$G(s) = \frac{0.2 \cdot \left(\dfrac{s}{2000 \cdot \pi}\right)}{\left(\left(\dfrac{s}{2000 \cdot \pi}\right)^2 + 0.2 \cdot \left(\dfrac{s}{2000 \cdot \pi}\right) + 1\right)} \cdot$$

$$\frac{0.2 \cdot \left(\dfrac{s}{2000 \cdot \pi}\right)}{\left(\left(\dfrac{s}{2000 \cdot \pi}\right)^2 + 0.1086 \cdot \left(\dfrac{s}{2000 \cdot \pi}\right) + 1.1891\right)} \cdot$$

$$\frac{0.2 \cdot \left(\dfrac{s}{2000 \cdot \pi}\right)}{\left(\left(\dfrac{s}{2000 \cdot \pi}\right)^2 + 0.0914 \cdot \left(\dfrac{s}{2000 \cdot \pi}\right) + 0.841\right)}$$

$$G(s) = \frac{1257 \cdot s}{(s^2 + 1257 \cdot s + 3.9478 \cdot 10^7)} \cdot$$

$$\frac{1257 \cdot s}{(s^2 + 682.6 \cdot s + 4.6944 \cdot 10^7)} \cdot$$

$$\frac{1257 \cdot s}{(s^2 + 574.04 \cdot s + 3.32 \cdot 10^7)}$$

Figure 6-20 is a computer generated bode magnitude plot of the filter transfer function with the specification superimposed in red lines. Note that the filter design meets the pass band specification and exceeds the stop band specification. Since the lower stop frequency determined the number of poles, it is closer to the filter response than the upper pass frequency.

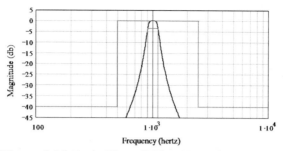

Figure 6-20 Bode Plot of Filter Transfer Function

6-8. Low Pass to Notch Transformation

Like the band pass filter, the notch filter also consists of a high pass and a low pass filter. However, the input is applied to both filters and their outputs are added. For the band pass filter the low pass filter had a higher pass band frequency than the high pass filter creating an overlap between the two filters which became the pass band. For the notch filter the low pass filter has a lower pass band frequency than the high pass filter creating a gap between the two filters. By adding the two filter outputs, low frequencies are passed on to the output as well as high frequencies but the frequencies in the gap are blocked.

Like the band pass filter, wide band notch filters are actually designed with a high pass design and a low pass design. With narrow band notch filters the transition bands of the two designs also interact and an approach similar to that used for the narrow band band pass filter must be used. This approach also uses a transformation of the low pass filter transfer function.

The concept of the notch filter notchwidth, which is similar to the band pass filter bandwidth, must first be introduced. From the notch filter specification graph of Figure 6-4, it can be determined that the filter has a lower pass frequency and an upper pass frequency. The difference between these two is the filter notchwidth (upper pass – lower pass). The center frequency of a notch filter also is not the arithmetic mean of the upper and lower pass frequencies but rather the geometric mean as shown in Equation (6-52).

$$\text{notchwidth} = N = \omega_{ph} - \omega_{pl} = 2 \cdot \pi \cdot \left(f_{ph} - f_{pl} \right)$$
$$\text{center frequency} = \omega_0 = \sqrt{\omega_{ph} \cdot \omega_{pl}} = 2 \cdot \pi \cdot f_0 = 2 \cdot \pi \cdot \sqrt{f_{ph} \cdot f_{pl}} \qquad (6\text{-}52)$$

The notchwidth of a notch filter is usually defined as the point where the magnitude is 3.0103 db down from the magnitude at the low or high frequencies. The quality factor, Q, of a notch filter is the ratio of the center frequency to the notchwidth as shown in Equation (6-53).

117

$$Q = \frac{\omega_0}{N} = \frac{2 \cdot \pi \cdot f_0}{2 \cdot \pi \cdot \left(f_{ph} - f_{pl}\right)} = \frac{f_0}{f_{ph} - f_{pl}} \tag{6-53}$$

Wide band notch filter designs are used when Q is less than or equal to 0.5 and narrow band notch filter designs are used when Q is greater than 0.5.

For the band pass filter the width of the pass band was smaller than the width of the stop band. For the notch filter the width of the pass band is larger than the width of the stop band. This resulted in the band pass filter specifications being transformed to a low pass equivalent. For the notch filter, the specifications need to be transformed to a high pass equivalent. The low pass to notch transformation must provide a low pass (s_N) and a high pass component ($1/s_N$) just like the low pass to band pass transformation but must also include and additional high pass transformation ($1/s_c$ where s_c represents the low and high pass components). Combining the low pass and high pass components as well as the additional high pass transformation is shown in Equation (6-54).

$$s_{LP} \Rightarrow \left(\frac{1}{s_c}\right) = \left(\frac{1}{s_N + \dfrac{1}{s_N}}\right) = \frac{s_N}{s_N^2 + 1} \tag{6-54}$$

Note that this is just the reciprocal of the band pass expression of Equation (6-42). The s^2 term in Equation (6-54) results in two notch frequencies for each low pass frequency which doubles the number of poles in the notch transfer function.

Just as the low pass to band pass transformation started with a normalized specification, the low pass to notch transformation also needs to start with a normalized specification. The pass and stop band frequencies must be normalized to the center frequency by dividing all frequencies by the center frequency. It doesn't matter whether the filter specification is in hertz or radians per second, just divide by the center frequency of the same units and the $2 \cdot \pi$ factors will cancel out.

In order to determine the low pass transfer function to start with, the normalized notch frequencies need to be transformed to the low pass equivalent. Going back to Equation (6-54) and replacing s with $j \cdot \omega$ yields the notch to low pass frequency transformation of Equation (6-55).

$$\frac{j \cdot \omega_N}{\left(j \cdot \omega_N\right)^2 + 1} \Rightarrow j \cdot \omega_{LP}$$

$$\frac{j \cdot \omega_N}{-\omega_N^2 + 1} = j \cdot \frac{\omega_N}{1 - \omega_N^2} \Rightarrow j \cdot \omega_{LP}$$

118

$$\frac{\omega_N}{1-\omega_N^2} \Rightarrow \omega_{LP} \qquad (6\text{-}55)$$

First transform the normalized notch lower pass frequency to the low pass equivalent as shown in Equation (6-56).

$$\frac{\omega_{pl}}{1-\omega_{pl}^2} = \frac{1}{\dfrac{1}{\omega_{pl}} - \omega_{pl}} \Rightarrow \omega_{PLP}$$

But from Equation (6-52) $\dfrac{1}{\omega_{pl}} = \omega_{ph} \left(\omega_0 = 1 = \sqrt{\omega_{ph} \cdot \omega_{pl}} \right)$

$$\frac{1}{\omega_{ph} - \omega_{pl}} = \frac{1}{N} \Rightarrow \omega_{PLP} \qquad (6\text{-}56)$$

Thus, the low pass equivalent pass band frequency is the reciprocal of the notch filter notchwidth. The geometric symmetry of the notch response can be extended to all points on the filter spectrum. That is Equation (6-57) must be satisfied for any pair of frequencies that have the same magnitude value.

$$\omega_0 = \sqrt{\omega_1 \cdot \omega_2}$$

And for the normalized notch function:

$$1 = \sqrt{\omega_1 \cdot \omega_2} = \omega_1 \cdot \omega_2 \text{ or } \omega_1 = \frac{1}{\omega_2} \text{ and } \omega_2 = \frac{1}{\omega_1} \qquad (6\text{-}57)$$

In transforming the stop band frequency back to the low pass equivalent there are two possibilities. Since the stop band specification is determined by a desired performance rather than the geometric symmetry like the pass band specification, either stop frequency could define the highest number of poles. Geometric symmetry can be used to determine which stop frequency to use. If the reciprocal of the lower stop frequency is higher than the upper stop frequency, then use the lower stop frequency for ω_{sl} and the reciprocal of the lower stop frequency for ω_{sh}. If the reciprocal of the lower stop frequency is lower than the upper stop frequency, then use the higher stop frequency for ω_{sh} and the reciprocal of the higher stop frequency for ω_{sl}. Equation (6-58) shows that the low pass equivalent stop band frequency is the reciprocal of the notch geometric stop width.

$$\frac{\omega_{sl}}{1-\omega_{sl}^2} = \frac{1}{\frac{1}{\omega_{sl}} - \omega_{sl}} \Rightarrow \omega_{SLP}$$

But from Equation (6-57) $\frac{1}{\omega_{sl}} = \omega_{sh}$

$$\frac{1}{\omega_{sh}-\omega_{sl}} = \frac{1}{\text{geometric stop width}} \Rightarrow \omega_{SLP} \qquad (6\text{-}58)$$

This defines the low pass equivalent specification. The Butterworth approximation is usually used for the narrow band design since the math is simpler and the narrow notch is smoother with no ripple. The equation for the number of poles for a Butterworth low pass filter (Equation (6-14)) can be used to determine the number of low pass equivalent poles required as shown in Equation (6-59).

$$n = \frac{\log_{10}\left(\frac{10^{(0.1 A_s)}-1}{10^{(0.1 A_p)}-1}\right)}{2\cdot\log_{10}\left(\frac{\omega_s}{\omega_p}\right)}$$

$$= \frac{\log_{10}\left(\frac{10^{(0.1 A_s)}-1}{10^{(0.1 A_p)}-1}\right)}{2\cdot\log_{10}\left(\frac{\frac{1}{\text{geometric stop width}}}{\frac{1}{\text{notchwidth}}}\right)}$$

$$= \frac{\log_{10}\left(\frac{10^{(0.1 A_s)}-1}{10^{(0.1 A_p)}-1}\right)}{2\cdot\log_{10}\left(\frac{\text{notchwidth}}{\text{geometric stop width}}\right)} \qquad (6\text{-}59)$$

Once the number of poles is determined the normalized transfer function for the low pass equivalent can be determined either from Table 6-1 or Equation (6-3). Instead of translating the transfer function to ω_c, It will be translated to 1/N. The transfer function for a real pole translated to 1/N and transformed to a notch transfer function using Equation (6-54) is shown in Equation (6-60).

$$G(s) = \frac{1}{\dfrac{s_{LP}}{\left(\dfrac{1}{N}\right)} + 1} \Bigg|_{s_{LP} = \frac{s}{s^2+1}} = \frac{1}{\left(\dfrac{N \cdot s}{s^2+1}\right) + 1} = \frac{s^2+1}{s^2 + N \cdot s + 1} \qquad (6\text{-}60)$$

Analyzing the bode plot of the notch transfer function of Equation (6-60) yields a plot consisting of a complex second order pole (the denominator) and a complex second order zero (the numerator). The pole has a Q of 1/N and is flat at zero db out to ω equal to 1 and then falls at a -40 db per decade rate. The zero has an infinite Q and is flat at zero db out to ω equal to 1 and then rises at a +40 db per decade rate. The + 40 db per decade rate of the zero cancels out the -40 db per decade rate of the pole yielding a flat response for all frequencies. However the infinite Q of the zero causes a notch at ω equal to 1 which goes to zero (when ω equals 1 the numerator is zero).

The transfer function for a low pass equivalent complex pole is of the form of Equation (6-61).

$$G(s_{LP}) = \frac{1}{s_{LP}^2 + \dfrac{1}{Q_{LP}} \cdot s_{LP} + 1} \qquad (6\text{-}61)$$

Translating this to 1/N and transforming to a notch transfer function using Equation (6-54) yields Equation (6-62).

$$G(s) = \frac{1}{\left(\dfrac{s_{LP}}{\left(\dfrac{1}{N}\right)}\right)^2 + \dfrac{1}{Q_{LP}} \cdot \left(\dfrac{s_{LP}}{\left(\dfrac{1}{N}\right)}\right) + 1} \Bigg|_{s_{LP} = \frac{s}{s^2+1}}$$

$$= \frac{1}{\left(\dfrac{N \cdot s}{s^2+1}\right)^2 + \dfrac{1}{Q_{LP}} \cdot \left(\dfrac{N \cdot s}{s^2+1}\right) + 1}$$

$$= \frac{\left(s^2+1\right)^2}{s^4 + \dfrac{N}{Q_{LP}} \cdot s^3 + \left(2 + N^2\right) \cdot s^2 + \dfrac{N}{Q_{LP}} \cdot s + 1} \qquad (6\text{-}62)$$

Equation (6-62) can be simplified to the product of two second order notch functions, one on either side of the normalized center frequency of 1 as shown in Equation (6-63).

$$G(s) = \frac{s^2 + 1}{\left(s^2 + \dfrac{\omega_b}{Q_b} \cdot s + \omega_b^2\right)} \cdot \frac{s^2 + 1}{\left(s^2 + \dfrac{1}{\omega_b \cdot Q_b} \cdot s + \left(\dfrac{1}{\omega_b}\right)^2\right)} \qquad (6\text{-}63)$$

The geometric symmetry of the notch function creates the ω_b and $1/\omega_b$ relationship between the two second order factors as well as both Q's being equal. Just as for the band pass filter, solving for ω_b and Q_b symbolically results in some very complex equations. The simpler solution is to use actual values before factoring the denominator of Equation (6-63).

Once the normalized transfer function of the notch filter is determined, it can be translated to any desired center frequency by replacing s with s/ω_0 where ω_0 is the desired center frequency.

Example 6-6

Determine the transfer function of a notch filter with a maximum pass band magnitude of 0 db. The notch center frequency is 1000 hertz with a 1500 hertz notchwidth at the -3.0103 db points. There shall be a maximum stop band magnitude of -40 db above 900 hertz and below 1100 hertz. Create a computer generated bode plot of the transfer function and compare it with the filter specification.

First determine the filter Q using Equation (6-53):

$$Q = \frac{\omega_0}{N} = \frac{2 \cdot \pi \cdot 1000}{2 \cdot \pi \cdot 1500} = 0.667$$

Since Q is greater than 0.5, use the narrow band approach.

Next, create the filter specification graph. Since the center frequency and notchwidth are given, they must be normalized and converted to the upper and lower pass frequencies. The normalized center frequency and notchwidth are:

$$\omega_0 = \frac{f_0}{1000} = 1 \qquad N = \frac{1500}{1000} = 1.5$$

The upper and lower pass frequencies can be calculated from Equation (6-52):

$$f_0 = 1000 = \sqrt{f_{ph} \cdot f_{pl}} \Rightarrow f_{pl} = \frac{1000^2}{f_{ph}}$$

$$N_f = 1500 = f_{ph} - f_{pl} = f_{ph} - \frac{1000^2}{f_{ph}}$$

$$f_{ph}^2 - 1500 \cdot f_{ph} - 1000^2 = 0 \Rightarrow f_{ph} = 2000 \quad \text{and} \quad f_{pl} = 500$$

The specification graph is shown in Figure 6-21:

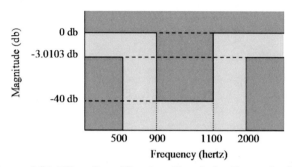

Figure 6-21 Filter Specification Graph for Example 6-6

The next step is to normalize the specifications. The magnitude axis is already normalized since the pass max value is 0 db. To normalize the frequency axis divide all of the specifications by f_0 or 1000. Since this is a frequency divided by a frequency, the normalized axis can be considered either hertz or radians per second. The normalized frequency values are:

$$\omega_{pl} = \frac{500}{1000} = 0.5 \quad \omega_{ph} = \frac{2000}{1000} = 2$$

$$\text{lower stop} = \frac{900}{1000} = 0.9 \quad \text{upper stop} = \frac{1100}{1000} = 1.1$$

The next step is to determine the low pass equivalent. To determine the equivalent low pass number of poles, determine the pass notchwidth N and the geometric stop notchwidth:

$$N = \frac{N_f}{f_0} = \frac{1500}{1000} = 1.5 \quad \text{or} \quad N = \omega_{ph} - \omega_{pl} = 2 - 0.5 = 1.5$$

$1/(\text{lower stop}) = 1/0.9 = 1.1111$ which is greater than the upper stop

Therefore: $\omega_{sl} = 0.9$ and $\omega_{sh} = \dfrac{1}{0.9} = 1.1111$

geometric stop width $= \omega_{sh} - \omega_{sl} = 1.1111 - 0.9 = 0.2111$

Determine the number of equivalent low pass poles using Equation (6-59):

$$n = \frac{\log_{10}\left(\dfrac{10^{(0.1 \cdot A_s)} - 1}{10^{(0.1 \cdot A_p)} - 1}\right)}{2 \cdot \log_{10}\left(\dfrac{\text{notchwidth}}{\text{geometric stop width}}\right)}$$

$$= \frac{\log_{10}\left(\dfrac{10^{(4)} - 1}{10^{(0.30103)} - 1}\right)}{2 \cdot \log_{10}\left(\dfrac{1.5}{0.2111}\right)} = 2.349$$

Rounding up to the next highest integer yields n equal to 3. Using Table 6-1 for the 3 pole entry, the normalized low pass equivalent transfer function is:

$$G(s)_{LP} = \frac{1}{(s+1)} \cdot \frac{1}{\left(s^2 + \dfrac{s}{1} + 1\right)}$$

Translating this to the pass band frequency of 1/N yields:

$$G(s)_{LP} = \frac{1}{(1.5 \cdot s + 1)} \cdot \frac{1}{\left((1.5 \cdot s)^2 + \dfrac{1}{1} \cdot (1.5 \cdot s) + 1\right)}$$

Transforming this to a notch using Equations (6-60) and (6-62) yields:

$$G(s) = \frac{s^2 + 1}{(s^2 + 1.5 \cdot s + 1)} \cdot \frac{(s^2 + 1)^2}{(s^4 + 1.5 \cdot s^3 + 4.25 \cdot s^2 + 1.5 \cdot s + 1)}$$

Factoring the 4th order denominator yields:

$$G(s) = \frac{s^2 + 1}{(s^2 + 1.5 \cdot s + 1)} \cdot$$

$$\frac{s^2 + 1}{(s^2 + 1.173 \cdot s + 3.5877)} \cdot \frac{s^2 + 1}{(s^2 + 0.327 \cdot s + 0..2787)}$$

Translating this out to the specification center frequency of 1000 hertz which is $2000 \cdot \pi$ radians per second yields:

$$G(s) = \dfrac{\left(\dfrac{s}{2000 \cdot \pi}\right)^2 + 1}{\left(\left(\dfrac{s}{2000 \cdot \pi}\right)^2 + 1.5 \cdot \left(\dfrac{s}{2000 \cdot \pi}\right) + 1\right)} \cdot$$

$$\dfrac{\left(\dfrac{s}{2000 \cdot \pi}\right)^2 + 1}{\left(\left(\dfrac{s}{2000 \cdot \pi}\right)^2 + 1.173 \cdot \left(\dfrac{s}{2000 \cdot \pi}\right) + 3.5877\right)} \cdot$$

$$\dfrac{\left(\dfrac{s}{2000 \cdot \pi}\right)^2 + 1}{\left(\left(\dfrac{s}{2000 \cdot \pi}\right)^2 + 0.327 \cdot \left(\dfrac{s}{2000 \cdot \pi}\right) + 0.2787\right)}$$

$$G(s) = \dfrac{s^2 + 3.9478 \cdot 10^7}{\left(s^2 + 9425 \cdot s + 3.9478 \cdot 10^7\right)} \cdot$$

$$\dfrac{s^2 + 3.9478 \cdot 10^7}{\left(s^2 + 7370 \cdot s + 1.4164 \cdot 10^8\right)} \cdot$$

$$\dfrac{s^2 + 3.9478 \cdot 10^7}{\left(s^2 + 2054 \cdot s + 1.1004 \cdot 10^7\right)}$$

Figure 6-22 is a computer generated bode magnitude plot of the filter transfer function with the specification superimposed in red lines.

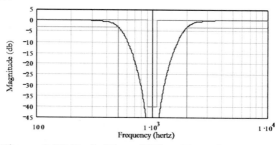

Figure 6-22 Bode Plot of Filter Transfer Function

Note that the filter design meets the pass band specification and exceeds the stop band specification.

The calculations involved in determining the filter transfer functions involve a lot of high level math, especially the band pass and notch filters.

There are several software programs available that can do the calculations. Some are just math software and others are developed specifically for filter design. It is important to understand how and why the filter transfer functions look the way they do and not just rely on software for the answer. Sometimes wrong assumptions are made that result in a software generated filter transfer function that doesn't meet the specification. Having the knowledge of how and why the filter transfer function should look is essential to understanding what went wrong with the software solution.

PROBLEMS

6-1. Determine the transfer function of a Butterworth low pass filter with a maximum pass band magnitude of 0 db, a minimum pass band magnitude of -2 db out to 1000 hertz, and a maximum stop band magnitude of -30 db from 5000 hertz on. Create a specification graph and a computer generated bode plot of the transfer function and compare it with the filter specification.

6-2. Determine the transfer function of a Butterworth low pass filter with a maximum pass band magnitude of 20 db, a minimum pass band magnitude of 19 db out to 2000 hertz, and a maximum stop band magnitude of -20 db from 10000 hertz on. Create a specification graph and a computer generated bode plot of the transfer function and compare it with the filter specification.

6-3. Determine the transfer function of a Chebyshev low pass filter with a maximum pass band magnitude of 0 db, a minimum pass band magnitude of -2 db out to 1000 hertz, and a maximum stop band magnitude of -40 db from 5000 hertz on. Create a specification graph and a computer generated bode plot of the transfer function and compare it with the filter specification.

6-4. Determine the transfer function of a Chebyshev low pass filter with a maximum pass band magnitude of 20 db, a minimum pass band magnitude of 19 db out to 2000 hertz, and a maximum stop band magnitude of -20 db from 5000 hertz on. Create a specification graph and a computer generated bode plot of the transfer function and compare it with the filter specification.

6-5. Determine the transfer function of a Butterworth high pass filter with a maximum pass band magnitude of 0 db, a minimum pass band magnitude of -2 db from 5000 hertz on out, and a maximum stop band magnitude of -30 db below 1000 hertz. Create a specification graph and a computer generated bode plot of the transfer function and compare it with the filter specification.

6-6. Determine the transfer function of a Butterworth high pass filter with a maximum pass band magnitude of 20 db, a minimum pass band magnitude of 19 db from 10000 hertz on out, and a maximum stop band magnitude of -20 db below 2000 hertz. Create a specification graph and a computer generated bode plot of the transfer function and compare it with the filter specification.

6-7. Determine the transfer function of a Chebyshev high pass filter with a maximum pass band magnitude of 0 db, a minimum pass band magnitude of -2 db from 5000 hertz on out, and a maximum stop band magnitude of -40 db below 1000 hertz. Create a specification graph and a computer generated bode plot of the transfer function and compare it with the filter specification.

6-8. Determine the transfer function of a Chebyshev high pass filter with a maximum pass band magnitude of 20 db, a minimum pass band magnitude of 19 db from 5000 hertz on out, and a maximum stop band magnitude of -20 db below 2000 hertz. Create a specification graph and a computer generated bode plot of the transfer function and compare it with the filter specification.

6-9. Determine the transfer function of a band pass filter with a maximum pass band magnitude of 0 db at a center frequency of 1000 hertz and a 250 hertz bandwidth at the -3.0103 db points. There shall be a maximum stop band magnitude of -40 db below 300 hertz and above 4,000 hertz. Create a specification graph and a computer generated bode plot of the transfer function and compare it with the filter specification.

6-10. Determine the transfer function of a band pass filter with a maximum pass band magnitude of 0 db at a center frequency of 3000 hertz and an 8000 hertz bandwidth at the -3.0103 db points. There shall be a maximum stop band magnitude of -40 db below 100 hertz and above 90,000 hertz. Create a specification graph and a computer generated bode plot of the transfer function and compare it with the filter specification.

6-11.   Determine the transfer function of a notch filter with a maximum pass band magnitude of 0 db. The notch center frequency is 2000 hertz with a 2100 hertz notchwidth at the -3.0103 db points. There shall be a maximum stop band magnitude of -40 db above 1900 hertz and below 2100 hertz. Create a specification graph and a computer generated bode plot of the transfer function and compare it with the filter specification.

6-12.   Determine the transfer function of a notch filter with a maximum pass band magnitude of 0 db. The notch center frequency is 1000 hertz with a 10,000 hertz notchwidth at the -3.0103 db points. There shall be a maximum stop band magnitude of -30 db above 700 hertz and below 1500 hertz. Create a specification graph and a computer generated bode plot of the transfer function and compare it with the filter specification.

CHAPTER 7
PASSIVE FILTER REALIZATIONS

7-1. Introduction

In Chapter 6 methods were developed to determine the transfer function that meets a given filter specification. This chapter will extend that process to the design of a passive circuit (a circuit that consists of resistors, inductors, and capacitors) that has the desired transfer function. Thus far the focus has been on analysis of passive circuits in which the transfer function of the circuit is determined. The focus will now shift to the design of a circuit that has a given transfer function.

The circuit design process, which is also referred to as circuit synthesis, is much more difficult than the circuit analysis process since there are an infinite number of component values that will give the same transfer function. A circuit design method will be developed for the low pass filter type and then, as was done in Chapter 6, transformation methods will be developed to convert the design to a high pass, band pass, or notch filter type.

Since resistive impedances are not frequency dependent, the passive filter designs will use primarily inductors and capacitors. Inductive impedances are directly proportional to inductance. Therefore, at low frequencies, large inductance values are needed to maintain realistic impedances.

The inductance is proportional to the physical size of the inductor and the square of the number of turns of wire. In order to keep the inductor size small a large number of turns of small diameter wire will be needed at low frequencies to have large inductance values. Long lengths of small diameter wire exhibit appreciable resistance which make the inductor non ideal. Therefore, passive designs are usually restricted to high frequency applications. Chapter 8 will focus on methods for the design of low frequency filters.

7-2. Low Pass Filter Realization

Low Pass filters can be implemented with a circuit called a ladder circuit. The circuit is named because of the configuration of the circuit elements which resemble a sideways ladder. There are many different ladder circuits, but the one shown in Figure 7-1 is a typical low pass ladder circuit.

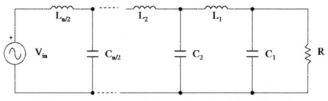

Figure 7-1 Typical Low Pass Ladder Circuit

The source, capacitors, and the resistor are the rungs of the ladder while the inductors are the upper rail. Each inductor and each capacitor correspond to a pole of the transfer function and the resistor represents the load on the filter circuit. The filter output is the voltage across the resistor. The value of n in Figure 7-1 represents the number of poles for the filter realization.

At low frequencies, the inductive impedance is low and the capacitive impedance is high resulting in the source voltage passing on to the output. At high frequencies, the inductive impedance is high and the capacitive impedance is low resulting in the source voltage being blocked from the output. This is the description of a low pass filter.

Since each inductor and each capacitor correspond to a pole of the transfer function, the number of inductors and capacitors equals the number of poles. The circuit of Figure 7-1 has the same number of inductors and capacitors and therefore corresponds to a filter realization with an even number of poles. Figure 7-2 has one more inductor than capacitor and represents a filter realization with an odd number of poles.

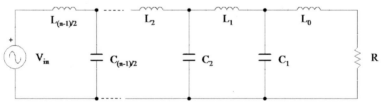

Figure 7-2 Typical Filter with Odd Number of Poles

The design task is to determine the value of the components that will yield the same transfer function for the circuit as the desired filter transfer function. The desired filter transfer function is determined using the methods of Chapter 6. In Chapter 6 the initial transfer function was derived in normalized form first and then transformed to the desired magnitude level and pass frequency.

The normalized form will be used to determine the circuit component values. The circuit will be designed in normalized form with the resistance value equal to 1 and then transformed to the desired frequency and impedance levels. Since this design is for a low pass filter, the numerator of the desired transfer function is 1 and the denominator is determined from Chapter 6 in normalized form. The normalized denominator must also be converted to an unfactored polynomial.

The circuit of either Figure 7-1 or 7-2 can be redrawn in a form that isolates the LC portion of the ladder circuit as shown in Figure 7-3. The transfer function of the LC ladder circuit will be called T(s). The Thévenin equivalent of the source and the LC ladder circuit will have a Thévenin voltage and Thévenin impedance as shown in Equation (7-1).

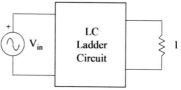

Figure 7-3 General Ladder Circuit

$$V_{TH} = V_{in} \cdot T(s) \quad \text{and} \quad Z_{TH} = \text{Ladder Impedance from R} \tag{7-1}$$

The recombined Thévenin equivalent circuit is shown in Figure 7-4. The output is the voltage across the resistor. The output voltage can be calculated from Figure 7-4 using the voltage divider rule as shown in Equation (7-2).

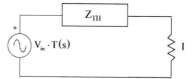

Figure 7-4 Thévenin Circuit

$$V_{out} = V_{in} \cdot T(s) \cdot \frac{1}{1 + Z_{TH}} \tag{7-2}$$

Equation (7-2) can be rearranged to solve for the circuit transfer function (which is the filter transfer function to be implemented) as shown in Equation (7-3).

$$\frac{V_{out}}{V_{in}} = \frac{T(s)}{1 + Z_{TH}} = \frac{1}{\text{Denominator}} \tag{7-3}$$

The denominator (which is in unfactored polynomial form) can be partitioned into terms of even and odd powers of s as shown in Equation (7-4).

$$\frac{V_{out}}{V_{in}} = \frac{T(s)}{1 + Z_{TH}} = \frac{1}{\text{Even} + \text{Odd}} = \frac{\frac{1}{\text{Even}}}{1 + \frac{\text{Odd}}{\text{Even}}} \tag{7-4}$$

From Equation (7-4), it can be determined that Z_{TH} can be represented by the filter transfer function denominator odd power of s terms divided by the filter transfer function denominator even power of s terms. The circuits of Figures 7-1 and 7-2 differ only by the L_0 inductor. The circuit of Figure 7-2 will be used for the filter design and if the number of filter poles is even L_0 will turn out to be 0 which is a short. Referring to Figure 7-2, the Z_{TH} is Z_{L0} (which equals $L_0 \cdot s$) plus the rest of the circuit. The rest of the circuit is Z_{C1} in parallel with the rest of the circuit as shown in Equation (7-5).

$$Zrest = \cfrac{1}{\cfrac{1}{Z_{C1}} + \cfrac{1}{Z_{rest}}} = \cfrac{1}{C_1 \cdot s + \cfrac{1}{Zrest}} \tag{7-5}$$

This process repeats creating the expression of Equation (7-6).

$$Z_{TH} = \frac{Odd}{Even} = L_0 \cdot s + \cfrac{1}{C_1 \cdot s + \cfrac{1}{L_1 \cdot s + \cfrac{1}{C_2 \cdot s + \cfrac{1}{L_2 \cdot s + \cdots}}}} \tag{7-6}$$

Equation (7-6) is a continued fraction expansion of Z_{TH} and will yield the values of the inductors and capacitors that will have the filter transfer function. If the filter transfer function has an odd number of poles, then the odd terms will have a higher power than the even terms and L_0 will be a finite value when the long division of Z_{TH} is performed. If the filter transfer function has an even number of poles, then the odd terms will have a lower power than the even terms and L_0 will be 0 when the long division of Z_{TH} is performed.

Example 7-1

Determine the normalized filter circuit for a 3 pole Butterworth -3.0103 db low pass filter.

From Table 6-1 for a 3 pole filter, the transfer function denominator is:

$$(s+1) \cdot (s^2 + s + 1) = s^3 + 2 \cdot s^2 + 2 \cdot s + 1$$

Then, using Equation (7-6):

$$Z_{TH} = \frac{s^3 + 2 \cdot s}{2 \cdot s^2 + 1} = 0.5 \cdot s + \cfrac{1}{\cfrac{1.5 \cdot s}{2 \cdot s^2 + 1}} \Rightarrow L_0 = 0.5$$

$$\frac{2 \cdot s^2 + 1}{1.5 \cdot s} = 1.3333 \cdot s + \frac{1}{\dfrac{1}{1.5 \cdot s}} \Rightarrow C_1 = 1.3333$$

$$\frac{1.5 \cdot s}{1} = 1.5 \cdot s \Rightarrow L_1 = 1.5$$

The normalized filter circuit is shown in Figure 7-5

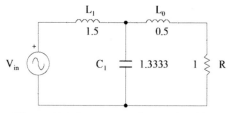

Figure 7-5 Butterworth 3 Pole Filter

The transfer function of Figure 7-5 can be verified by applying the Thévenin equivalent and the voltage divider rule. First, the Thévenin equivalent of V_{in}, L_1, and C_1 is:

$$V_{TH} = \frac{V_{in} \cdot \dfrac{1}{1.3333 \cdot s}}{1.5 \cdot s + \dfrac{1}{1.3333 \cdot s}} = \frac{V_{in}}{2 \cdot s^2 + 1}$$

$$Z_{TH} = \frac{(1.5 \cdot s) \cdot \left(\dfrac{1}{1.3333 \cdot s} \right)}{1.5 \cdot s + \dfrac{1}{1.3333 \cdot s}} = \frac{1.5 \cdot s}{2 \cdot s^2 + 1}$$

V_{out} can then be calculated using the voltage divider rule:

$$V_{out} = V_{TH} \cdot \frac{1}{1 + 0.5 \cdot s + Z_{TH}} = \frac{V_{in}}{2 \cdot s^2 + 1} \cdot \frac{1}{1 + 0.5 \cdot s + \dfrac{1.5 \cdot s}{2 \cdot s^2 + 1}}$$

$$= \frac{V_{in}}{s^3 + 2 \cdot s^2 + 2 \cdot s + 1}$$

The transfer function is then:

$$\frac{V_{out}}{V_{in}} = \frac{1}{s^3 + 2 \cdot s^2 + 2 \cdot s + 1} = \frac{1}{(s+1) \cdot (s^2 + s + 1)}$$

which agrees with Table 6-1.

Example 7-2

Determine the normalized filter circuit for a 4 pole Butterworth -3.0103 db low pass filter.

From Table 6-1 for a 4 pole filter, the transfer function denominator is:

$$\left(s^2 + \frac{s}{1.3066} + 1\right) \cdot \left(s^2 + \frac{s}{0.5412} + 1\right)$$
$$= s^4 + 2.6131 \cdot s^3 + 3.4142 \cdot s^2 + 26131 \cdot s + 1$$

Then, using Equation (7-6):

$$Z_{TH} = \frac{2.6131 \cdot s^3 + 2.6131 \cdot s}{s^4 + 3.4142 \cdot s^2 + 1} = 0 +$$

$$\cfrac{1}{\cfrac{2.6131 \cdot s^3 + 2.6131 \cdot s}{s^4 + 3.4142 \cdot s^2 + 1}} \Rightarrow L_0 = 0$$

$$\cfrac{s^4 + 3.4142 \cdot s^2 + 1}{2.6131 \cdot s^3 + 2.6131 \cdot s} = 0.3827 \cdot s +$$

$$\cfrac{1}{\cfrac{2.4142 \cdot s^2 + 1}{2.6131 \cdot s^3 + 2.6131 \cdot s}} \Rightarrow C_1 = 0.3827$$

$$\cfrac{2.6131 \cdot s^3 + 2.6131 \cdot s}{2.4142 \cdot s^2 + 1} = 1.0824 \cdot s +$$

$$\cfrac{1}{\cfrac{1.5307 \cdot s}{2.4142 \cdot s^2 + 1}} \Rightarrow L_1 = 1.0824$$

$$\frac{2.4142 \cdot s^2 + 1}{1.5307 \cdot s} = 1.5772 \cdot s + \cfrac{1}{\cfrac{1}{1.5307 \cdot s}} \Rightarrow C_2 = 1.5772$$

$$\frac{1.5307 \cdot s}{1} = 1.5307 \cdot s \Rightarrow L_2 = 1.5307$$

The normalized filter circuit is shown in Figure 7-5

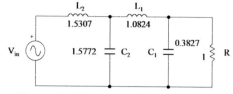

Figure 7-6 Butterworth 4 Pole Filter

This process can be completed for any Butterworth low pass filter with any number of poles. Table 7-1 lists the component values for the first 8 poles.

Table 7-1 Butterworth Low Pass Filter Component Values

n	L_0	C1	L1	C2	L2	C3	L3	C4	L4
1	1.0								
2	0	0.7071	1.4142						
3	0.50	1.3333	1.50						
4	0	0.3827	1.0824	1.5772	1.5307				
5	0.3090	0.8944	1.3820	1.6944	1.5451				
6	0	0.2588	0.7579	1.2016	1.5529	1.7593	1.5529		
7	0.2225	0.6560	1.0550	1.3972	1.6588	1.7988	1.5576		
8	0	0.1951	0.5776	0.9370	1.2588	1.5283	1.7287	1.8246	1.5607

A similar table could be constructed for Chebyshev low pass filters but a separate table would be required for each value of ripple factor because the pole break frequencies are different with each pole and each ripple factor. The Butterworth ripple is constant at -3.0103 db and the filter ripple is achieved by adjusting the cutoff frequency, so only 1 table is needed. The same process used in Equation (7-6) works just as well for Chebyshev transfer functions where the denominator poles are Chebyshev poles rather than Butterworth poles.

It is a good design practice to tighten the design ripple from the specification ripple to allow for component value tolerance variation. Since this would yield design ripple factors that were not round numbers, a reasonable set of Chebyshev tables would not be of much use.

7-3. Frequency and Impedance Translation

The circuits of Figures 7-5 and 7-6 are normalized to a cutoff frequency of 1 radian/sec and a 1 ohm resistor value. The component values need to be translated to the desired cutoff frequency and a realistic resistor value. This process is called frequency and impedance translation. In Chapter 6 the transfer functions were translated to a different frequency by replacing s with s/ω. The passive filter circuits can also be translated to a different frequency by replacing the s in the inductive and capacitive impedances with s/ω as shown in Equation (7-7).

$$Z_L = L \cdot s_n \Big|_{s_n = \frac{s}{\omega}} \Rightarrow \frac{L}{\omega} \cdot s$$

$$Z_C = \frac{1}{C \cdot s_n} \Big|_{s_n = \frac{s}{\omega}} \Rightarrow \frac{1}{\frac{C}{\omega} \cdot s} \tag{7-7}$$

To translate the filter design to the cutoff frequency for a Butterworth filter or the pass frequency for a Chebyshev filter, simply divide the inductor and capacitor values by the desired frequency.

It is also desirable to change the resistor value to something more realistic than 1 ohm. If all of the impedances in a circuit are multiplied by the same factor, the frequency response of the circuit is unchanged. Therefore, what ever factor the resistance is multiplied by (K_R) will also need to be multiplied times the inductive impedance and the capacitive impedance as shown in Equation (7-8).

$$\text{Translated } Z_L = K_R \cdot L \cdot s$$

$$\text{Translated } Z_C = K_R \cdot \frac{1}{C \cdot s} = \frac{1}{\frac{C}{K_R} \cdot s} \tag{7-8}$$

To translate the resistor value to a more realistic value, multiply the resistor and inductor values by K_R and divide the capacitor values by K_R. Equation (7-9) combines Equations (7-7) and (7-8).

$$\text{Translated Resistance} = K_R$$

$$\text{Translated Inductance} = \frac{K_R}{K_\omega} \cdot L \tag{7-9}$$

$$\text{Translated Capacitance} = \frac{C}{K_R \cdot K_\omega}$$

Example 7-3

Design a passive Butterworth low pass filter with a maximum pass band magnitude of 0 db, a minimum pass band magnitude of -2 db out to 1 megahertz, and a maximum stop band magnitude of -40 db from 4 megahertz on. The filter load resistance will be 1000 ohm.

First create the filter specification graph shown in Figure 7-7.

136

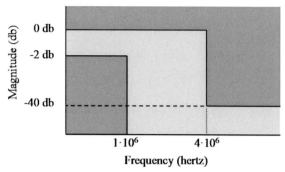

Figure 7-7 Filter Specification Graph for Example 7-3

Note that the frequencies are in hertz and not radians/sec. The frequencies must be converted from hertz to radians/sec:

pass band frequency $= \omega_p = 2 \cdot \pi \cdot 1 \cdot 10^6 = 6.283 \cdot 10^6$

stop band frequency $= \omega_s = 2 \cdot \pi \cdot 4 \cdot 10^6 = 25.130 \cdot 10^6$

The magnitude axis is already normalized to pass max equal to 0 db. In order to allow for component variations, the design minimum pass band magnitude needs to be tightened by 10%. Since the magnitude values are in db, the 10% cannot be applied to the db values. Converting -2 db from db to magnitude yields:

$$-2 \text{ db} \Rightarrow 10^{(-2/20)} = 0.7943$$

Tightening this value by 10% yields 0.8738. Converting this back to db yields -1.172 db.

The pass min value (A_p) is 1.172 db and the stop max value (A_s) is 40 db. Use Equation (6-14) to determine the number of poles:

$$n = \frac{\log_{10}\left(\dfrac{10^{(0.1 \cdot A_s)} - 1}{10^{(0.1 \cdot A_p)} - 1}\right)}{2 \cdot \log_{10}\left(\dfrac{\omega_s}{\omega_p}\right)} = \frac{\log_{10}\left(\dfrac{10^4 - 1}{10^{0.1172} - 1}\right)}{2 \cdot \log_{10}\left(\dfrac{25.130 \cdot 10^6}{6.283 \cdot 10^6}\right)} = 3.745$$

Rounding n up to the next highest integer means that 4 poles will be required. Next use Equation (6-15) to determine the filter cutoff frequency ω_c:

$$\omega_c = \frac{\omega_p}{\left(10^{(0.1 \cdot A_p)} - 1\right)^{1/2 \cdot n}} = \frac{6.283 \cdot 10^6}{\left(10^{0.1172} - 1\right)^{1/8}} = 7.274 \cdot 10^6$$

Using the Table 7-1 for the 4 pole entry yields:

$$C_1 = 0.3827 \quad L_1 = 1.0824 \quad C_2 = 1.5772 \quad L_2 = 1.5307$$

Use Equation (7-9) to translate the component values:

$$\text{Translated Resis} \tan \text{ce} = K_R = 1000$$

$$\text{Translated Induc} \tan \text{ce} = \frac{K_R}{K_\omega} \cdot L = \frac{1000}{7.274 \cdot 10^6} \cdot L$$

$$\text{Translated Capaci} \tan \text{ce} = \frac{C}{K_R \cdot K_\omega} = \frac{C}{1000 \cdot 7.274 \cdot 10^6}$$

The final circuit is shown in Figure 7-8:

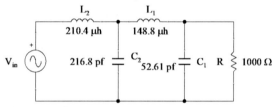

Figure 7-8 Example 7-3 Circuit

The component values on the circuit schematic of Figure 7-8 are actual calculated values. Standard values will need to be selected as close as possible to those values. If necessary, inductors can be implemented with two inductors in series and capacitors can be implemented with two capacitors in parallel (one a large value and one a small value) to get closer to the calculated value. The specification was intentionally tightened to allow for component value variances.

Verification of the filter frequency response can be accomplished using AC circuit analysis methods. However, this method is difficult for higher number of poles. Several computer software circuit analysis programs are available that will make this task much easier.

Figure 7-9 is the frequency response of the circuit of Figure 7-8 using a computer software circuit analysis program. The analysis was run with the calculated component values. The specification is superimposed in red lines. Note that the filter response exceeds both the pass band and stop band specifications.

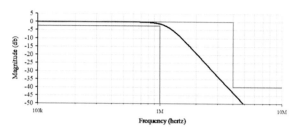

Figure 7-9 Filter Frequency Response

Example 7-4

Design a passive Chebyshev low pass filter with a maximum pass band magnitude of 0 db, a minimum pass band magnitude of -2 db out to 1 megahertz, and a maximum stop band magnitude of -40 db from 4 megahertz on. The filter load resistance will be 1000 ohm.

First create the filter specification graph shown in Figure 7-10.

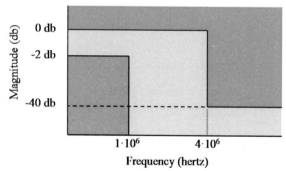

Figure 7-10 Filter Specification Graph for Example 7-4

Note that the frequencies are in hertz and not radians/sec. The frequencies must be converted from hertz to radians/sec:

$$\text{pass band frequency} = \omega_p = 2 \cdot \pi \cdot 1 \cdot 10^6 = 6.283 \cdot 10^6$$

$$\text{stop band frequency} = \omega_s = 2 \cdot \pi \cdot 4 \cdot 10^6 = 25.130 \cdot 10^6$$

The magnitude axis is already normalized to pass max equal to 0 db. In order to allow for component variations, the design minimum pass band magnitude needs to be tightened by 10%.

Since the magnitude values are in db, the 10% cannot be applied to the db values. Converting -2 db from db to magnitude yields:

$$-2 \text{ db} \Rightarrow 10^{(-2/20)} = 0.7943$$

Tightening this value by 10% yields 0.8738. Converting this back to db yields -1.172 db.

The pass min value (A_p) is 1.172 db and the stop max value (A_s) is 40 db. Use Equation (6-28) to determine the number of poles:

$$n = \frac{\cosh^{-1}\left(\sqrt{\dfrac{10^{(0.1 \cdot A_s)} - 1}{10^{(0.1 \cdot A_p)} - 1}}\right)}{\cosh^{-1}\left(\dfrac{\omega_s}{\omega_p}\right)} = \frac{\cosh^{-1}\left(\sqrt{\dfrac{10^4 - 1}{10^{0.1172} - 1}}\right)}{\cosh^{-1}\left(\dfrac{25.13 \cdot 10^6}{6.283 \cdot 10^6}\right)} = 2.852$$

Rounding n up to the next highest integer means that 3 poles will be required. Using Equation (6-19), determine ε:

$$\varepsilon = \sqrt{10^{(0.1 \cdot A_p)} - 1} = \sqrt{10^{0.1172} - 1} = 0.5566$$

Using the Table 6-1 for the 3 pole entry, determine the Butterworth pole values:

$$\alpha_1 = 0.5 \quad \omega_1 = 0.866$$
$$\alpha_2 = 1.0$$

The Chebyshev transformation factors using Equation (6-21) are:

$$K_\alpha = \sinh\left(\frac{1}{n} \cdot \sinh^{-1}\left(\frac{1}{\varepsilon}\right)\right)$$

$$= \sinh\left(\frac{1}{3} \cdot \sinh^{-1}\left(\frac{1}{0.5566}\right)\right) = 0.4649$$

$$K_\omega = \cosh\left(\frac{1}{n} \cdot \sinh^{-1}\left(\frac{1}{\varepsilon}\right)\right)$$

$$= \cosh\left(\frac{1}{3} \cdot \sinh^{-1}\left(\frac{1}{0.5566}\right)\right) = 1.1028$$

The Chebyshev pole values are then:

$$\alpha_1 = 0.5 \cdot 0.4649 = 0.2325$$
$$\omega_1 = 0.866 \cdot 1.1028 = 0.9550$$
$$\alpha_2 = 1.0 \cdot 0.4649 = 0.4649$$

The normalized Chebyshev poles are then:

$$(s + 0.4649) \cdot \left((s + 0.2325)^2 + 0.955^2 \right)$$

Converting this to unfactored polynomial form yields:

$$s^3 + 0.9299 \cdot s^2 + 1.1823 \cdot s + 0.4491$$

Then, using Equation (7-6):

$$Z_{TH} = \frac{s^3 + 1.1823 \cdot s}{0.9299 \cdot s^2 + 0.4491} = 1.0754 \cdot s +$$

$$\frac{1}{\dfrac{0.6993 \cdot s}{0.9299 \cdot s^2 + 0.4491}} \Rightarrow L_0 = 1.0754$$

$$\frac{0.9299 \cdot s^2 + 0.4491}{0.6993 \cdot s} = 1.3298 \cdot s +$$

$$\frac{1}{\dfrac{0.4491}{0.6993 \cdot s}} \Rightarrow C_1 = 1.3298$$

$$\frac{0.6993 \cdot s}{0.4491} = 1.5571 \cdot s \Rightarrow L_1 = 1.5571$$

Use Equation (7-9) to translate the component values:

$$\text{Translated Resistance} = K_R = 1000$$

$$\text{Translated Inductance} = \frac{K_R}{K_\omega} \cdot L = \frac{1000}{6.283 \cdot 10^6} \cdot L$$

$$\text{Translated Capacitance} = \frac{C}{K_R \cdot K_\omega} = \frac{C}{1000 \cdot 6.283 \cdot 10^6}$$

The final circuit is shown in Figure 7-11:

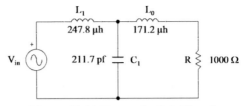

Figure 7-11 Example 7-4 Circuit

The component values on the circuit schematic of Figure 7-8 are actual calculated values. Standard values will need to be selected as close as possible to those values. If necessary, inductors can be implemented with two inductors in series and capacitors can be implemented with two capacitors in parallel (one a large value and one a small value) to get closer to the calculated value. The specification was intentionally tightened to allow for component value variances.

Figure 7-12 is the frequency response of the circuit of Figure 7-11 using a computer software circuit analysis program. The analysis was run with the calculated component values. The specification is superimposed in red lines. Note that the filter response exceeds both the pass band and stop band specifications.

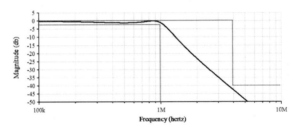

Figure 7-12 Filter Frequency Response

7-4. Low Pass to High Pass Transformation

In Chapter 6 the transformation of a low pass transfer function to a high pass transfer function was accomplished by replacing s with 1/s. Since inductive and capacitive impedances are a function of s, this same process can be used with the low pass filter components. As with the method used in Chapter 6, the substitution will be done on the normalized filter circuit. Equation (7-10) shows the results of the substitution. From Equation (7-10)

142

it can be seen that inductors are replaced by a capacitor with a value equal to $1/L$ and capacitors are replaced by an inductor with a value of $1/C$.

$$L \cdot s_{LP}\big|_{s_{LP}=\frac{1}{s}} = \frac{L}{s} = \frac{1}{\frac{1}{L} \cdot s} \Rightarrow L_{LP} \to C_{HP} = \frac{1}{L_{LP}}$$

$$\frac{1}{C \cdot s_{LP}}\bigg|_{s_{LP}=\frac{1}{s}} = \frac{s}{C} = \frac{1}{C} \cdot s \Rightarrow C_{LP} \to L_{HP} = \frac{1}{C_{LP}}$$

(7-10)

Once the circuit has been transformed using Equation (7-10), it can be translated to the proper frequency and impedance using Equation (7-9).

The design of a high pass filter then consists of designing a normalized low pass filter, using Equation (7-10) to transform the circuit, and then translating the design to the proper frequency and impedance.

Example 7-5

Design a passive Butterworth high pass filter with a maximum pass band magnitude of 0 db, a minimum pass band magnitude of -2 db from 1 megahertz on, and a maximum stop band magnitude of -40 db out to 250 kilohertz. The filter load resistance will be 1000 ohm.

First create the filter specification graph shown in Figure 7-13.

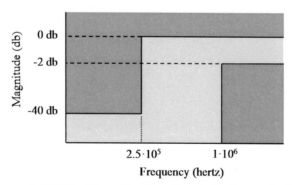

Figure 7-13 Filter Specification Graph for Example 7-5

Note that the frequencies are in hertz and not radians/sec. The frequencies must be converted from hertz to radians/sec:

$$\text{pass band frequency} = \omega_p = 2 \cdot \pi \cdot 1 \cdot 10^6 = 6.283 \cdot 10^6$$

$$\text{stop band frequency} = \omega_s = 2 \cdot \pi \cdot 2.5 \cdot 10^5 = 1.571 \cdot 10^6$$

143

The magnitude axis is already normalized to pass max equal to 0 db. In order to allow for component variations, the design minimum pass band magnitude needs to be tightened by 10%. Since the magnitude values are in db, the 10% cannot be applied to the db values. Converting -2 db from db to magnitude yields:

$$-2 \text{ db} \Rightarrow 10^{(-2/20)} = 0.7943$$

Tightening this value by 10% yields 0.8738. Converting this back to db yields -1.172 db.

The pass min value (A_p) is 1.172 db and the stop max value (A_s) is 40 db. Use Equation (6-31) to determine the number of poles:

$$n = \frac{\log_{10}\left(\dfrac{10^{(0.1 \cdot A_s)} - 1}{10^{(0.1 \cdot A_p)} - 1}\right)}{2 \cdot \log_{10}\left(\dfrac{\omega_p}{\omega_s}\right)} = \frac{\log_{10}\left(\dfrac{10^4 - 1}{10^{0.1172} - 1}\right)}{2 \cdot \log_{10}\left(\dfrac{6.283 \cdot 10^6}{1.571 \cdot 10^6}\right)} = 3.745$$

Rounding n up to the next highest integer means that 4 poles will be required. Next use Equation (6-32) to determine the filter cutoff frequency ω_c:

$$\omega_c = \omega_p \cdot \left(10^{(0.1 \cdot A_p)} - 1\right)^{1/2 \cdot n}$$

$$= 6.283 \cdot 10^6 \cdot \left(10^{(0.1172)} - 1\right)^{1/8} = 5.427 \cdot 10^6$$

Using the Table 7-1 for the 4 pole entry yields:

$$C_{1LP} = 0.3827 \quad L_{1LP} = 1.0824 \quad C_{2LP} = 1.5772 \quad L_{2LP} = 1.5307$$

Use Equation (7-10) to transform the components to a high pass configuration.

$$L_{1HP} = \frac{1}{C_{1LP}} = \frac{1}{0.3827} = 2.613$$

$$L_{2HP} = \frac{1}{C_{2LP}} = \frac{1}{1.5772} = 0.634$$

$$C_{1HP} = \frac{1}{L_{1LP}} = \frac{1}{1.0824} = 0.9239$$

$$C_{2HP} = \frac{1}{L_{2LP}} = \frac{1}{1.5307} = 0.6533$$

Use Equation (7-9) to translate the component values:

Translated Resistance $= K_R = 1000$

$$\text{Translated Inductance} = \frac{K_R}{K_\omega} \cdot L = \frac{1000}{5.427 \cdot 10^6} \cdot L$$

$$\text{Translated Capacitance} = \frac{C}{K_R \cdot K_\omega} = \frac{C}{1000 \cdot 5.427 \cdot 10^6}$$

The final circuit is shown in Figure 7-14:

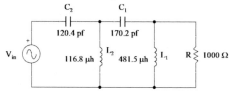

Figure 7-14 Example 7-5 Circuit

The component values on the circuit schematic of Figure 7-14 are actual calculated values. Standard values will need to be selected as close as possible to those values. If necessary, inductors can be implemented with two inductors in series and capacitors can be implemented with two capacitors in parallel (one a large value and one a small value) to get closer to the calculated value. The specification was intentionally tightened to allow for component value variances.

Figure 7-15 is the frequency response of the circuit of Figure 7-14 using a computer software circuit analysis program. The analysis was run with the calculated component values. The specification is superimposed in red lines. Note that the filter response exceeds both the pass band and stop band specifications.

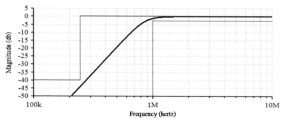

Figure 7-15 Filter Frequency Response

7-5. Low Pass to Band Pass Transformation

In Chapter 6 the transformation of a low pass transfer function to a band pass transfer function was accomplished by replacing s with $s + 1/s$. Prior to performing the transformation the low pass transfer function needed to be translated to the band pass bandwidth, B. Applying this same process to the inductive and capacitive impedances of the low pass filter circuit results in Equation (7-11).

$$\left. L \cdot \frac{s_{LP}}{B} \right|_{s_{LP} = s + \frac{1}{s}} = \frac{L}{B} \cdot s + \frac{1}{\dfrac{B}{L} \cdot s} \Rightarrow L_{LP} \rightarrow L_{BP} \text{ in series with } C_{BP}$$

$$\text{where: } L_{BP} = \frac{L}{B} \text{ and } C_{BP} = \frac{B}{L}$$

$$\left. \frac{1}{C \cdot \dfrac{s_{LP}}{B}} \right|_{s_{LP} = s + \frac{1}{s}} = \frac{1}{\dfrac{1}{\dfrac{C}{B} \cdot s} + \dfrac{1}{\dfrac{B}{C} \cdot s}} \Rightarrow C_{LP} \rightarrow C_{BP} \parallel L_{BP} \qquad (7\text{-}11)$$

$$\text{where: } C_{BP} = \frac{C}{B} \text{ and } L_{BP} = \frac{B}{C}$$

Note that each inductor transforms into an inductor and a capacitor and each capacitor transforms into an inductor and a capacitor. This doubles the number of capacitors and inductors which is required to double the number of poles as was the case for the transfer function transformation.

Once the circuit has been transformed using Equation (7-11), it can be translated to the proper frequency and impedance using Equation (7-9). The design of a band pass filter then consists of designing a normalized low pass filter, using Equation (7-11) to transform the circuit, and then translating the design to the proper frequency and impedance.

Example 7-6

Design a passive Butterworth band pass filter with a maximum pass band magnitude of 0 db at a center frequency of 1 megahertz and a 250 kilohertz bandwidth at the -3.0103 db points. There shall be a maximum stop band magnitude of -40 db below 300 kilohertz and above 4 megahertz. The filter load resistance will be 1000 ohm.

First determine the filter Q using Equation (6-41):

$$Q = \frac{\omega_0}{B} = \frac{2 \cdot \pi \cdot 1 \cdot 10^6}{2 \cdot \pi \cdot 2.5 \cdot 10^5} = 4$$

Since Q is greater than 0.5, use the narrow band approach.

Next, create the filter specification graph. Since the center frequency and bandwidth are given, they must be normalized and converted to the upper and lower pass frequencies. The normalized center frequency and bandwidth are:

$$\omega_0 = \frac{f_0}{1 \cdot 10^6} = 1 \quad B = \frac{2.5 \cdot 10^5}{1 \cdot 10^6} = 0.25$$

The upper and lower pass frequencies can be calculated from Equation (6-40):

$$f_0 = 1 \cdot 10^6 = \sqrt{f_{ph} \cdot f_{pl}} \Rightarrow f_{pl} = \frac{1 \cdot 10^{12}}{f_{ph}}$$

$$B_f = 2.5 \cdot 10^5 = f_{ph} - f_{pl} = f_{ph} - \frac{1 \cdot 10^{12}}{f_{ph}}$$

$$f_{ph}^2 - 2.5 \cdot 10^5 \cdot f_{ph} - 1 \cdot 10^{12} = 0$$

$$\Rightarrow f_{ph} = 1.133 \cdot 10^6 \quad \text{and} \quad f_{pl} = 8.83 \cdot 10^5$$

The specification graph is shown in Figure 7-16:

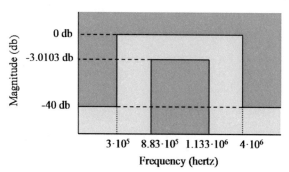

Figure 7-16 Filter Specification Graph for Example 7-6

The next step is to normalize the specifications. The magnitude axis is already normalized since the pass max value is 0 db. In order to allow for component variations, the design bandwidth needs to be increased by 10%. The design bandwidth then becomes 275 kilohertz. The design upper and lower pass frequencies can be calculated from Equation (6-40):

$$f_0 = 1 \cdot 10^6 = \sqrt{f_{ph} \cdot f_{pl}} \Rightarrow f_{pl} = \frac{1 \cdot 10^{12}}{f_{ph}}$$

$$B_f = 2.75 \cdot 10^5 = f_{ph} - f_{pl} = f_{ph} - \frac{1 \cdot 10^{12}}{f_{ph}}$$

$$f_{ph}^2 - 2.75 \cdot 10^5 \cdot f_{ph} - 1 \cdot 10^{12} = 0$$

$$\Rightarrow f_{ph} = 1.1469 \cdot 10^6 \quad \text{and} \quad f_{pl} = 8.719 \cdot 10^5$$

To normalize the frequency axis divide all of the specifications by f_0 or 1 megahertz. Since this is a frequency divided by a frequency, the normalized axis can be considered either hertz or radians per second. The normalized frequency values are:

$$\omega_{pl} = \frac{8.719 \cdot 10^5}{1 \cdot 10^6} = 0.8719 \quad \omega_{ph} = \frac{1.1469 \cdot 10^6}{1 \cdot 10^6} = 1.1469$$

$$\text{lower stop} = \frac{3 \cdot 10^5}{1 \cdot 10^6} = 0.3 \quad \text{upper stop} = \frac{4 \cdot 10^6}{1 \cdot 10^6} = 4$$

The next step is to determine the low pass equivalent. To determine the low pass equivalent number of poles, determine the pass bandwidth B and the geometric stop bandwidth:

$$B = \frac{B_f}{f_0} = \frac{2.75 \cdot 10^5}{1 \cdot 10^6} = 0.275$$

or $B = \omega_{ph} - \omega_{pl} = 1.1469 - 0.8719 = 0.275$

$1/(\text{lower stop}) = 1/0.3 = 3.333$

which is less than the upper stop

Therefore: $\omega_{sl} = 0.3$ and $\omega_{sh} = \frac{1}{0.3} = 3.333$

geometric stop width $= \omega_{sh} - \omega_{sl} = 3.333 - 0.3 = 3.033$

Determine the number of low pass equivalent poles using Equation (6-47):

$$n = \frac{\log_{10}\left(\frac{10^{(0.1 \cdot A_s)} - 1}{10^{(0.1 \cdot A_p)} - 1}\right)}{2 \cdot \log_{10}\left(\frac{\text{geometric stop width}}{\text{bandwidth}}\right)}$$

148

$$n = \frac{\log_{10}\left(\frac{10^{(4)}-1}{10^{(0.30103)}-1}\right)}{2 \cdot \log_{10}\left(\frac{3.033}{0.275}\right)} = 1.918$$

Rounding up to the next highest integer yields n equal to 2. Using the Table 7-1 for the 2 pole entry yields:

$$C_{1LP} = 0.7071 \quad L_{1LP} = 1.4142$$

Use Equation (7-11) to transform the components to a band pass configuration.

$$L_{1BP} = \frac{B}{C_{1LP}} = \frac{0.275}{0.7071} = 0.3889$$

$$C_{1BP} = \frac{C_{1LP}}{B} = \frac{0.7071}{0.275} = 2.5713$$

$$L_{2BP} = \frac{L_{1LP}}{B} = \frac{1.4142}{0.275} = 5.1425$$

$$C_{2BP} = \frac{B}{L_{1LP}} = \frac{0.275}{1.4142} = 0.1945$$

Use Equation (7-9) to translate the component values:

$$\text{Translated Resistance} = K_R = 1000$$

$$\text{Translated Inductance} = \frac{K_R}{K_\omega} \cdot L = \frac{1000}{2 \cdot \pi \cdot 10^6} \cdot L$$

$$\text{Translated Capacitance} = \frac{C}{K_R \cdot K_\omega} = \frac{C}{1000 \cdot 2 \cdot \pi \cdot 10^6}$$

The final circuit is shown in Figure 7-17:

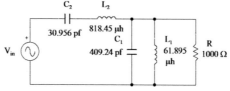

Figure 7-17 Example 7-6 Circuit

The component values on the circuit schematic of Figure 7-17 are actual calculated values. Standard values will need to be selected as close as possible to those values. If necessary, inductors can be implemented with two inductors in series and capacitors can be implemented with two capacitors in parallel (one a large value and one a small value) to get closer to the calculated value. The specification was intentionally tightened to allow for component value variances.

Figure 7-18 is the frequency response of the circuit of Figure 7-17 using a computer software circuit analysis program. The analysis was run with the calculated component values. The specification is superimposed in red lines. Note that the filter response exceeds both the pass band and stop band specifications.

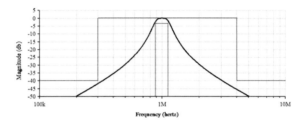

Figure 7-18 Filter Frequency Response

7-6. Low Pass to Notch Transformation

In Chapter 6 the transformation of a low pass transfer function to a notch transfer function was accomplished by replacing s with $1/(s+1/s)$. Prior to performing the transformation the low pass transfer function needed to be translated to the reciprocal of the notchwidth, $1/N$. Applying this same process to the inductive and capacitive impedances of the low pass filter circuit results in Equation (7-12).

$$L \cdot \frac{s_{LP}}{1/N}\bigg|_{s_{LP}=\frac{1}{s+\frac{1}{s}}} = \cfrac{1}{\cfrac{1}{\cfrac{1}{\cfrac{1}{L \cdot N} \cdot s}} + \cfrac{1}{L \cdot N \cdot s}} \Rightarrow L_{LP} \rightarrow C_N \parallel L_N$$

$$\text{where: } L_N = L \cdot N \text{ and } C_N = \frac{1}{L \cdot N}$$

(7-12)

150

$$\frac{1}{C \cdot \dfrac{S_{LP}}{1/N}}\Bigg|_{S_{LP}=\frac{1}{s+\frac{1}{s}}} = \frac{1}{C \cdot N} \cdot s + \frac{1}{C \cdot N \cdot s}$$

$$\Rightarrow C_{LP} \rightarrow L_N \text{ in series with } C_N \qquad (7\text{-}12)$$

$$\text{where: } L_N = \frac{1}{C \cdot N} \text{ and } C_N = C \cdot N$$

Note that each inductor transforms into an inductor and a capacitor and each capacitor transforms into an inductor and a capacitor. This doubles the number of capacitors and inductors which is required to double the number of poles as was the case for the transfer function transformation.

Once the circuit has been transformed using Equation (7-12), it can be translated to the proper frequency and impedance using Equation (7-9). The design of a notch filter then consists of designing a normalized low pass filter, using Equation (7-12) to transform the circuit, and then translating the design to the proper frequency and impedance.

Example 7-7

Design a passive Butterworth notch filter with a maximum pass band magnitude of 0 db. The notch center frequency is 1 megahertz with a 1.5 megahertz notchwidth at the -3.0103 db points. There shall be a maximum stop band magnitude of -40 db above 0.94 megahertz and below 1.05 megahertz. The filter load resistance will be 1000 ohm.

First determine the filter Q using Equation (6-53):

$$Q = \frac{\omega_0}{N} = \frac{2 \cdot \pi \cdot 1 \cdot 10^6}{2 \cdot \pi \cdot 1.5 \cdot 10^6} = 0.667$$

Since Q is greater than 0.5, use the narrow band approach.

Next, create the filter specification graph. Since the center frequency and notchwidth are given, they must be normalized and converted to the upper and lower pass frequencies. The normalized center frequency and notchwidth are:

$$\omega_0 = \frac{f_0}{1 \cdot 10^6} = 1 \qquad N = \frac{1.5 \cdot 10^6}{1 \cdot 10^6} = 1.5$$

The upper and lower pass frequencies can be calculated from Equation (6-52):

$$f_0 = 1 \cdot 10^6 = \sqrt{f_{ph} \cdot f_{pl}} \Rightarrow f_{pl} = \frac{1 \cdot 10^{12}}{f_{ph}}$$

$$N_f = 1.5 \cdot 10^6 = f_{ph} - f_{pl} = f_{ph} - \frac{1 \cdot 10^{12}}{f_{ph}}$$

$$f_{ph}^2 - 1.5 \cdot 10^6 \cdot f_{ph} - 1 \cdot 10^{12} = 0 \Rightarrow f_{ph} = 2 \cdot 10^6 \text{ and } f_{pl} = 0.5 \cdot 10^6$$

The specification graph is shown in Figure 7-19:

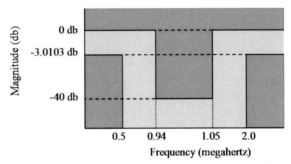

Figure 7-19 Filter Specification Graph for Example 7-7

The next step is to normalize the specifications. The magnitude axis is already normalized since the pass max value is 0 db. In order to allow for component variations, the design notchwidth needs to be decreased by 10%. The design notchwidth then becomes 1.35 megahertz. The design upper and lower pass frequencies can be calculated from Equation (6-52):

$$f_0 = 1 \cdot 10^6 = \sqrt{f_{ph} \cdot f_{pl}} \Rightarrow f_{pl} = \frac{1 \cdot 10^{12}}{f_{ph}}$$

$$N_f = 1.35 \cdot 10^6 = f_{ph} - f_{pl} = f_{ph} - \frac{1 \cdot 10^{12}}{f_{ph}}$$

$$f_{ph}^2 - 1.35 \cdot 10^6 \cdot f_{ph} - 1 \cdot 10^{12} = 0$$

$$\Rightarrow f_{ph} = 1.8815 \cdot 10^6 \quad \text{and} \quad f_{pl} = 0.5315 \cdot 10^6$$

To normalize the frequency axis divide all of the specifications by f_0 or 1 megahertz. Since this is a frequency divided by a frequency, the normalized axis can be considered either hertz or radians per second. The normalized frequency values are:

$$\omega_{pl} = \frac{0.5315 \cdot 10^6}{1 \cdot 10^6} = 0.5315 \quad \omega_{ph} = \frac{1.8815 \cdot 10^6}{1 \cdot 10^6} = 1.8815$$

$$\text{lower stop} = \frac{0.94 \cdot 10^6}{1 \cdot 10^6} = 0.94 \quad \text{upper stop} = \frac{1.05 \cdot 10^6}{1 \cdot 10^6} = 1.05$$

The next step is to determine the low pass equivalent. To determine the equivalent low pass number of poles, determine the pass notchwidth N and the geometric stop notchwidth:

$$N = \frac{N_f}{f_0} = \frac{1.35 \cdot 10^6}{1 \cdot 10^6} = 1.35$$

$$\text{or } N = \omega_{ph} - \omega_{pl} = 1.8815 - 0.5315 = 1.35$$

$1/(\text{lower stop}) = 1/0.94 = 1.0638$ which is greater than

the upper stop

Therefore: $\omega_{sl} = 0.94$ and $\omega_{sh} = \dfrac{1}{0.94} = 1.0638$

geometric stop width $= \omega_{sh} - \omega_{sl} = 1.0638 - 0.94 = 0.1238$

Determine the number of equivalent low pass poles using Equation (6-59):

$$n = \frac{\log_{10}\left(\dfrac{10^{(0.1 \cdot A_s)} - 1}{10^{(0.1 \cdot A_p)} - 1}\right)}{2 \cdot \log_{10}\left(\dfrac{\text{notchwidth}}{\text{geometric stop width}}\right)}$$

$$= \frac{\log_{10}\left(\dfrac{10^{(4)} - 1}{10^{(0.30103)} - 1}\right)}{2 \cdot \log_{10}\left(\dfrac{1.35}{0.1238}\right)} = 1.927$$

Rounding up to the next highest integer yields n equal to 2. Using the Table 7-1 for the 2 pole entry yields:

$$C_{1LP} = 0.7071 \quad L_{1LP} = 1.4142$$

Use Equation (7-12) to transform the components to a notch configuration.

$$L_{1N} = \frac{1}{C_{1LP} \cdot N} = \frac{1}{0.7071 \cdot 1.35} = 1.0476$$

$$C_{1N} = C_{1LP} \cdot N = 0.7071 \cdot 1.35 = 0.9546$$

$$L_{2N} = L_{1LP} \cdot N = 1.4142 \cdot 1.35 = 1.9092$$

$$C_{2N} = \frac{1}{L_{1LP} \cdot N} = \frac{1}{1.4142 \cdot 1.35} = 0.5238$$

Use Equation (7-9) to translate the component values:

$$\text{Translated Resis tan ce} = K_R = 1000$$

$$\text{Translated Induc tan ce} = \frac{K_R}{K_\omega} \cdot L = \frac{1000}{2 \cdot \pi \cdot 10^6} \cdot L$$

$$\text{Translated Capaci tan ce} = \frac{C}{K_R \cdot K_\omega} = \frac{C}{1000 \cdot 2 \cdot \pi \cdot 10^6}$$

The final circuit is shown in Figure 7-20:

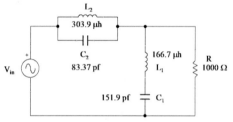

Figure 7-20 Example 7-7 Circuit

The component values on the circuit schematic of Figure 7-20 are actual calculated values. Standard values will need to be selected as close as possible to those values. If necessary, inductors can be implemented with two inductors in series and capacitors can be implemented with two capacitors in parallel (one a large value and one a small value) to get closer to the calculated value. The specification was intentionally tightened to allow for component value variances.

Figure 7-21 is the frequency response of the circuit of Figure 7-20 using a computer software circuit analysis program. The analysis was run with the calculated component values. The specification is superimposed in red lines. Note that the filter response exceeds both the pass band and stop band specifications.

154

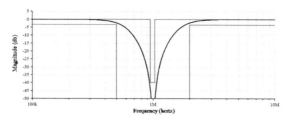

Figure 7-21 Filter Frequency Response

PROBLEMS

7-1. Design a passive Butterworth low pass filter with a maximum pass band magnitude of 0 db, a minimum pass band magnitude of -2 db out to 1 megahertz, and a maximum stop band magnitude of -30 db from 5 megahertz on. The filter load resistance will be 1000 ohm.

7-2. Design a passive Chebyshev low pass filter with a maximum pass band magnitude of 0 db, a minimum pass band magnitude of -2 db out to 1 megahertz, and a maximum stop band magnitude of -40 db from 3 megahertz on. The filter load resistance will be 1000 ohm.

7-3. Design a passive Butterworth high pass filter with a maximum pass band magnitude of 0 db, a minimum pass band magnitude of -1 db from 1 megahertz on, and a maximum stop band magnitude of -40 db out to 200 kilohertz. The filter load resistance will be 1000 ohm.

7-4. Design a passive Chebyshev high pass filter with a maximum pass band magnitude of 0 db, a minimum pass band magnitude of -2 db from 1 megahertz on, and a maximum stop band magnitude of -40 db out to 200 kilohertz. The filter load resistance will be 1000 ohm.

7-5. Design a passive Butterworth band pass filter with a maximum pass band magnitude of 0 db at a center frequency of 1 megahertz and a 200 kilohertz bandwidth at the -3.0103 db points. There shall be a maximum stop band magnitude of -40 db below 500 kilohertz and above 2.5 megahertz. The filter load resistance will be 1000 ohm.

7-6. Design a passive Butterworth notch filter with a maximum pass band magnitude of 0 db. The notch center frequency is 1 megahertz with a 1.1 megahertz notchwidth at the -3.0103 db points. There shall be a maximum stop band magnitude of -40 db above 0.95 megahertz and below 1.04 megahertz. The filter load resistance will be 1000 ohm.

CHAPTER 8
ACTIVE FILTER REALIZATIONS

8-1. Introduction

Chapter 7 focused on implementing filter designs using only passive circuit elements (resistors, capacitors, and inductors). One of the problems with passive designs was the significant resistance component of inductors at low frequencies which limited the designs to high frequency applications. This chapter will focus on what is referred to as active realizations which incorporate an amplifier (usually an operational amplifier). The addition of the amplifier allows the filter design to be implemented using only resistors, capacitors, and amplifiers. The elimination of the inductor allows for low frequency filter designs but limits the high frequency design applications due to the bandwidth of the amplifier.

Another advantage of active realizations is that the filter design can be broken down into stages which make the design much simpler. The amplifier low output impedance makes this possible since the load on each amplifier stage has no effect on previous stages. Passive realizations must be designed for the entire filter since each stage of the ladder circuit loads the previous stages. This was evidenced in Chapter 7 where the design required transforming the transfer function developed in Chapter 6 from factored form to un-factored form. The active designs will work directly with the factored form of the transfer function developed in Chapter 6.

8-2. The Operational Amplifier

The operational amplifier (or op-amp) is a high voltage gain differential amplifier. The schematic symbol for the op-amp is shown in Figure 8-1. There are actually two more functions that are typically not shown since they are 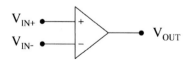 not involved in a circuit analysis. They are the DC power inputs – one a positive DC voltage and the other a negative DC voltage. Since energy must be conserved, the op-amp uses energy from the DC power sources

Figure 8-1 Op-Amp Symbol

157

rather than the input signal to increase the input voltage and/or current levels as required by the circuit load.

The equation for the op-amp is shown in Equation (8-1).

$$V_{OUT} = A \cdot (V_{IN+} - V_{IN-})$$ (8-1)

A is the gain of the op-amp. Since the op-amp has a high gain (typically 200,000 to 300,000), it has few direct applications. The op-amp circuit usually employs feedback to make it more useful. Figure 8-2 shows a simple non-inverting amplifier application that uses feedback.

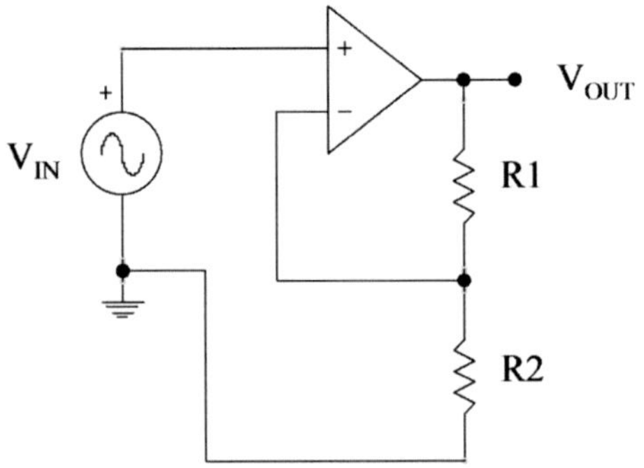

Figure 8-2 Non-Inverting Amplifier

Apply Equation (8-1) to the circuit of Figure 8-2. V_{IN+} is equal to V_{IN} and V_{IN-} (the feedback voltage) is determined using the voltage divider rule as shown in Equation (8-2).

$$V_{IN-} = V_{OUT} \cdot \frac{R2}{R1 + R2}$$ (8-2)

Applying Equation (8-1) yields Equation (8-3).

$$V_{OUT} = A \cdot (V_{IN+} - V_{IN-}) = A \cdot \left(V_{IN} - V_{OUT} \cdot \frac{R2}{R1 + R2} \right)$$

$$V_{OUT} + A \cdot V_{OUT} \cdot \frac{R2}{R1 + R2} = A \cdot V_{IN}$$

$$V_{OUT} \cdot \left(1 + A \cdot \frac{R2}{R1 + R2} \right) = A \cdot V_{IN}$$

$$V_{OUT} = V_{IN} \cdot \left(\frac{A}{1 + A \cdot \dfrac{R2}{R1 + R2}} \right) = V_{IN} \cdot \left(\frac{1}{\dfrac{1}{A} + \dfrac{1}{1 + \dfrac{R1}{R2}}} \right)$$ (8-3)

For A much much greater than 1 and much much greater than the circuit gain, Equation (8-3) reduces to ($V_{OUT} = V_{IN}(1+R1/R2)$). In other words the gain of the circuit is determined by R1 and R2 and not the op-amp gain.

The op-amp gain is not the only specification that must be considered. An op-amp has input impedance, output impedance, offset voltage and current, and a limited frequency response. The availability of field effect transistor (FET) input operational amplifiers has virtually eliminated the effects of input impedance (10^{12} ohms) and input and offset current (10^{-11} amps). The output impedance for any op-amp will be made negligible as long as the circuit is operating within the frequency response of the operational amplifier. Thus, the only other parameters to be concerned about are frequency response and offset voltage.

The offset voltage at the circuit output is a function of circuit gain. The offset voltage specification is typically a few millivolts but when translated to the output it must be multiplied by the circuit gain. If output offset voltage becomes a problem, it can be calibrated out with a DC bias network.

The frequency response of most op-amps can be characterized as a two-pole transfer function with the first break at a relatively low frequency value and the second break at a much higher frequency. The gain of the op-amp will be the magnitude of the frequency response at DC. These values are shown in the frequency response plot of Figure 8-3.

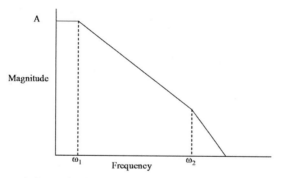

Figure 8-3 Typical Op-Amp Frequency Specification

A is the op-amp gain, ω_1 is the value of the first break frequency and ω_2 is the value of the second break frequency. For a typical op-amp, A is 200,000, ω_1 is 300 rad/sec, and ω_2 is $3 \cdot 10^8$ rad/sec. For those typical values the transfer function for the op-amp gain is shown in Equation (8-4).

$$\text{Gain} = \frac{A}{\left(\dfrac{s}{\omega_1}+1\right)\cdot\left(\dfrac{s}{\omega_2}+1\right)} = \frac{2 \cdot 10^5}{\left(\dfrac{s}{300}+1\right)\cdot\left(\dfrac{s}{3\cdot 10^8}+1\right)} \qquad (8\text{-}4)$$

Adding this transfer function to a filter circuit greatly complicates the math. Therefore, the calculations are typically done using an ideal op-amp with infinite gain, infinite frequency response, infinite input impedance, zero input current, and zero offset voltage. If the ideal op-amp circuit analysis is not adequate, circuit simulation software can be used to determine the filter response as was done for passive filter realizations.

The op-amp frequency response needs to be considered during the design of the filter circuit. The op-amp gain needs to be at least 25 db above the circuit gain at any frequency. When designing a low pass or band pass filter, as long as the filter response falls within the op-amp response by 25 db, the op-amp response should have negligible effect. When designing a high pass or notch filter, the op-amp is going to act as a cascaded low pass filter at the high frequency response limit of the op-amp. This essentially makes the high pass filter a wide band band pass filter and the notch filter a low pass filter with a notch.

8-3. Low Pass Filter Realization

In Chapter 6 the transfer function for a low pass filter was developed in factored form. When the number of filter poles (n) is even, all of the factors are second order. When the number of filter poles (n) is odd, a single first order factor plus second order factors are needed. The active realizations will be implemented by designing an op-amp circuit for each factor and then cascading the circuits (the input is applied to the first circuit whose output is the input to the second circuit, etc.). A design will be needed for a first order factor (single pole design) and one for a second order factor (two pole design).

The non-inverting amplifier circuit of Figure 8-2 can be modified as shown in Figure 8-4 to implement the design of a one pole low pass filter circuit.

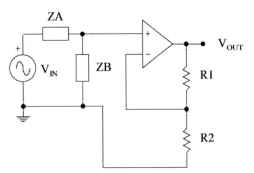

Figure 8-4 One Pole Filter Stage

ZA and ZB can be either a resistor or a capacitor. Since the ideal non-inverting amplifier gain is (1+R1/R2) times the voltage at the + input of the op-amp, the gain of the circuit of Figure 8-4 can be determined using the voltage divider rule as shown in Equation (8-5). The transfer function of the circuit of Figure 8-5 is determined by rearranging Equation (8-5) as shown in Equation (8-6).

160

$$V_{OUT} = V_{IN} \cdot \frac{ZB}{ZA + ZB} \cdot \left(1 + \frac{R1}{R2}\right) \qquad (8\text{-}5)$$

$$\frac{V_{OUT}}{V_{IN}} = \frac{1 + \dfrac{R1}{R2}}{\dfrac{ZA}{ZB} + 1} \qquad (8\text{-}6)$$

From Chapter 6, a first order pole factor will be of the form shown in Equation (8-7).

$$\frac{V_{OUT}}{V_{IN}} = \frac{K}{\dfrac{s}{\omega_c} + 1} \qquad (8\text{-}7)$$

The K is the gain factor which for passive realizations was always 1 since other values would require an amplifier. For Equation (8-6) K is equal to $(1+R1/R2)$. The design task then becomes the selection of ZA and ZB such that the denominator of Equation (8-6) matches the denominator of Equation (8-7). Choosing ZA to be a resistor RA and ZB to be a capacitor $1/(s \cdot CB)$ will accomplish this as shown in Equation (8-8).

$$\frac{V_{OUT}}{V_{IN}} = \frac{1 + \dfrac{R1}{R2}}{\dfrac{RA}{\dfrac{1}{s \cdot CB}} + 1} = \frac{1 + \dfrac{R1}{R2}}{s \cdot RA \cdot CB + 1} = \frac{1 + \dfrac{R1}{R2}}{\dfrac{s}{\dfrac{1}{RA \cdot CB}} + 1} \qquad (8\text{-}8)$$

Comparing Equations (8-7) and (8-8) yields, K equal to $(1+R1/R2)$ and ω_c equal to $1/(RA \cdot CB)$. Figure 8-5 is the circuit for a one pole low pass filter stage.

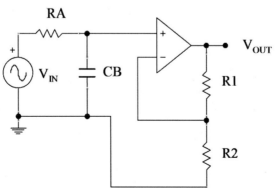

Figure 8-5 One Pole Low Pass Filter Stage

The design procedure for a one pole low pass stage is to first select a standard value for R2 and then determine R1 using Equation (8-9).

$$K = 1 + \frac{R1}{R2} \Rightarrow R1 = R2 \cdot (K - 1) \tag{8-9}$$

Next, select a standard value for CB and then determine RA from Equation (8-10).

$$\omega_c = \frac{1}{RA \cdot CB} \Rightarrow RA = \frac{1}{CB \cdot \omega_c} \tag{8-10}$$

For the special case where K is 1, R1 is zero or a short and R2 is infinite or an open. If the calculated component values are not reasonable, reselect either R2 or CB and recalculate until reasonable values are obtained.

The non-inverting amplifier circuit of Figure 8-2 can be modified as shown in Figure 8-6 to implement the design of a two pole low pass filter circuit. This circuit is also referred to as the Sallen-Key filter circuit, named after R. P. Sallen and E. L. Key of MIT Lincoln Laboratory who introduced the circuit in 1955 (their circuit had unity gain feedback).

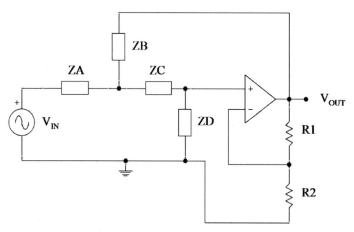

Figure 8-6 Two Pole Filter Stage

In order to simplify the analysis of the circuit of Figure 8-6, the amplifier gain will be defined as K. The amplifier gain of the circuit of Figure 8-6 is the same as the amplifier gain of the circuit of Figure 8-5 which is shown in Equation (8-11).

$$K = 1 + \frac{R1}{R2} \tag{8-11}$$

Analysis of the circuit of Figure 8-6 will require establishing branch currents and node voltages. Figure 8-7 shows the branch currents and node voltages.

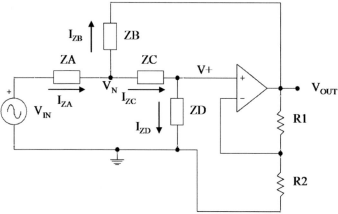

Figure 8-7 Two pole Stage with Branch Currents and Node Voltages

Summing the branch currents at node V_N yields Equation (8-12).

$$I_{ZA} = I_{ZB} + I_{ZC} \tag{8-12}$$

Equation (8-13) is Equation (8-12) converted to node voltages.

$$\frac{V_{IN} - V_N}{ZA} = \frac{V_N - V_{OUT}}{ZB} + \frac{V_N - V+}{ZC} \tag{8-13}$$

The amplifier output can also be expressed as the gain times V+ which can be solved for V+ as shown in Equation (8-14).

$$V_{OUT} = K \cdot V+ \quad \Rightarrow \quad V+ = \frac{V_{OUT}}{K} \tag{8-14}$$

Summing the branch currents at node V+ and converting to node voltages yields Equation (8-15).

$$I_{ZC} = I_{ZD}$$
$$\frac{V_N - V+}{ZC} = \frac{V+}{ZD} \tag{8-15}$$

Substituting Equation (8-14) into Equation (8-15) and solving for V_N yields Equation (8-16).

$$\frac{V_N - \dfrac{V_{OUT}}{K}}{ZC} = \frac{\dfrac{V_{OUT}}{K}}{ZD}$$

$$\frac{V_N}{ZC} = \frac{\frac{V_{OUT}}{K}}{ZD} + \frac{\frac{V_{OUT}}{K}}{ZC}$$

$$V_N = \frac{V_{OUT}}{K} \cdot \left(\frac{ZC}{ZD} + 1\right) \tag{8-16}$$

Substituting Equations (8-14) and (8-16) into Equation (8-13) yields Equation (8-17).

$$\frac{V_{IN} - \frac{V_{OUT}}{K} \cdot \left(\frac{ZC}{ZD} + 1\right)}{ZA} = \frac{\frac{V_{OUT}}{K} \cdot \left(\frac{ZC}{ZD} + 1\right) - V_{OUT}}{ZB} + \frac{\frac{V_{OUT}}{K} \cdot \left(\frac{ZC}{ZD} + 1\right) - \frac{V_{OUT}}{K}}{ZC} \tag{8-17}$$

Simplifying Equation (8-17) yields Equation (8-18).

$$\frac{V_{IN}}{ZA} = \frac{V_{OUT} \cdot ZC}{K \cdot ZA \cdot ZD} + \frac{V_{OUT}}{K \cdot ZA} + \frac{V_{OUT} \cdot ZC}{K \cdot ZB \cdot ZD} + \frac{V_{OUT}}{K \cdot ZB} - \frac{V_{OUT}}{ZB} + \frac{V_{OUT}}{K \cdot ZD} + \frac{V_{OUT}}{K \cdot ZC} - \frac{V_{OUT}}{K \cdot ZC} \tag{8-18}$$

Simplifying Equation (8-18) yields Equation (8-19).

$$V_{IN} = \frac{V_{OUT}}{K} \cdot \left(\frac{ZC}{ZD} + 1 + \frac{ZA \cdot ZC}{ZB \cdot ZD} + \frac{ZA}{ZB} - \frac{K \cdot ZA}{ZB} + \frac{ZA}{ZD}\right) \tag{8-19}$$

Solving Equation (819) for V_{OUT}/V_{IN} (the circuit transfer function) yields Equation (8-20).

$$\frac{V_{OUT}}{V_{IN}} = \frac{K}{\frac{ZA \cdot ZC}{ZB \cdot ZD} + \frac{ZA}{ZD} + \frac{ZC}{ZD} + \frac{ZA \cdot (1 - K)}{ZB} + 1} \tag{8-20}$$

From Chapter 6, a second order pole factor will be of the form shown in Equation (8-21).

$$\frac{V_{OUT}}{V_{IN}} = \frac{K}{\frac{s^2}{\omega_c^2} + \frac{1}{Q} \cdot \frac{s}{\omega_c} + 1} \tag{8-21}$$

For Equation (8-20) to be of this form, ZA and ZC must be resistors RA and RC and ZB and ZD must be capacitors $1/(s \cdot CB)$ and $1/(s \cdot CD)$. Equation (8-22) is Equation (8-20) with those substitutions made.

$$\frac{V_{OUT}}{V_{IN}} = \frac{K}{(RA \cdot RC \cdot CB \cdot CD) \cdot s^2 + (RA \cdot CD + RC \cdot CD + RA \cdot CB \cdot (1-K)) \cdot s + 1} \quad (8\text{-}22)$$

Equating like coefficients of the denominators of Equations (8-21) and (8-22) yields Equations (8-23) and (8-24).

$$RA \cdot RC \cdot CB \cdot CD = \frac{1}{\omega_c^2} \quad (8\text{-}23)$$

$$RA \cdot CD + RC \cdot CD + RA \cdot CB \cdot (1-K) = \frac{1}{\omega_c \cdot Q} \quad (8\text{-}24)$$

Since there are only two equations and four unknowns, standard values will be selected for CB and CD and the equations solved for RA and RC. Equation (8-23) can be solved for RA as shown in Equation (8-25).

$$RA = \frac{1}{RC \cdot CB \cdot CD \cdot \omega_c^2} \quad (8\text{-}25)$$

Substituting Equation (8-25) into Equation (8-24) yields Equation (8-26).

$$\frac{1}{RC \cdot CB \cdot \omega_c^2} + RC \cdot CD + \frac{1-K}{RC \cdot CD \cdot \omega_c^2} = \frac{1}{\omega_c \cdot Q} \quad (8\text{-}26)$$

Multiplying both sides of Equation (8-26) by RC and then solving for RC yields Equation (8-27).

$$\frac{1}{CB \cdot \omega_c^2} + RC^2 \cdot CD + \frac{1-K}{CD \cdot \omega_c^2} = \frac{RC}{\omega_c \cdot Q}$$

$$RC^2 \cdot CD - \frac{RC}{\omega_c \cdot Q} + \frac{1}{CB \cdot \omega_c^2} + \frac{1-K}{CD \cdot \omega_c^2} = 0$$

$$RC = \frac{\dfrac{1}{\omega_c \cdot Q} + \sqrt{\dfrac{1}{\omega_c^2 \cdot Q^2} - 4 \cdot CD \cdot \left(\dfrac{1}{CB \cdot \omega_c^2} + \dfrac{1-K}{CD \cdot \omega_c^2}\right)}}{2 \cdot CD}$$

$$RC = \frac{1}{2 \cdot \omega_c \cdot Q \cdot CD} + \frac{1}{\omega_c} \cdot \sqrt{\frac{1}{4 \cdot Q^2 \cdot CD^2} - \frac{1}{CD} \cdot \left(\frac{1}{CB} + \frac{1-K}{CD}\right)} \quad (8\text{-}27)$$

Only the positive sign for the radical was used since RC must be a positive real number. Also the portion of the equation under the radical must be positive for RC to be real and not imaginary. This places a constraint on the selection of CB and CD. Equation (8-28) establishes the maximum value for CD for the given CB selection.

$$\frac{1}{4 \cdot Q^2 \cdot CD_{max}^2} = \frac{1}{CD_{max}} \cdot \left(\frac{1}{CB} + \frac{1-K}{CD_{max}} \right)$$

$$\frac{1}{4 \cdot Q^2 \cdot CD_{max}} = \left(\frac{1}{CB} + \frac{1-K}{CD_{max}} \right)$$

$$1 = 4 \cdot Q^2 \cdot CD_{max} \cdot \left(\frac{1}{CB} + \frac{1-K}{CD_{max}} \right)$$

$$1 = 4 \cdot Q^2 \cdot \frac{CD_{max}}{CB} + 4 \cdot Q^2 \cdot (1-K)$$

$$4 \cdot Q^2 \cdot \frac{CD_{max}}{CB} = 1 - 4 \cdot Q^2 \cdot (1-K)$$

$$\frac{CD_{max}}{CB} = \frac{1}{4 \cdot Q^2} + K - 1$$

$$CD_{max} = CB \cdot \left(\frac{1}{4 \cdot Q^2} + K - 1 \right) \tag{8-28}$$

Figure 8-8 is the circuit for a two pole low pass filter stage.

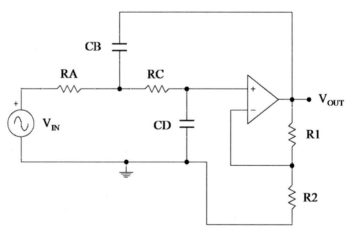

Figure 8-8 Two Pole Low Pass Filter Stage

The design procedure starts by determining the filter transfer function in factored form using the methods of Chapter 6. The transfer function can then be partitioned into first and second order stages. If the number of poles is odd there will be one first order stage and the rest second order stages. The first order stage can be implemented using the one pole filter stage of Figure 8-5 and the second order stages can be implemented using the two pole filter stage of Figure 8-8.

Next, select standard capacitor values and calculate the resistor values. If the resistor values are less than desirable, select another capacitor standard value and recalculate the resistor values. Repeat this process until acceptable resistor values are obtained. Since the two pole filter calculations are rather lengthy, an EXCEL spreadsheet or math software can be used to speed the process and minimize errors.

The filter stages can then be cascaded to form the final filter design. In order to keep the individual filter stages from saturating, always place the one pole circuit closest to the input and follow in the cascading operation with the two pole circuits in order of increasing Q.

If the filter gain is greater than unity (0 dB), the gain will need to be distributed among the number of stages used to implement the filter. This is best accomplished by distributing the gain based on the Q of each stage (to keep the individual filter stages from saturating) with the lower Q stages having a higher gain than the higher Q stages. Since the value of Q determines the amount of peaking, Q values less than 1 should be defaulted to 1. Solution of Equations (8-29) will define the gain for each stage.

$$K_1 \cdot K_2 \cdot K_3 \cdots K_n = K$$
$$K_1 \cdot Q_1 = K_2 \cdot Q_2 = K_3 \cdot Q_3 \cdots = K_n \cdot Q_n \qquad (8-29)$$

Example 8-1

Design an active Butterworth low pass filter with a maximum pass band magnitude of 0 db, a minimum pass band magnitude of -2 db out to 1000 hertz, and a maximum stop band magnitude of -40 db from 4000 hertz on.

First create the filter specification graph shown in Figure 8-9.

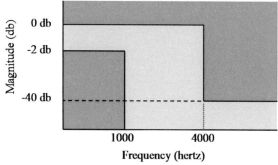

Figure 8-9 Filter Specification Graph for Example 8-1

167

Note that the frequencies are in hertz and not radians/sec. This is typical in filter designs and the frequencies must be converted from hertz to radians/sec to determine the filter transfer function:

$$\text{pass band frequency} = \omega_p = 2 \cdot \pi \cdot 1000 = 6,283$$

$$\text{stop band frequency} = \omega_s = 2 \cdot \pi \cdot 4000 = 25,130$$

The magnitude axis is already normalized to pass max equal to 0 db. In order to allow for component variations, the design minimum pass band magnitude needs to be tightened by 10%. Since the magnitude values are in db, the 10% cannot be applied to the db values. Converting -2 db from db to magnitude yields:

$$-2 \text{ db} \Rightarrow 10^{(-2/20)} = 0.7943$$

Tightening this value by 10% yields 0.8738. Converting this back to db yields -1.172 db.

The pass min value (A_p) is 1.172 db and the stop max value (A_s) is 40 db. Use Equation (6-14) to determine the number of poles:

$$n = \frac{\log_{10}\left(\dfrac{10^{(0.1 \cdot A_s)} - 1}{10^{(0.1 \cdot A_p)} - 1}\right)}{2 \cdot \log_{10}\left(\dfrac{\omega_s}{\omega_p}\right)} = \frac{\log_{10}\left(\dfrac{10^4 - 1}{10^{0.1172} - 1}\right)}{2 \cdot \log_{10}\left(\dfrac{25,130}{6,283}\right)} = 3.745$$

Rounding n up to the next highest integer means that 4 poles will be required. Next use Equation (6-15) to determine the filter cutoff frequency ω_c:

$$\omega_c = \frac{\omega_p}{\left(10^{(0.1 \cdot A_p)} - 1\right)^{1/2 \cdot n}} = \frac{6,283}{\left(10^{0.1172} - 1\right)^{1/8}} = 7,274$$

Using the Table 6-1 for the 4 pole entry and ω_c from above determine the filter transfer function:

$$G(s) = \cfrac{1}{\left(\left(\cfrac{s}{7{,}274}\right)^2 + \cfrac{1}{0.5412}\cdot\left(\cfrac{s}{7{,}274}\right)+1\right)} \cdot$$

$$\cfrac{1}{\left(\left(\cfrac{s}{7{,}274}\right)^2 + \cfrac{1}{1.3066}\cdot\left(\cfrac{s}{7{,}274}\right)+1\right)}$$

This is a 4 pole filter that can be implemented as an active filter with 2 two pole stages of the type in Figure 8-8. The value of ω_c for both stages is 7,274 rad/sec and the Q for the first stage (the input stage) is 0.5412 and the Q for the second stage (the output stage) is 1.3066.

The stage 1 design has K_1 equal to 1 which makes $R1_1$ a short and $R2_1$ an open. Select CB_1 equal to 0.068 microfarads and use Equation (8-28) To determine CD_{1max} :

$$CD_{1max} = CB_1 \cdot \left(\frac{1}{4\cdot Q_1^2}+K_1-1\right) = 0.058\cdot 10^{-6}$$

Select CD_1 equal to 0.01 microfarads. Use Equation (8-27) to calculate RC_1:

$$RC_1 = \frac{1}{2\cdot\omega_c\cdot Q_1\cdot CD_1}+\frac{1}{\omega_c}\cdot\sqrt{\frac{1}{4\cdot Q_1^2\cdot CD_1^2}-\frac{1}{CD_1}\cdot\left(\frac{1}{CB_1}+\frac{1-K_1}{CD_1}\right)}$$

$$RC_1 = 24.256 \text{ kohms}$$

Use Equation (8-25) to calculate RA_1:

$$RA_1 = \frac{1}{RC_1\cdot CB_1\cdot CD_1\cdot\omega_c^2} = 1.1458 \text{ kohms}$$

Select standard values of 24.3 kohms and 1.15 kohms for RC_1 and RA_1 respectively.

The stage 2 design also has K_2 equal to 1 which makes $R1_2$ a short and $R2_2$ an open. Select CB_2 equal to 0.082 microfarads and use Equation (8-28) To determine CD_{2max} :

$$CD_{2max} = CB_2 \cdot \left(\frac{1}{4\cdot Q_2^2}+K_2-1\right) = 0.012\cdot 10^{-6}$$

Select CD_2 equal to 0.0082 microfarads. Use Equation (8-27) to calculate RC_2:

$$RC_2 = \frac{1}{2 \cdot \omega_c \cdot Q_2 \cdot CD_2} + \frac{1}{\omega_c} \cdot \sqrt{\frac{1}{4 \cdot Q_2^2 \cdot CD_2^2} - \frac{1}{CD_2} \left(\frac{1}{CB_2} + \frac{1-K_2}{CD_2} \right)}$$

$RC_2 = 10.03$ kohms

Use Equation (8-25) to calculate RA_2:

$$RA_2 = \frac{1}{RC_2 \cdot CB_2 \cdot CD_2 \cdot \omega_c^2} = 2.803 \text{ kohms}$$

Select standard values of 10.0 kohms and 2.80 kohms for RC_2 and RA_2 respectively. The final circuit is shown in Figure 8-10.

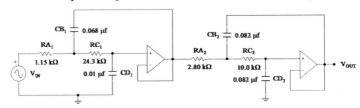

Figure 8-10 Example 8-1 Circuit

Figure 8-11 is the frequency response of the circuit of Figure 8-10 using a computer software circuit analysis program. The specification is superimposed in red lines. Note that the filter response exceeds both the pass band and stop band specifications.

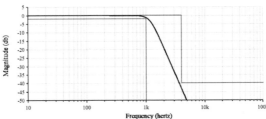

Figure 8-11 Filter Frequency Response

Example 8-2

Design an active Chebyshev low pass filter with a maximum pass band magnitude of 20 db, a minimum pass band magnitude of 18 db out to 1000 hertz, and a maximum stop band magnitude of -20 db from 3000 hertz on.

170

First create the filter specification graph shown in Figure 8-12.

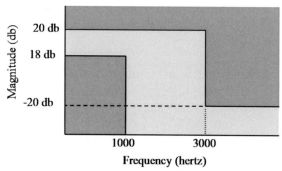

Figure 8-12 Filter Specification Graph for Example 8-2

Note that the frequencies are in hertz and not radians/sec. This is typical in filter designs and the frequencies must be converted from hertz to radians/sec to determine the filter transfer function:

$$\text{pass band frequency} = \omega_p = 2 \cdot \pi \cdot 1000 = 6{,}283$$

$$\text{stop band frequency} = \omega_s = 2 \cdot \pi \cdot 3000 = 18{,}850$$

The magnitude axis needs to be normalized to pass max equal to 0 db.

$$\text{Pass Max} = 20 - 20 = 0 \text{ db}$$

$$\text{Pass Min} = 18 - 20 = -2 \text{ db}$$

$$\text{Stop Max} = -20 - 20 = -40 \text{ db}$$

In order to allow for component variations, the design minimum pass band magnitude needs to be tightened by 10%. Since the magnitude values are in db, the 10% cannot be applied to the db values. Converting -2 db from db to magnitude yields:

$$-2 \text{ db} \Rightarrow 10^{(-2/20)} = 0.7943$$

Tightening this value by 10% yields 0.8738. Converting this back to db yields -1.172 db.

The pass min value (A_p) is 1.172 db and the stop max value (A_s) is 40 db. Use Equation (6-28) to determine the number of poles:

$$n = \cfrac{\cosh^{-1}\left(\sqrt{\cfrac{10^{(0.1 \cdot A_s)} - 1}{10^{(0.1 \cdot A_p)} - 1}}\right)}{\cosh^{-1}\left(\cfrac{\omega_s}{\omega_p}\right)} = \cfrac{\cosh^{-1}\left(\sqrt{\cfrac{10^4 - 1}{10^{0.1172} - 1}}\right)}{\cosh^{-1}\left(\cfrac{18,850}{6,283}\right)} = 3.338$$

Rounding n up to the next highest integer means that 4 poles will be required. Using Equation (6-19), determine ε:

$$\varepsilon = \sqrt{10^{(0.1 \cdot A_p)} - 1} = \sqrt{10^{0.1172} - 1} = 0.5566$$

Using the Table 6-1 for the 4 pole entry, determine the Butterworth pole values:

$$\alpha_1 = 0.3827 \qquad \omega_1 = 0.9239$$
$$\alpha_2 = 0.9239 \qquad \omega_2 = 0.3827$$

The Chebyshev transformation factors using Equation (6-21) are:

$$K_\alpha = \sinh\left(\frac{1}{n} \cdot \sinh^{-1}\left(\frac{1}{\varepsilon}\right)\right)$$

$$= \sinh\left(\frac{1}{4} \cdot \sinh^{-1}\left(\frac{1}{0.5566}\right)\right) = 0.3436$$

$$K_\omega = \cosh\left(\frac{1}{n} \cdot \sinh^{-1}\left(\frac{1}{\varepsilon}\right)\right)$$

$$= \cosh\left(\frac{1}{4} \cdot \sinh^{-1}\left(\frac{1}{0.5566}\right)\right) = 1.0574$$

The Chebyshev pole values are then:

$$\alpha_1 = 0.3827 \cdot 0.3436 = 0.1315$$
$$\omega_1 = 0.9239 \cdot 1.0574 = 0.9769$$
$$\alpha_2 = 0.9239 \cdot 0.3426 = 0.3175$$
$$\omega_2 = 0.3827 \cdot 1.0574 = 0.4046$$

The normalized Chebyshev poles are then:

$$\left((s + 0.1315)^2 + 0.9769^2\right) \cdot \left((s + 0.3175)^2 + 0.4046^2\right)$$

172

Converting this to a normalized transfer function in break frequency form yields:

$$G_N(s) = \cfrac{1}{\left(\cfrac{s}{0.5143}\right)^2 + \cfrac{1}{0.81}\cdot\left(\cfrac{s}{0.5143}\right)+1} \cdot$$
$$\cfrac{1}{\left(\cfrac{s}{0.9857}\right)^2 + \cfrac{1}{3.7479}\cdot\left(\cfrac{s}{0.9857}\right)+1}$$

Translating this to the pass frequency ω_p yields:

$$G_T(s) = \cfrac{1}{\left(\cfrac{s}{3,231}\right)^2 + \cfrac{1}{0.81}\cdot\left(\cfrac{s}{3,231}\right)+1} \cdot$$
$$\cfrac{1}{\left(\cfrac{s}{6,193}\right)^2 + \cfrac{1}{3.7479}\cdot\left(\cfrac{s}{6,193}\right)+1}$$

The gain of a Chebyshev filter with an even number of poles must be the pass min value. From above the normalized value is 0.8738 which when denormalized must be multiplied by 10 (20 db). The gain of 8.738 must be distributed between the two stages using Equations (8-29):

$$K_1 \cdot K_2 = 8.738 \Rightarrow K_2 = \frac{8.738}{K_1}$$

$$K_1 \cdot Q_1 = K_2 \cdot Q_2 \Rightarrow K_1 \cdot 1.0 = K_2 \cdot 3.7479$$

$$K_1^2 = 32.749 \Rightarrow K_1 = 5.723 \text{ and } K_2 = 1.527$$

The final transfer function then becomes:

$$G(s) = \cfrac{5.723}{\left(\cfrac{s}{3,231}\right)^2 + \cfrac{1}{0.81}\cdot\left(\cfrac{s}{3,231}\right)+1} \cdot$$
$$\cfrac{1.527}{\left(\cfrac{s}{6,193}\right)^2 + \cfrac{1}{3.7479}\cdot\left(\cfrac{s}{6,193}\right)+1}$$

This also is a 4 pole filter that can be implemented as an active filter with 2 two pole stages of the type in Figure 8-8. The value of ω_c for stage 1 (the input stage) is 3,231 rad/sec and the Q for stage 1 is 0.81. The value of ω_c for stage 2 (the output stage) is 6,193 rad/sec and the Q for stage 2 is 3.7479.

The stage 1 design has K_1 equal to 5.723. Choose $R2_1$ equal to 2,150 ohms and calculate $R1_1$ using Equation (8-9):

$$RI_1 = R2_1 \cdot (K_1 - 1) = 2,150 \cdot (5.723 - 1) = 10,154$$

Select $R1_1$ equal to 10,200 ohms which gives a gain of 5.744. Next select CB_1 equal to 0.01 microfarads and use Equation (8-28) To determine CD_{1max} :

$$CD_{1max} = CB_1 \cdot \left(\frac{1}{4 \cdot Q_1^2} + K_1 - 1 \right) = 0.0513 \cdot 10^{-6}$$

Select CD_1 equal to 0.039 microfarads. Use Equation (8-27) to calculate RC_1:

$$RC_1 = \frac{1}{2 \cdot \omega_{c1} \cdot Q_1 \cdot CD_1} + \frac{1}{\omega_{c1}} \cdot \sqrt{\frac{1}{4 \cdot Q_1^2 \cdot CD_1^2} - \frac{1}{CD_1} \cdot \left(\frac{1}{CB_1} + \frac{1 - K_1}{CD_1} \right)}$$

$$RC_1 = 13.68 \text{ kohms}$$

Use Equation (8-25) to calculate RA_1:

$$RA_1 = \frac{1}{RC_1 \cdot CB_1 \cdot CD_1 \cdot \omega_{c1}^2} = 17.95 \text{ kohms}$$

Select standard values of 13.7 kohms and 17.8 kohms for RC_1 and RA_1 respectively.

The stage 2 design has K_2 equal to 1.527. Choose $R2_2$ equal to 2,260 ohms and calculate $R1_2$ using Equation (8-29):

$$RI_2 = R2_2 \cdot (K_2 - 1) = 2,260 \cdot (1.527 - 1) = 1,191$$

Select $R1_2$ equal to 1,180 ohms which gives a gain of 1.522. Next select CB_2 equal to 0.01 microfarads and use Equation (8-28) To determine CD_{2max} :

$$CD_{2max} = CB_2 \cdot \left(\frac{1}{4 \cdot Q_2^2} + K_2 - 1 \right) = 0.0054 \cdot 10^{-6}$$

Select CD_2 equal to 0.0033 microfarads. Use Equation (8-27) to calculate RC_2:

$$RC_2 = \frac{1}{2 \cdot \omega_{c2} \cdot Q_2 \cdot CD_2} + \frac{1}{\omega_{c2}} \cdot \sqrt{\frac{1}{4 \cdot Q_2^2 \cdot CD_2^2} - \frac{1}{CD_2} \cdot \left(\frac{1}{CB_2} + \frac{1 - K_2}{CD_2}\right)}$$

$RC_2 = 28.94$ kohms

Use Equation (8-25) to calculate RA_2:

$$RA_2 = \frac{1}{RC_2 \cdot CB_2 \cdot CD_2 \cdot \omega_{c2}^2} = 27.3 \text{ kohms}$$

Select standard values of 28.7 kohms and 27.4 kohms for RC_2 and RA_2 respectively. The final circuit is shown in Figure 8-13.

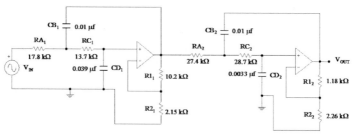

Figure 8-13 Example 8-2 Circuit

Figure 8-14 is the frequency response of the circuit of Figure 8-13 using a computer software circuit analysis program. The specification is superimposed in red lines. Note that the filter response exceeds both the pass band and stop band specifications.

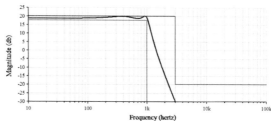

Figure 8-14 Filter Frequency Response

8-4. High Pass Filter Realization

In Chapter 6 the transfer function for a low pass filter was converted to a high pass transfer function by replacing s with 1/s. The high pass transfer function was then represented in polynomial form as shown in Equation (8-30).

$$G(s) = \frac{V_{OUT}}{V_{IN}} = \frac{K \cdot s}{s + \omega_c} \quad \text{for a first order factor}$$

$$G(s) = \frac{V_{OUT}}{V_{IN}} = \frac{K \cdot s^2}{s^2 + \frac{\omega_c}{Q} \cdot s + \omega_c^2} \quad \text{for a second order factor} \qquad (8\text{-}30)$$

The one pole filter stage of Figure 8-4 can also be used to design a first order high pass filter stage by proper selection of ZA and ZB. Choosing ZA to be a capacitor CA and ZB to be a resistor RB will accomplish this. Substituting $1/(s \cdot CA)$ for ZA and RB for ZB into Equation (8-6) yields Equation (8-31).

$$\frac{V_{OUT}}{V_{IN}} = \frac{1 + \frac{R1}{R2}}{\frac{ZA}{ZB} + 1} = \frac{K}{\frac{1}{s \cdot CA} + 1} = \frac{K \cdot s}{s + \frac{1}{RB \cdot CA}} \qquad (8\text{-}31)$$

Where: $K = 1 + \frac{R1}{R2}$ and $\omega_c = \frac{1}{RB \cdot CA}$

Figure 8-55 is the circuit for a one pole high pass filter stage.

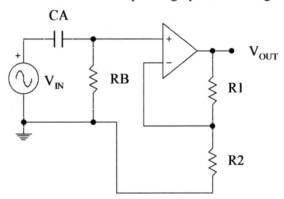

Figure 8-15 One Pole High Pass Filter Stage

The design procedure for a one pole high pass stage is to first select a standard value for R2 and then determine R1 using Equation (8-32).

$$K = 1 + \frac{R1}{R2} \Rightarrow R1 = R2 \cdot (K - 1) \qquad (8\text{-}32)$$

Next, select a standard value for CA and then determine RB from Equation (8-33).

$$\omega_c = \frac{1}{RB \cdot CA} \Rightarrow RB = \frac{1}{CA \cdot \omega_c} \tag{8-33}$$

For the special case where K is 1, R1 is zero or a short and R2 is infinite or an open. If the calculated component values are not reasonable, reselect either R2 or CA and recalculate until reasonable values are obtained.

The second order factor high pass transfer function of Equation (8-30) can be implemented with the circuit of Figure 8-6 to form a two pole high pass filter stage by proper selection of ZA, ZB, ZC, and ZD. Choosing ZA and ZC as capacitors and ZB and ZD as resistors will accomplish this. Substituting $1/(s \cdot CA)$ for ZA, RB for ZB, $1/(s \cdot CC)$ for ZC, and RD for ZD into Equation (8-20) yields Equation (8-34).

$$
\begin{aligned}
\frac{V_{OUT}}{V_{IN}} &= \frac{K}{\dfrac{ZA \cdot ZC}{ZB \cdot ZD} + \dfrac{ZA}{ZD} + \dfrac{ZC}{ZD} + \dfrac{ZA \cdot (1-K)}{ZB} + 1} \\[2ex]
&= \frac{K}{\dfrac{\frac{1}{s^2 \cdot CA \cdot CC}}{RB \cdot RD} + \dfrac{\frac{1}{s \cdot CA}}{RD} + \dfrac{\frac{1}{s \cdot CC}}{RD} + \dfrac{\frac{1-K}{s \cdot CA}}{RB} + 1} \\[2ex]
&= \frac{K}{\dfrac{1}{s^2 \cdot RB \cdot RD \cdot CA \cdot CC} + \dfrac{1}{s \cdot RD \cdot CA} + \dfrac{1}{s \cdot RD \cdot CC} + \dfrac{1-K}{s \cdot RB \cdot CA} + 1} \\[2ex]
&= \frac{K \cdot s^2}{s^2 + s \cdot \left(\dfrac{1}{RD \cdot CA} + \dfrac{1}{RD \cdot CC} + \dfrac{1-K}{RB \cdot CA} \right) + \dfrac{1}{RB \cdot RD \cdot CA \cdot CC}}
\end{aligned}
\tag{8-34}
$$

Equating like coefficients of Equations (8-34) and (8-30) yields Equations (8-35) and (8-36).

$$\frac{1}{RB \cdot RD \cdot CA \cdot CC} = \omega_c^2 \tag{8-35}$$

$$\frac{1}{RD \cdot CA} + \frac{1}{RD \cdot CC} + \frac{1-K}{RB \cdot CA} = \frac{\omega_c}{Q} \tag{8-36}$$

Since there are only two equations and four unknowns, standard values will be selected for CA and CC and the equations solved for RB and RD. Equation (8-35) can be solved for RD as shown in Equation (8-37).

$$RD = \frac{1}{RB \cdot CA \cdot CC \cdot \omega_c^2} \tag{8-37}$$

Substituting Equation (8-37) into Equation (8-36) yields Equation (8-38).

$$RB \cdot CC \cdot \omega_c^2 + RB \cdot CA \cdot \omega_c^2 + \frac{1-K}{RB \cdot CA} = \frac{\omega_c}{Q} \tag{8-38}$$

Multiplying both sides of Equation (8-38) by RB/ω_c^2 and solving for RB yields Equation (8-39).

$$RB^2 \cdot CC + RB^2 \cdot CA + \frac{1-K}{CA \cdot \omega_c^2} = \frac{RB}{Q \cdot \omega_c}$$

$$RB^2 \cdot (CA + CC) - \frac{RB}{Q \cdot \omega_c} + \frac{1-K}{CA \cdot \omega_c^2} = 0$$

$$RB = \frac{1}{2 \cdot (CA + CC) \cdot Q \cdot \omega_c} + \frac{1}{\omega_c} \cdot \sqrt{\frac{1}{4 \cdot (CA + CC)^2 \cdot Q^2} - \frac{1-K}{CA \cdot (CA + CC)}} \tag{8-39}$$

Just as was done for the low pass filter, only the positive radical was used. Unlike the low pass design the radical for the high pass design is always positive since K must be greater than or equal to 1. Since there is no restriction on the capacitor value choice, CA and CC will be chosen to be equal. Substituting C for CA and CC in Equation (8-39) yields Equation (8-40).

$$RB = \frac{1}{4 \cdot C \cdot Q \cdot \omega_c} + \frac{1}{C \cdot \omega_c} \cdot \sqrt{\frac{1}{16 \cdot Q^2} - \frac{1-K}{2}} \tag{8-40}$$

Figure 8-16 is the circuit for a two pole low pass filter stage.

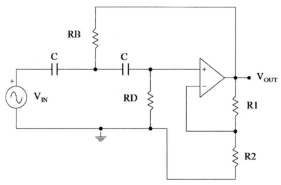

Figure 8-16 Two Pole High Pass Filter Stage

The design procedure starts by determining the filter transfer function in factored form using the methods of Chapter 6. The transfer function can then be partitioned into first and second order stages. If the number of poles is odd there will be one first order stage and the rest second order stages. The first order stage can be implemented using the one pole filter stage of Figure 8-15 and the second order stages can be implemented using the two pole filter stage of Figure 8-16.

Next, select a standard capacitor value for C and calculate the resistor values. If the resistor values are less than desirable, select another capacitor standard value and recalculate the resistor values. Repeat this process until acceptable resistor values are obtained. Since the two pole filter calculations are rather lengthy, an EXCEL spreadsheet or math software can be used to speed the process and minimize errors.

The filter stages can then be cascaded to form the final filter design. In order to keep the individual filter stages from saturating, always place the one pole circuit closest to the input and follow in the cascading operation with the two pole circuits in order of increasing Q.

If the filter gain is greater than unity (0 dB), the gain will need to be distributed among the number of stages used to implement the filter. This is best accomplished by distributing the gain based on the Q of each stage (to keep the individual filter stages from saturating) with the lower Q stages having a higher gain than the higher Q stages. Since the value of Q determines the amount of peaking, Q values less than 1 should be defaulted to 1. Solution of Equations (8-29) will define the gain for each stage.

Example 8-3

Design an active Butterworth high pass filter with a maximum pass band magnitude of 0 db, a minimum pass band magnitude of -2 db from 2000 hertz on out, and a maximum stop band magnitude of -40 db below 500 hertz.

First create the filter specification graph shown in Figure 8-17. The frequencies must be converted from hertz to radians/sec to determine the filter transfer function:

$$\text{pass band frequency} = \omega_p = 2 \cdot \pi \cdot 2000 = 12,566$$

$$\text{stop band frequency} = \omega_s = 2 \cdot \pi \cdot 500 = 3,142$$

The magnitude axis is already normalized to pass max equal to 0 db. In order to allow for component variations, the design minimum pass band magnitude needs to be tightened by 10%.

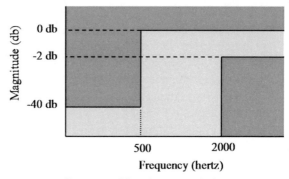

Figure 8-17 Filter Specification Graph for Example 8-3

Since the magnitude values are in db, the 10% cannot be applied to the db values. Converting -2 db from db to magnitude yields:

$$-2 \text{ db} \Rightarrow 10^{(-2/20)} = 0.7943$$

Tightening this value by 10% yields 0.8738. Converting this back to db yields -1.172 db. The pass min value (A_p) is 1.172 db and the stop max value (A_s) is 40 db. Use Equation (6-31) to determine the number of poles:

$$n = \frac{\log_{10}\left(\dfrac{10^{(0.1 \cdot A_s)} - 1}{10^{(0.1 \cdot A_p)} - 1}\right)}{2 \cdot \log_{10}\left(\dfrac{\omega_p}{\omega_s}\right)} = \frac{\log_{10}\left(\dfrac{10^4 - 1}{10^{0.1172} - 1}\right)}{2 \cdot \log_{10}\left(\dfrac{12,566}{3,142}\right)} = 3.745$$

Rounding n up to the next highest integer means that 4 poles will be required. Next use Equation (6-32) to determine the filter cutoff frequency ω_c:

$$\omega_c = \omega_p \cdot \left(10^{(0.1 \cdot A_p)} - 1\right)^{1/2 \cdot n} = 12,566 \cdot \left(10^{(0.1172)} - 1\right)^{1/8} = 10,854$$

Using the Table 6-1 for the 4 pole entry, Equation (6-34), and ω_c from above determine the filter transfer function:

$$G(s) = \frac{s^2}{\left(s^2 + \dfrac{10,854}{0.5412} \cdot s + 10,854^2\right)} \cdot \frac{s^2}{\left(s^2 + \dfrac{10,854}{1.3066} \cdot s + 10,854^2\right)}$$

This is a 4 pole filter that can be implemented as an active filter with 2 two pole stages of the type in Figure 8-16. The value of

180

ω_c for both stages is 10,854 rad/sec and the Q for the first stage (the input stage) is 0.5412 and the Q for the second stage (the output stage) is 1.3066.

The stage 1 design has K_1 equal to 1 which makes $R1_1$ a short and $R2_1$ an open. Select C_1 equal to 0.0068 microfarads and use Equation (8-40) to determine RB_1:

$$RB_1 = \frac{1}{4 \cdot C_1 \cdot Q_1 \cdot \omega_c} + \frac{1}{C_1 \cdot \omega_c} \cdot \sqrt{\frac{1}{16 \cdot Q_1^2} - \frac{1 - K_1}{2}}$$

$$RB_1 = 12.517 \text{ kohms}$$

Use Equation (8-37) to calculate RD_1:

$$RD_1 = \frac{1}{RB_1 \cdot C_1^2 \cdot \omega_c^2} = 14.665 \text{ kohms}$$

Select standard values of 12.4 kohms and 14.7 kohms for RB_1 and RD_1 respectively.

The stage 2 design has K_2 equal to 1 which makes $R1_2$ a short and $R2_2$ an open. Select C_2 equal to 0.018 microfarads and use Equation (8-40) to determine RB_2:

$$RB_2 = \frac{1}{4 \cdot C_2 \cdot Q_2 \cdot \omega_c} + \frac{1}{C_2 \cdot \omega_c} \cdot \sqrt{\frac{1}{16 \cdot Q_2^2} - \frac{1 - K_2}{2}}$$

$$RB_2 = 1.9587 \text{ kohms}$$

Use Equation (8-37) to calculate RD_2:

$$RD_2 = \frac{1}{RB_2 \cdot C_2^2 \cdot \omega_c^2} = 13.376 \text{ kohms}$$

Select standard values of 1.96 kohms and 13.3 kohms for RB_2 and RD_2 respectively. The final circuit is shown in Figure 8-18.

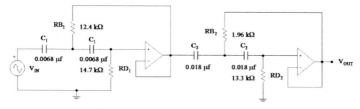

Figure 8-18 Example 8-3 Circuit

Figure 8-19 is the frequency response of the circuit of Figure 8-18 using a computer software circuit analysis program. The specification is superimposed in red lines. Note that the filter response exceeds both the pass band and stop band specifications.

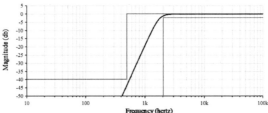

Figure 8-19 Filter Frequency Response

Example 8-4

Design an active Chebyshev high pass filter with a maximum pass band magnitude of 20 db, a minimum pass band magnitude of 18 db from 5000 hertz on out, and a maximum stop band magnitude of -20 db below 2000 hertz.

First create the filter specification graph shown in Figure 8-20:

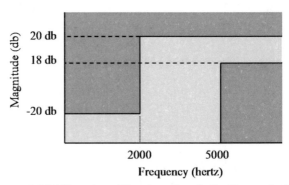

Figure 8-20 Filter Specification Graph for Example 8-4

The frequencies must be converted from hertz to radians/sec to determine the filter transfer function:

$$\text{pass band frequency} = \omega_p = 2 \cdot \pi \cdot 5000 = 31,416$$

$$\text{stop band frequency} = \omega_s = 2 \cdot \pi \cdot 2000 = 12,566$$

The magnitude axis needs to be normalized to pass max equal to 0 db.

$$\text{Pass Max} = 20 - 20 = 0 \text{ db}$$
$$\text{Pass Min} = 18 - 20 = -2 \text{ db}$$
$$\text{Stop Max} = -20 - 20 = -40 \text{ db}$$

In order to allow for component variations, the design minimum pass band magnitude needs to be tightened by 10%. Since the magnitude values are in db, the 10% cannot be applied to the db values. Converting -2 db from db to magnitude yields:

$$-2 \text{ db} \Rightarrow 10^{(-2/20)} = 0.7943$$

Tightening this value by 10% yields 0.8738. Converting this back to db yields -1.172 db. The pass min value (A_p) is 1.172 db and the stop max value (A_s) is 40 db. Use Equation (6-39) to determine the number of poles:

$$n = \frac{\cosh^{-1}\left(\sqrt{\frac{10^{(0.1 A_s)} - 1}{10^{(0.1 A_p)} - 1}}\right)}{\cosh^{-1}\left(\frac{\omega_p}{\omega_s}\right)} = \frac{\cosh^{-1}\left(\sqrt{\frac{10^4 - 1}{10^{0.1172} - 1}}\right)}{\cosh^{-1}\left(\frac{31,416}{12,566}\right)} = 3.756$$

Rounding n up to the next highest integer means that 4 poles will be required. Using Equation (6-19), determine ε:

$$\varepsilon = \sqrt{10^{(0.1 A_p)} - 1} = \sqrt{10^{0.1172} - 1} = 0.5566$$

Using the Table 6-1 for the 4 pole entry, determine the Butterworth pole values:

$$\alpha_1 = 0.3827 \quad \omega_1 = 0.9239$$
$$\alpha_2 = 0.9239 \quad \omega_2 = 0.3827$$

The Chebyshev transformation factors are:

$$K_\alpha = \sinh\left(\frac{1}{n} \cdot \sinh^{-1}\left(\frac{1}{\varepsilon}\right)\right)$$
$$= \sinh\left(\frac{1}{4} \cdot \sinh^{-1}\left(\frac{1}{0.5566}\right)\right) = 0.3436$$

$$K_\omega = \cosh\left(\frac{1}{n} \cdot \sinh^{-1}\left(\frac{1}{\varepsilon}\right)\right)$$

$$= \cosh\left(\frac{1}{4} \cdot \sinh^{-1}\left(\frac{1}{0.5566}\right)\right) = 1.0574$$

The Chebyshev pole values are then:

$$\alpha_1 = 0.3827 \cdot 0.3436 = 0.1315$$
$$\omega_1 = 0.9239 \cdot 1.0574 = 0.9769$$
$$\alpha_2 = 0.9239 \cdot 0.3436 = 0.3175$$
$$\omega_2 = 0.3827 \cdot 1.0574 = 0.4046$$

The normalized Chebyshev poles are then:

$$\left((s+0.1315)^2 + 0.9769^2\right) \cdot \left((s+0.3175)^2 + 0.4046^2\right)$$

Converting the poles to break frequency format yields:

$$\left(\left(\frac{s}{0.9857}\right)^2 + \frac{1}{3.7479}\cdot\left(\frac{s}{0.9857}\right)+1\right)\cdot$$

$$\left(\left(\frac{s}{0.5143}\right)^2 + \frac{1}{0.81}\cdot\left(\frac{s}{0.5143}\right)+1\right)$$

Using Equation (6-38), convert these poles to the high pass form translated to ω_p.

$$s^2 + \frac{1}{3.7479}\cdot\left(\frac{31,416}{0.9857}\right)\cdot s+\left(\frac{31,416}{0.9857}\right)^2 = s^2 + \frac{31,872}{3.7479}\cdot s + 31,872^2$$

$$s^2 + \frac{1}{0.81}\cdot\left(\frac{31,416}{0.5143}\right)\cdot s+\left(\frac{31,416}{0.5143}\right)^2 = s^2 + \frac{61,085}{0.81}\cdot s + 61,085^2$$

Since n is even the numerator constant of the transfer function must be the pass min value converted using Equation (6-6):

$$\text{numerator} = \frac{1}{10^{(A_p/20)}} = \frac{1}{10^{0.0586}} = 0.8738$$

The transfer function magnitudes must also be denormalized by adding a multiplying factor of 10 $(20 \cdot \log_{10} 10 = 20 \text{ db})$. The total numerator gain factor is then 8.738 which must be distributed between the two stages using Equations (8-29). Note that Q values less than 1 default to 1.

184

$$K_1 \cdot K_2 = 8.738 \Rightarrow K_2 = \frac{8.738}{K_1}$$

$$K_1 \cdot Q_1 = K_2 \cdot Q_2 \Rightarrow K_1 \cdot 1.0 = K_2 \cdot 3.7479$$

$$K_1^2 = 32.749 \Rightarrow K_1 = 5.723 \text{ and } K_2 = 1.527$$

The Chebyshev high pass transfer function then becomes:

$$G(s) = \frac{5.723 \cdot s^2}{s^2 + \dfrac{61,085}{0.81} \cdot s + 61,085^2} \cdot \frac{1.527 \cdot s^2}{s^2 + \dfrac{31,872}{3.7479} \cdot s + 31,872^2}$$

This is a 4 pole filter that can be implemented as an active filter with 2 two pole stages of the type in Figure 8-16. The value of ω_p for stage 1 (the input stage) is 61,085 rad/sec and the Q for stage 1 is 0.81. The value of ω_p for stage 2 (the output stage) is 31,872 rad/sec and the Q for stage 2 is 3.7479.

The stage 1 design has K_1 equal to 5.723. Choose $R2_1$ equal to 2,150 ohms and calculate $R1_1$ using Equation (8-9):

$$R1_1 = R2_1 \cdot (K_1 - 1) = 2,150 \cdot (5.723 - 1) = 10,154$$

Select $R1_1$ equal to 10,200 ohms which gives a gain of 5.744. Select C_1 equal to 820 picofarads and use Equation (8-40) to determine RB_1:

$$RB_1 = \frac{1}{4 \cdot C_1 \cdot Q_1 \cdot \omega_{c1}} + \frac{1}{C_1 \cdot \omega_{c1}} \cdot \sqrt{\frac{1}{16 \cdot Q_1^2} - \frac{1 - K_1}{2}}$$

$$RB_1 = 37.52 \text{ kohms}$$

Use Equation (8-37) to calculate RD_1:

$$RD_1 = \frac{1}{RB_1 \cdot C_1^2 \cdot \omega_{c1}^2} = 10.623 \text{ kohms}$$

Select standard values of 37.4 kohms and 10.7 kohms for RB_1 and RD_1 respectively.

The stage 2 design has K_2 equal to 1.527. Choose $R2_2$ equal to 2,260 ohms and calculate $R1_2$ using Equation (8-29):

$$R1_2 = R2_2 \cdot (K_2 - 1) = 2,260 \cdot (1.527 - 1) = 1,191$$

Select $R1_2$ equal to 1,180 ohms which gives a gain of 1.522. Select C_2 equal to 0.0018 microfarads and use Equation (8-40) to determine RB_2:

$$RB_2 = \frac{1}{4 \cdot C_2 \cdot Q_2 \cdot \omega_{c2}} + \frac{1}{C_2 \cdot \omega_{c2}} \cdot \sqrt{\frac{1}{16 \cdot Q_2^2} - \frac{1-K_2}{2}}$$

$$RB_2 = 10.143 \text{ kohms}$$

Use Equation (8-37) to calculate RD_2:

$$RD_2 = \frac{1}{RB_2 \cdot C_2^2 \cdot \omega_{c2}^2} = 29.954 \text{ kohms}$$

Select standard values of 10.2 kohms and 30.1 kohms for RB_2 and RD_2 respectively. The final circuit is shown in Figure 8-21.

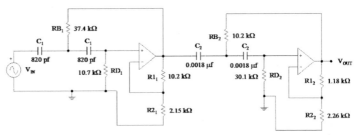

Figure 8-21 Example 8-4 Circuit

Figure 8-22 is the frequency response of the circuit of Figure 8-21 using a computer software circuit analysis program. The specification is superimposed in red lines. Note that the filter response exceeds both the pass band and stop band specifications.

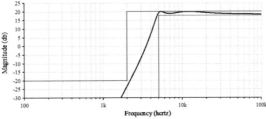

Figure 8-22 Filter Frequency Response

8-5. Band Pass Filter Realization

In Chapter 6 the transfer function for a low pass filter was converted to a narrow band band pass filter by replacing s with $(s + 1/s)$. This resulted in doubling the number of poles which means only two pole stages will be required. This only applies to the narrow band implementation $(Q > 0.5)$. Wide band implementations $(Q \le 0.5)$ are accomplished by separate low pass and high pass designs as was discussed in Chapter 6. From Chapter 6, a 2 pole band pass transfer function is as shown in Equation (8-41).

$$G(s) = \frac{V_{OUT}}{V_{IN}} = \frac{K \cdot B \cdot s}{s^2 + B \cdot s + \omega_0^2} \tag{8-41}$$

In Equation (8-41), K is the pass max magnitude at the center frequency, B is the bandwidth (upper pass frequency - lower pass frequency), and ω_0 is the center frequency. From Chapter 6 $Q = \omega_0 / B$.

The two pole filter stage of Figure 8-6 is not the best choice for a band pass filter implementation. The multiple feedback circuit of Figure 8-23 is the best choice.

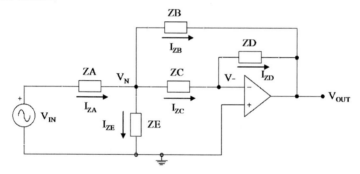

Figure 8-23 Two Pole Multiple Feedback Circuit

Summing branch currents at node V_N yields Equation (8-42).

$$I_{ZA} = I_{ZB} + I_{ZC} + I_{ZE} \tag{8-42}$$

Converting Equation (8-42) to node voltages yields Equation (8-43).

$$\frac{V_{IN} - V_N}{ZA} = \frac{V_N - V_{OUT}}{ZB} + \frac{V_N - V-}{ZC} + \frac{V_N - 0}{ZE} \tag{8-43}$$

The op-amp output is the op-amp gain times the difference between the op-amp inputs. Since the op-amp gain is very large (infinite for the ideal op-amp), the difference between the op-amp inputs must be very small (zero for the ideal op-amp) for the output to be a finite value. From Figure 8-23, the V+ input is zero, therefore the V- input must also be zero. Applying this to Equation (8-43) yields Equation (8-44).

$$\frac{V_{IN} - V_N}{ZA} = \frac{V_N - V_{OUT}}{ZB} + \frac{V_N}{ZC} + \frac{V_N}{ZE} \tag{8-44}$$

Summing the branch currents at the V- node (note that for the ideal op-amp the input current is zero) yields Equation (8-45).

$$I_{ZC} = I_{ZD} \tag{8-45}$$

Converting Equation (8-45) to node voltages (remember that V- is zero) yields Equation (8-46).

$$\frac{V_N}{ZC} = \frac{-V_{OUT}}{ZD} \tag{8-46}$$

Solving Equation (8-46) for V_N yields Equation (8-47).

$$V_N = -\frac{V_{OUT} \cdot ZC}{ZD} \tag{8-47}$$

Substituting Equation (8-47) into Equation (8-44) yields Equation (8-48).

$$\frac{V_{IN}}{ZA} + \frac{V_{OUT} \cdot ZC}{ZA \cdot ZD} = -\frac{V_{OUT} \cdot ZC}{ZB \cdot ZD} - \frac{V_{OUT}}{ZB} - \frac{V_{OUT}}{ZD} - \frac{V_{OUT} \cdot ZC}{ZD \cdot ZE} \tag{8-48}$$

Rearranging Equation (8-48) yields Equation (8-49).

$$\frac{V_{IN}}{ZA} = -V_{OUT} \cdot \left(\frac{ZC}{ZA \cdot ZD} + \frac{ZC}{ZB \cdot ZD} + \frac{1}{ZB} + \frac{1}{ZD} + \frac{ZC}{ZD \cdot ZE} \right) \tag{8-49}$$

Solving Equation (8-49) for V_{OUT}/V_{IN} yields Equation (8-50).

$$V_{IN} = -V_{OUT} \cdot \left(\frac{ZC}{ZD} + \frac{ZA \cdot ZC}{ZB \cdot ZD} + \frac{ZA}{ZB} + \frac{ZA}{ZD} + \frac{ZA \cdot ZC}{ZD \cdot ZE} \right)$$

$$\frac{V_{OUT}}{V_{IN}} = -\frac{1}{\dfrac{ZC}{ZD} + \dfrac{ZA \cdot ZC}{ZB \cdot ZD} + \dfrac{ZA}{ZB} + \dfrac{ZA}{ZD} + \dfrac{ZA \cdot ZC}{ZD \cdot ZE}} \tag{8-50}$$

Note that Equation (8-50) has a negative sign as compared to the band pass transfer function of Equation (8-41). This merely indicates a 180 degree phase shift which in some situations must be considered.

In order for Equation (8-50) to be a band pass filter transfer function like Equation (8-41), ZA, ZD, and ZE must be resistors RA, RD, and RE and ZB and ZC must be capacitors $1/(s \cdot CB)$ and $1/(s \cdot CB)$. Equation (8-51) is Equation (8-50) with those substitutions made.

$$\frac{V_{OUT}}{V_{IN}} = -\frac{\dfrac{1}{RA \cdot CB} \cdot s}{s^2 + \left(\dfrac{1}{RD \cdot CC} + \dfrac{1}{RD \cdot CB}\right) \cdot s + \dfrac{1}{RD \cdot CB \cdot CC} \cdot \left(\dfrac{1}{RA} + \dfrac{1}{RE}\right)} \qquad (8\text{-}51)$$

Equating like coefficients of Equation (8-41 and Equation (8-51) yields Equations (8-52), (8-53) and (8-54).

$$K \cdot B = \frac{1}{RA \cdot CB} \qquad (8\text{-}52)$$

$$B = \frac{1}{RD \cdot CC} + \frac{1}{RD \cdot CB} \qquad (8\text{-}53)$$

$$\omega_0^2 = \frac{1}{RD \cdot CB \cdot CC} \cdot \left(\frac{1}{RA} + \frac{1}{RE}\right) \qquad (8\text{-}54)$$

Since there are only three equations and five unknowns, standard values will be selected for CB and CC and the equations solved for RA, RD and RE. Equation (8-52) can be solved for RA yielding Equation (8-55).

$$K \cdot B = \frac{1}{RA \cdot CB} \Rightarrow RA = \frac{1}{K \cdot B \cdot CB} \qquad (8\text{-}55)$$

Equation (8-53) can be solved for RD as shown in Equation (8-56).

$$B = \frac{1}{RD \cdot CC} + \frac{1}{RD \cdot CB} \Rightarrow RD = \frac{1}{B} \cdot \left(\frac{1}{CC} + \frac{1}{CB}\right) \qquad (8\text{-}56)$$

Equations (8-55) and (8-56) can be substituted into Equation (8-54) to solve for RE as shown in Equation (8-57).

$$\omega_0^2 = \frac{1}{RD \cdot CB \cdot CC} \cdot \left(\frac{1}{RA} + \frac{1}{RE}\right)$$

$$\omega_0^2 = \frac{1}{\dfrac{1}{B} \cdot \left(\dfrac{1}{CC} + \dfrac{1}{CB}\right) \cdot CB \cdot CC} \cdot \left(\frac{1}{\dfrac{1}{K \cdot B \cdot CB}} + \frac{1}{RE}\right)$$

$$\omega_0^2 = \frac{K \cdot B^2 \cdot CB}{CB + CC} + \frac{B}{(CB + CC) \cdot RE}$$

$$\frac{B}{(CB + CC) \cdot RE} = \omega_0^2 - \frac{K \cdot B^2 \cdot CB}{CB + CC}$$

$$\frac{1}{RE} = \frac{\omega_0^2 \cdot (CB + CC)}{B} - K \cdot B \cdot CB$$

$$RE = \frac{1}{\dfrac{\omega_0^2 \cdot (CB + CC)}{B} - K \cdot B \cdot CB} \tag{8-57}$$

Since RE must be positive, restrictions are placed on CB and CC. In order for the denominator of Equation (8-57) to be positive, Equation (8-58) must be met.

$$\frac{\omega_0^2 \cdot (CB + CC)}{B} > K \cdot B \cdot CB \tag{8-58}$$

Rearranging Equation (8-58) yields Equation (8-59).

$$\frac{\omega_0^2 \cdot (CB + CC)}{B} > K \cdot B \cdot CB$$

$$\frac{(CB + CC)}{CB} > \frac{K \cdot B^2}{\omega_0^2}$$

$$\frac{CC}{CB} + 1 > \frac{K \cdot B^2}{\omega_0^2}$$

$$CC > \left(\frac{K \cdot B^2}{\omega_0^2} - 1 \right) \cdot CB$$

$$CC_{min} = \left(\frac{K \cdot B^2}{\omega_0^2} - 1 \right) \cdot CB \tag{8-59}$$

Equation (8-59) places a restriction on CC relative to CB unless the right side of Equation (8-59) is negative. In that case there is no restriction. Figure 8-24 is the circuit for a two pole band pass filter stage.

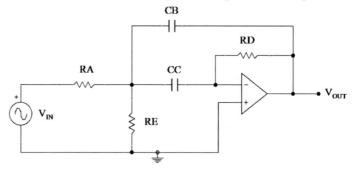

Figure 8-24 Two Pole Band Pass Filter Stage

The design procedure starts by determining the filter transfer function partitioned in the form of Equation (8-41) using the methods of Chapter 6.

The stages can be implemented using the two pole filter stage of Figure 8-24. Next, select a standard capacitor value for CB, determine the minimum value for CC, and select a standard capacitor value for CC. Then calculate the resistor values. If the resistor values are less than desirable, select other standard capacitor values and recalculate the resistor values. Repeat this process until acceptable resistor values are obtained. Since the filter calculations are rather lengthy, an EXCEL spreadsheet or math software can be used to speed the process and minimize errors.

The filter stages can then be cascaded to form the final filter design. In order to keep the individual filter stages from saturating, always place the circuit with the lowest value of K closest to the input and follow in the cascading operation in order of increasing K.

Example 8-5

Design an active band pass Butterworth filter with a maximum pass band magnitude of 0 db at a center frequency of 1000 hertz and a 200 hertz bandwidth at the -3.0103 db points. There shall be a maximum stop band magnitude of -40 db below 350 hertz and above 3000 hertz.

First determine the filter Q using Equation (6-41):

$$Q = \frac{\omega_0}{B} = \frac{2 \cdot \pi \cdot 1000}{2 \cdot \pi \cdot 200} = 5$$

Since Q is greater than 0.5, use the narrow band approach.

Next, create the filter specification graph. Since the center frequency and bandwidth are given, they must be normalized and converted to the upper and lower pass frequencies. The normalized center frequency and bandwidth are:

$$\omega_0 = \frac{f_0}{1000} = 1 \quad B = \frac{200}{1000} = 0.2$$

The upper and lower pass frequencies can be calculated from Equation (6-40):

$$f_0 = 1000 = \sqrt{f_{ph} \cdot f_{pl}} \Rightarrow f_{pl} = \frac{1000^2}{f_{ph}}$$

$$B_f = 200 = f_{ph} - f_{pl} = f_{ph} - \frac{1000^2}{f_{ph}}$$

$$f_{ph}^2 - 200 \cdot f_{ph} - 1000^2 = 0 \Rightarrow f_{ph} = 1105 \quad \text{and} \quad f_{pl} = 905$$

The specification graph is shown in Figure 8-25.

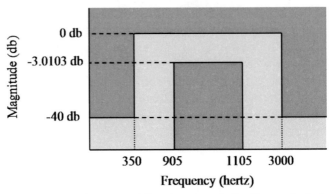

Figure 8-25 Filter Specification Graph for Example 8-5

The next step is to normalize the specifications. The magnitude axis is already normalized since the pass max value is 0 db. In order to allow for component variations, the design bandwidth needs to be increased by 10%. The design bandwidth then becomes 220 hertz. The design upper and lower pass frequencies can be calculated from Equation (6-40):

$$f_0 = 1000 = \sqrt{f_{ph} \cdot f_{pl}} \Rightarrow f_{pl} = \frac{1000^2}{f_{ph}}$$

$$B_f = 220 = f_{ph} - f_{pl} = f_{ph} - \frac{1000^2}{f_{ph}}$$

$$f_{ph}^2 - 220 \cdot f_{ph} - 1000^2 = 0 \Rightarrow f_{ph} = 1116 \quad \text{and} \quad f_{pl} = 896$$

To normalize the frequency axis divide all of the specifications by f_0 or 1000. Since this is a frequency divided by a frequency, the normalized axis can be considered either hertz or radians per second. The normalized frequency values are:

$$\omega_{pl} = \frac{896}{1000} = 0.896 \quad \omega_{ph} = \frac{1116}{1000} = 1.116$$

$$\text{lower stop} = \frac{350}{1000} = 0.35 \quad \text{upper stop} = \frac{3000}{1000} = 3.0$$

192

The next step is to determine the low pass equivalent. To determine the low pass equivalent number of poles, determine the pass bandwidth B and the geometric stop bandwidth:

$$B = \frac{B_f}{f_0} = \frac{220}{1000} = 0.22$$

$$\text{or } B = \omega_{ph} - \omega_{pl} = 1.116 - 0.896 = 0.22$$

$$1/(\text{lower stop}) = 1/0.35 = 2.857$$

which is less than the upper stop frequency

Therefore: $\omega_{sl} = 0.35$ and $\omega_{sh} = 2.857$

geometric stop width $= \omega_{sh} - \omega_{sl} = 2.857 - 0.35 = 2.51$

Determine the number of low pass equivalent poles using Equation (6-47):

$$n = \frac{\log_{10}\left(\frac{10^{(0.1 \cdot A_s)} - 1}{10^{(0.1 \cdot A_p)} - 1}\right)}{2 \cdot \log_{10}\left(\frac{\text{geometric stop width}}{\text{bandwidth}}\right)}$$

$$= \frac{\log_{10}\left(\frac{10^{(4)} - 1}{10^{(0.30103)} - 1}\right)}{2 \cdot \log_{10}\left(\frac{2.51}{0.22}\right)} = 1.893$$

Rounding up to the next highest integer yields n equal to 2. Using Table 6-1 for the 2 pole entry, the normalized low pass equivalent transfer function is:

$$G(s)_{LP} = \frac{1}{\left(s^2 + \dfrac{s}{0.7071} + 1\right)}$$

Translating this to the pass band frequency of B yields:

$$G(s)_{LP} = \frac{1}{\left(\dfrac{s}{0.22}\right)^2 + \dfrac{1}{0.7071} \cdot \left(\dfrac{s}{0.22}\right) + 1}$$

Transforming this to band pass using Equation (6-50) yields:

$$G(s) = \cfrac{B^2 \cdot s^2}{s^4 + \cfrac{B}{Q_{LP}} \cdot s^3 + (2 + B^2) \cdot s^2 + \cfrac{B}{Q_{LP}} \cdot s + 1}$$

$$G(s) = \frac{0.0484 \cdot s^2}{\left(s^4 + 0.3111 \cdot s^3 + 2.0484 \cdot s^2 + 0.3111 \cdot s + 1\right)}$$

Factoring the 4th order denominator yields:

$$G(s) = \frac{0.22 \cdot s}{\left(s^2 + 0.1676 \cdot s + 1.1687\right)} \cdot$$

$$\frac{0.22 \cdot s}{\left(s^2 + 0.1435 \cdot s + 0.8557\right)}$$

Translating this out to the specification center frequency of 1000 hertz which is $2000 \cdot \pi$ radians per second yields:

$$G(s) = \frac{0.22 \cdot \left(\dfrac{s}{2000 \cdot \pi}\right)}{\left(\left(\dfrac{s}{2000 \cdot \pi}\right)^2 + 0.1676 \cdot \left(\dfrac{s}{2000 \cdot \pi}\right) + 1.1687\right)} \cdot$$

$$\frac{0.22 \cdot \left(\dfrac{s}{2000 \cdot \pi}\right)}{\left(\left(\dfrac{s}{2000 \cdot \pi}\right)^2 + 0.1435 \cdot \left(\dfrac{s}{2000 \cdot \pi}\right) + 0.8557\right)}$$

$$G(s) = \frac{1382 \cdot s}{\left(s^2 + 1053.4 \cdot s + 4.61383 \cdot 10^7\right)} \cdot$$

$$\frac{1382 \cdot s}{\left(s^2 + 901.3 \cdot s + 3.378 \cdot 10^7\right)}$$

This is a 4 pole filter that can be implemented as an active filter with 2 two pole stages of the type in Figure 8-24. By comparing the desired transfer function for each stage with Equation (8-41) the following requirements can be defined:

Stage 1:

$$\omega_{01} = \sqrt{4.61383 \cdot 10^7} = 6793$$

$$B_1 = 1053.4$$

$$K_1 = \frac{1382}{1053.4} = 1.3119$$

Stage 2:

$$\omega_{02} = \sqrt{3.378 \cdot 10^7} = 5812$$

$$B_2 = 901.3$$

$$K_2 = \frac{1382}{901.3} = 1.5334$$

Choose CB_1 equal to 0.027 microfarads and calculate RA_1 using Equation (8-55).

$$RA_1 = \frac{1}{K_1 \cdot B_1 \cdot CB_1}$$

$$RA_1 = 26.80 \text{ kohms}$$

Choose RA_1 equal to 26.7 kohms. Calculate CC_{1min} using Equation (8-59).

$$CC_{1min} = \left(\frac{K_1 \cdot B_1^2}{\omega_{01}^2} - 1 \right) \cdot CB_1$$

Since CC_{1min} is negative there is no restriction

Choose CC_1 equal to 0.033 microfarads and Calculate RD_1 using Equation (8-56).

$$RD_1 = \frac{1}{B_1} \cdot \left(\frac{1}{CC_1} + \frac{1}{CB_1} \right)$$

$$RD_1 = 63.93 \text{ kohms}$$

Choose RD_1 equal to 63.4 kohms. Calculate RE_1 using Equation (8-57).

$$RE_1 = \frac{1}{\dfrac{\omega_{01}^2 \cdot (CB_1 + CC_1)}{B_1} - K_1 \cdot B_1 \cdot CB_1}$$

$$RE_1 = 385.9 \text{ ohms}$$

Choose RE_1 equal to 383 ohms.

Choose CB_2 equal to 0.027 microfarads and calculate RA_2 using Equation (8-55).

$$RA_2 = \frac{1}{K_2 \cdot B_2 \cdot CB_2}$$

$$RA_2 = 26.80 \text{ kohms}$$

Choose RA_2 equal to 26.7 kohms. Calculate CC_{2min} using Equation (8-59).

$$CC_{2min} = \left(\frac{K_2 \cdot B_2^2}{\omega_{02}^2} - 1 \right) \cdot CB_2$$

Since CC_{2min} is negative there is no restriction

Choose CC_2 equal to 0.047 microfarads and Calculate RD_2 using Equation (8-56).

$$RD_2 = \frac{1}{B_2} \cdot \left(\frac{1}{CC_2} + \frac{1}{CB_2} \right)$$

$$RD_2 = 64.70 \text{ kohms}$$

Choose RD_2 equal to 64.9 kohms. Calculate RE_2 using Equation (8-57).

$$RE_2 = \frac{1}{\dfrac{\omega_{02}^2 \cdot (CB_2 + CC_2)}{B_2} - K_2 \cdot B_2 \cdot CB_2}$$

$$RE_2 = 365.5 \text{ ohms}$$

Choose RE_2 equal to 365 ohms. The final circuit is shown in Figure 8-26.

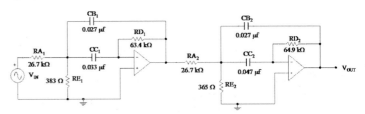

Figure 8-26 Example 8-5 Circuit

Figure 8-27 is the frequency response of the circuit of Figure 8-26 using a computer software circuit analysis program. The specification is superimposed in red lines. Note that the filter response exceeds both the pass band and stop band specifications.

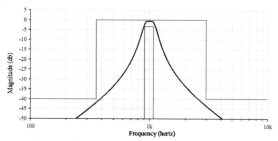

Figure 8-27 Filter Frequency Response

8-6. Notch Filter Realization

In Chapter 6 the transfer function for a low pass filter was converted to a narrow band notch filter by replacing s with $(1/(s+1/s))$. This resulted in doubling the number of poles which means only two pole stages will be required. This only applies to the narrow band implementation $(Q > 0.5)$. Wide band implementations $(Q \leq 0.5)$ are accomplished by separate low pass and high pass designs as was discussed in Chapter 6. From Chapter 6, a 2 pole notch transfer function is as shown in Equation (8-60).

$$G(s) = \frac{V_{OUT}}{V_{IN}} = \frac{K \cdot (s^2 + \omega_0^2)}{s^2 + N \cdot s + \omega_b^2}$$

(8-60)

In Equation (8-60), K is the pass max magnitude at low and high frequencies, N is the notchwidth (upper pass frequency - lower pass frequency), ω_0 is the center frequency, and ω_b is the denominator break frequency which can be greater than, less than, or equal to ω_0. Also from Chapter 6, $Q = \omega_0 / N$.

Although two pole notch filter circuits exist they are more complex than the low pass, high pass, and band pass circuits discussed previously. A simpler approach is an indirect approach using high pass and low pass circuits. The transfer function of a notch filter from Equation (8-60) can be manipulated into high pass and low pass functions as shown in Equation (8-61).

$$G(s) = \frac{K \cdot (s^2 + \omega_0^2)}{s^2 + N \cdot s + \omega_b^2}$$

$$G(s) = \left(\frac{K \cdot s^2}{s^2 + N \cdot s + \omega_b^2}\right) + \left(\frac{K \cdot \left(\frac{\omega_0^2}{\omega_b^2}\right) \cdot \omega_b^2}{s^2 + N \cdot s + \omega_b^2}\right)$$

(8-61)

From Equation (8-61) it can be seen that the high pass section can be implemented by Equation (8-34) and Figure 8-16 with K equal to K, ω_c/Q equal to N, and ω_c equal to ω_b. Also the low pass section can be implemented by Equation (8-22) and Figure 8-8 with K equal to $K \cdot (\omega_0^2/\omega_b^2)$, ω_c/Q equal to N, and ω_c equal to ω_b. Figure 8-28 shows the two pole notch filter circuit. The third op-amp and its 3 resistors provides the summation of the high pass and low pass sections as well as the numerator gain expressions. Also be aware that the output is negated as was the case for the band pass filter. The verification of the third op-amp summing function is left as an exercise in Problem 8-11.

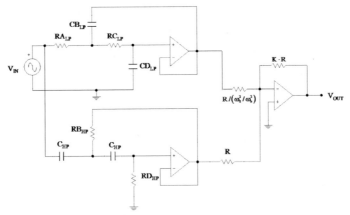

Figure 8-28 Two Pole Notch Filter Circuit

The design procedure starts by determining the filter transfer function partitioned in the form of Equation (8-60) using the methods of Chapter 6. The stages can be implemented using the two pole filter stage of Figure 8-28. Use the high pass procedure of Section 8-4 and the low pass procedure of Section 8-3 with K equal to 1. Then select a value for R for the third op-amp circuit and calculate the other two resistors from Equations (8-62).

Resistor $(K \cdot R) = K \cdot R$

$$\text{Resistor } \left(R/\left(\omega_0^2/\omega_b^2\right)\right) = \frac{R}{\omega_0^2/\omega_b^2} \tag{8-62}$$

The filter stages can then be cascaded to form the final filter design. In order to keep the individual filter stages from saturating, always place the circuit with the lowest value of K for the low pass section closest to the input and follow in the cascading operation in order of increasing K.

Example 8-6

Design an active notch Butterworth filter with a maximum pass band magnitude of 0 db. The notch center frequency is 1000 hertz with a 1500 hertz notchwidth at the -3.0103 db points. There shall be a maximum stop band magnitude of -40 db above 950 hertz and below 1050 hertz.

First determine the filter Q using Equation (6-53):

$$Q = \frac{\omega_0}{N} = \frac{2 \cdot \pi \cdot 1000}{2 \cdot \pi \cdot 1500} = 0.667$$

Since Q is greater than 0.5, use the narrow band approach.
Next, create the filter specification graph. Since the center frequency and notchwidth are given, they must be normalized and converted to the upper and lower pass frequencies. The normalized center frequency and notchwidth are:

$$\omega_0 = \frac{f_0}{1000} = 1 \quad N = \frac{1500}{1000} = 1.5$$

The upper and lower pass frequencies can be calculated from Equation (6-52):

$$f_0 = 1000 = \sqrt{f_{ph} \cdot f_{pl}} \Rightarrow f_{pl} = \frac{1000^2}{f_{ph}}$$

$$N_f = 1500 = f_{ph} - f_{pl} = f_{ph} - \frac{1000^2}{f_{ph}}$$

$$f_{ph}^2 - 1500 \cdot f_{ph} - 1000^2 = 0 \Rightarrow f_{ph} = 2000 \quad \text{and} \quad f_{pl} = 500$$

The specification graph is shown in Figure 8-29:

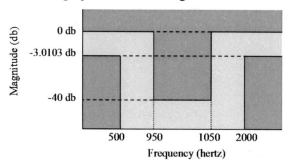

Figure 8-29 Filter Specification Graph for Example 8-6

The next step is to normalize the specifications. The magnitude axis is already normalized since the pass max value is 0 db. In order to allow for component variations, the design notchwidth needs to be decreased by 10%. The design notchwidth then becomes 1350 hertz. The design upper and lower pass frequencies can be calculated from Equation (6-52):

$$f_0 = 1000 = \sqrt{f_{ph} \cdot f_{pl}} \Rightarrow f_{pl} = \frac{1000^2}{f_{ph}}$$

$$N_f = 1350 = f_{ph} - f_{pl} = f_{ph} - \frac{1000^2}{f_{ph}}$$

$$f_{ph}^2 - 1350 \cdot f_{ph} - 1000^2 = 0 \Rightarrow f_{ph} = 1881 \quad \text{and} \quad f_{pl} = 531$$

To normalize the frequency axis divide all of the specifications by f_0 or 1000. Since this is a frequency divided by a frequency, the normalized axis can be considered either hertz or radians per second. The normalized frequency values are:

$$\omega_{pl} = \frac{531}{1000} = 0.531 \quad \omega_{ph} = \frac{1881}{1000} = 1.881$$

$$\text{lower stop} = \frac{950}{1000} = 0.95 \quad \text{upper stop} = \frac{1050}{1000} = 1.05$$

The next step is to determine the low pass equivalent. To determine the equivalent low pass number of poles, determine the pass notchwidth N and the geometric stop notchwidth:

$$N = \frac{N_f}{f_0} = \frac{1350}{1000} = 1.35$$

$$\text{or } N = \omega_{ph} - \omega_{pl} = 1.881 - 0.531 = 1.35$$

$1/(\text{lower stop}) = 1/0.95 = 1.0526$ which is greater than
the upper stop

Therefore: $\omega_{sl} = 0.95$ and $\omega_{sh} = \dfrac{1}{0.95} = 1.0526$

geometric stop width $= \omega_{sh} - \omega_{sl} = 1.0526 - 0.95 = 0.1026$

Determine the number of equivalent low pass poles using Equation (6-59):

$$n = \frac{\log_{10}\left(\frac{10^{(0.1 \cdot A_s)} - 1}{10^{(0.1 \cdot A_p)} - 1}\right)}{2 \cdot \log_{10}\left(\frac{\text{notchwidth}}{\text{geometric stop width}}\right)}$$

$$= \frac{\log_{10}\left(\frac{10^{(4)} - 1}{10^{(0.30103)} - 1}\right)}{2 \cdot \log_{10}\left(\frac{1.35}{0.1026}\right)} = 1.787$$

Rounding up to the next highest integer yields n equal to 2. Using Table 6-1 for the 2 pole entry, the normalized low pass equivalent transfer function is:

$$G(s)_{LP} = \frac{1}{s^2 + \dfrac{s}{0.7071} + 1}$$

Translating this to the pass band frequency of 1/N yields:

$$G(s)_{LP} = \frac{1}{\left((1.35 \cdot s)^2 + \dfrac{1}{0.7071} \cdot (1.35 \cdot s) + 1\right)}$$

Transforming this to a notch using Equations (6-60) and (6-62) yields:

$$G(s) = \frac{\left(s^2 + 1\right)^2}{\left(s^4 + \dfrac{1.35}{0.7071} \cdot s^3 + 3.8225 \cdot s^2 + \dfrac{1.35}{0.7071} \cdot s + 1\right)}$$

Factoring the 4th order denominator yields:

$$G(s) = \frac{s^2 + 1}{\left(s^2 + 1.3994 \cdot s + 2.7447\right)} \cdot \frac{s^2 + 1}{\left(s^2 + 0.5098 \cdot s + 0.3643\right)}$$

Translating this out to the specification center frequency of 1000 hertz which is $2000 \cdot \pi$ radians per second yields:

$$G(s) = \frac{\left(\dfrac{s}{2000 \cdot \pi}\right)^2 + 1}{\left(\left(\dfrac{s}{2000 \cdot \pi}\right)^2 + 1.3994 \cdot \left(\dfrac{s}{2000 \cdot \pi}\right) + 2.7447\right)} \cdot$$

$$\frac{\left(\dfrac{s}{2000 \cdot \pi}\right)^2 + 1}{\left(\left(\dfrac{s}{2000 \cdot \pi}\right)^2 + 0.5098 \cdot \left(\dfrac{s}{2000 \cdot \pi}\right) + 0.3643\right)}$$

$$G(s) = \frac{s^2 + 3.9478 \cdot 10^7}{\left(s^2 + 8793 \cdot s + 1.0836 \cdot 10^8\right)} \cdot$$

$$\frac{s^2 + 3.9478 \cdot 10^7}{\left(s^2 + 3203 \cdot s + 1.4382 \cdot 10^7\right)}$$

Placing stage 1 in the form of Equation (8-61) yields:

$$G_1(s) = \frac{s^2 + 3.9478 \cdot 10^7}{\left(s^2 + 8793 \cdot s + 1.0836 \cdot 10^8\right)}$$

$$= \frac{s^2}{\left(s^2 + 8793 \cdot s + 1.0836 \cdot 10^8\right)}$$

$$+ \frac{0.3643 \cdot 1.0836 \cdot 10^8}{\left(s^2 + 8793 \cdot s + 1.0836 \cdot 10^8\right)}$$

For stage 1 ω_b, ω_0, and Q are:

$$\omega_{b1} = \sqrt{1.0836 \cdot 10^8} = 10,410$$

$$\omega_0 = \sqrt{3.9478 \cdot 10^7} = 6,283$$

$$\frac{\omega_{b1}}{Q_1} = 8,793 \Rightarrow Q_1 = \frac{10,410}{8,793} = 1.1839$$

For the stage 1 high pass section design, select CA_1 and CC_1 equal to 0.018 microfarads and use Equation (8-40) to determine RB_1:

$$RB_1 = \frac{1}{4 \cdot C_1 \cdot Q_1 \cdot \omega_{c1}} + \frac{1}{C_1 \cdot \omega_{c1}} \cdot \sqrt{\frac{1}{16 \cdot Q_1^2} - \frac{1 - K_1}{2}}$$

$$RB_1 = 2.254 \text{ kohms}$$

Use Equation (8-37) to calculate RD_1:

$$RD_1 = \frac{1}{RB_1 \cdot C_1^2 \cdot \omega_{c1}^2} = 12.64 \text{ kohms}$$

Select standard values of 2.26 kohms and 12.7 kohms for RB_1 and RD_1 respectively.

For the stage 1 low pass section design, select CB_1 equal to 0.082 microfarads and use Equation (8-28) To determine CD_{1max}:

$$CD_{1max} = CB_1 \cdot \left(\frac{1}{4 \cdot Q_1^2} + K_1 - 1\right) = 0.0146 \cdot 10^{-6}$$

Select CD_1 equal to 0.0056 microfarads. Use Equation (8-27) to calculate RC_1:

$$RC_1 = \frac{1}{2 \cdot \omega_{c1} \cdot Q_1 \cdot CD_1} + \frac{1}{\omega_{c1}} \cdot \sqrt{\frac{1}{4 \cdot Q_1^2 \cdot CD_1^2} - \frac{1}{CD_1} \cdot \left(\frac{1}{CB_1} + \frac{1 - K_1}{CD_1}\right)}$$

$$RC_1 = 12.94 \text{ kohms}$$

Use Equation (8-25) to calculate RA_1:

$$RA_1 = \frac{1}{RC_1 \cdot CB_1 \cdot CD_1 \cdot \omega_{c1}^2} = 1.553 \text{ kohms}$$

Select standard values of 13.0 kohms and 1.54 kohms for RC_1 and RA_1 respectively.

For the stage 1 summing op-amp section design, select R_1 equal to 1000 ohms and use Equations (8-62) to calculate the remaining resistors:

Resistor $(K \cdot R) = K_1 \cdot R_1 = 1 \cdot 1,000 = 1,000$ ohms

Resistor $\left(R / \left(\omega_0^2 / \omega_b^2\right)\right) = \frac{R_1}{\omega_0^2 / \omega_{b1}^2} = \frac{1,000}{0.3643} = 2,745$ ohms

Select $K_1 R_1$ equal to 1.00 kohms and $R_1 / \left(\omega_0^2 / \omega_{b1}^2\right)$ equal to 2.74 kohms.

Placing stage 2 in the form of Equation (8-61) yields:

$$G_2(s) = \frac{s^2 + 3.9478 \cdot 10^7}{\left(s^2 + 3203 \cdot s + 1.4382 \cdot 10^7\right)}$$

$$G_2(s) = \frac{s^2}{(s^2 + 3203 \cdot s + 1.4382 \cdot 10^7)}$$
$$+ \frac{2.745 \cdot 1.4382 \cdot 10^7}{(s^2 + 3203 \cdot s + 1.4382 \cdot 10^7)}$$

For stage 2 ω_b, ω_0, and Q are:

$$\omega_{b2} = \sqrt{1.4382 \cdot 10^7} = 3,792$$

$$\omega_0 = \sqrt{3.9478 \cdot 10^7} = 6,283$$

$$\frac{\omega_{b2}}{Q_2} = 3,203 \Rightarrow Q_2 = \frac{3,792}{3,203} = 1.1839$$

For the stage 2 high pass section design, select CA_2 and CC_2 equal to 0.047 microfarads and use Equation (8-40) to determine RB_2:

$$RB_2 = \frac{1}{4 \cdot C_2 \cdot Q_2 \cdot \omega_{c2}} + \frac{1}{C_2 \cdot \omega_{c2}} \cdot \sqrt{\frac{1}{16 \cdot Q_2^2} - \frac{1 - K_2}{2}}$$

$$RB_2 = 2.3697 \text{ kohms}$$

Use Equation (8-37) to calculate RD_2:

$$RD_2 = \frac{1}{RB_2 \cdot C_2^2 \cdot \omega_{c2}^2} = 13.29 \text{ kohms}$$

Select standard values of 2.37 kohms and 13.3 kohms for RB_2 and RD_2 respectively.

For the stage 2 low pass section design, select CB_2 equal to 0.039 microfarads and use Equation (8-28) To determine CD_{2max}:

$$CD_{2\max} = CB_2 \cdot \left(\frac{1}{4 \cdot Q_2^2} + K_2 - 1 \right) = 0.00696 \cdot 10^{-6}$$

Select CD_2 equal to 0.0068 microfarads. Use Equation (8-27) to calculate RC_2 and Equation (8-25) to calculate RA_2:

$$RC_2 = \frac{1}{2 \cdot \omega_{c2} \cdot Q_2 \cdot CD_2} + \frac{1}{\omega_{c2}} \cdot \sqrt{\frac{1}{4 \cdot Q_2^2 \cdot CD_2^2} - \frac{1}{CD_2} \left(\frac{1}{CB_2} + \frac{1 - K_2}{CD_2} \right)}$$

$$RC_2 = 18.83 \text{ kohms}$$

$$RA_2 = \frac{1}{RC_2 \cdot CB_2 \cdot CD_2 \cdot \omega_{c2}^2} = 13.92 \text{ kohms}$$

Select standard values of 18.7 kohms and 14.0 kohms for RC_2 and RA_2 respectively.

For the stage 2 summing op-amp section design, select R_2 equal to 4990 ohms and use Equations (8-62) to calculate the remaining resistors:

$$\text{Resistor } (K \cdot R) = K_2 \cdot R_2 = 1 \cdot 4,990 = 4,990 \text{ ohms}$$

$$\text{Resistor } \left(R / \left(\omega_0^2 / \omega_b^2 \right) \right) = \frac{R_2}{\omega_0^2 / \omega_{b2}^2} = \frac{4,990}{2.745} = 1,818 \text{ ohms}$$

Select $K_2 R_2$ equal to 4.99 kohms and $R_2 / \left(\omega_0^2 / \omega_{b2}^2 \right)$ equal to 1.82 kohms. The final circuit is shown in Figure 8-30.

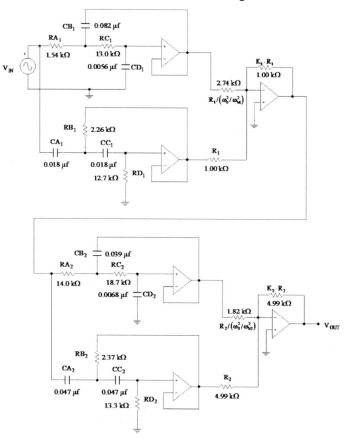

Figure 8-30 Example 8-6 Circuit

Figure 8-31 is the frequency response of the circuit of Figure 8-30 using a computer software circuit analysis program. The specification is superimposed in red lines. Note that the filter response exceeds both the pass band and stop band specifications.

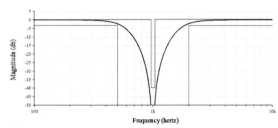

Figure 8-31 Filter Frequency Response

8-7. State Variable Filter Realization

There is another popular filter realization circuit called the state variable filter. Before the digital computer was invented all circuits and systems were analog, even the computers. The analog computer, which is now extinct, consisted of an array of operational amplifiers on a plug board. The op-amps could be configured as inverters, gain functions, integrators, or differentiators by plugging in the appropriate resistors and capacitors and interconnecting them with wire jumpers.

The analog computer was used primarily to solve complex integral-differential equations of systems. Actually, using today's terminology, the system of equations was being simulated by the analog computer. The process of rearranging a differential equation for simulation on the analog computer is referred to as the state variable process and can also be applied to defining a filter circuit which has been adopted by several manufactures of analog filter modules.

Since the filter transfer functions are in the Laplace domain, it is necessary to realize that multiplication and division by s is equivalent to differentiation and integration in the time domain as was discussed in Chapter 4. Thus, the filter transfer function is essentially an integral-differential equation.

To start the rearranging process, consider the 2 pole low pass filter transfer function of Equation (8-21). Change the form of the equation from break frequency form to polynomial form as shown in Equation (8-63).

$$\frac{V_{OLP}}{V_{IN}} = \frac{K}{\dfrac{s^2}{\omega_c^2} + \dfrac{1}{Q} \cdot \dfrac{s}{\omega_c} + 1} = \frac{K \cdot \omega_c^2}{s^2 + \dfrac{\omega_c}{Q} \cdot s + \omega_c^2} \tag{8-63}$$

206

The denominators can be eliminated by cross multiplying as show in Equation (8-64).

$$\frac{V_{OLP}}{V_{IN}} = \frac{K \cdot \omega_c^2}{s^2 + \frac{\omega_c}{Q} \cdot s + \omega_c^2}$$

$$\Rightarrow V_{OLP} \cdot \left(s^2 + \frac{\omega_c}{Q} \cdot s + \omega_c^2 \right) = V_{IN} \cdot K \cdot \omega_c^2 \qquad (8\text{-}64)$$

In Chapter 5, bode plots for the function s and $1/s$ were shown to be a positive slope for s and a negative slope for $1/s$. The important factor here is that for the s function, the magnitude increases as frequency is increased and for the $1/s$ function, the magnitude decreases as frequency is increased. Consequently, the s function (differentiation) results in a much noisier signal than the $1/s$ function (integration). Therefore, the differential equation solutions using analog computer methods are implemented with integrators rather than differentiators.

Equation (8-64) has s and s^2 terms which is the equivalent of differentiators. Dividing both sides of Equation (8-64) by s^2 will convert the equation to integrators as shown in Equation (8-65).

$$V_{OLP} \cdot \left(s^2 + \frac{\omega_c}{Q} \cdot s + \omega_c^2 \right) = V_{IN} \cdot K \cdot \omega_c^2$$

$$\Rightarrow V_{OLP} \cdot \left(1 + \frac{\omega_c}{Q \cdot s} + \frac{\omega_c^2}{s^2} \right) = \frac{V_{IN} \cdot K \cdot \omega_c^2}{s^2} \qquad (8\text{-}65)$$

Rearranging Equation (8-65) yields Equation (8-66).

$$V_{OLP} = \frac{V_{IN} \cdot K \cdot \omega_c^2}{s^2} - \frac{V_{OLP} \cdot \omega_c^2}{s^2} - \frac{V_{OLP} \cdot \omega_c}{Q \cdot s} \qquad (8\text{-}66)$$

Equation (8-66) can be modified slightly to achieve a common denominator of s^2 on the right side of the equation. This modification will generate a much more versatile filter circuit. The modified equation is shown in Equation (8-67).

$$V_{OLP} = \frac{V_{IN} \cdot K \cdot \omega_c^2}{s^2} - \frac{V_{OLP} \cdot \omega_c^2}{s^2} - \frac{V_{OLP} \cdot \frac{\omega_c}{Q} \cdot s}{s^2} \qquad (8\text{-}67)$$

Using analog computer methods, Equation (8-67) is first converted to block diagram form using integrators, summers, and gain blocks as shown in Figure 8-32.

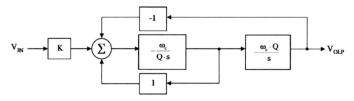

Figure 8-32 State Variable Filter Block Diagram

The block diagram blocks represent transfer functions that when properly combined will yield Equation (8-67). The summer has three inputs which all contribute to the output. V_{OLP1} gets its flow from V_{IN} as shown in Equation (8-68) which corresponds to the first term on the right side of Equation (8-67).

$$V_{OLP1} = V_{IN} \cdot K \cdot \left(-\frac{\omega_c}{Q \cdot s} \right) \cdot \left(-\frac{\omega_c \cdot Q}{s} \right) = \frac{V_{IN} \cdot K \cdot \omega_c^2}{s^2} \qquad (8\text{-}68)$$

V_{OLP2} gets its flow from V_{OLP} as shown in Equation (8-69) which corresponds to the second term on the right side of Equation (8-67).

$$V_{OLP2} = V_{OLP} \cdot (-1) \cdot \left(-\frac{\omega_c}{Q \cdot s} \right) \cdot \left(-\frac{\omega_c \cdot Q}{s} \right) = -\frac{V_{OLP} \cdot \omega_c^2}{s^2} \qquad (8\text{-}69)$$

V_{OLP3} gets its flow from the first integrator block output. In order for the output to be V_{OLP}, the input to the second integrator block must be $V_{OLP} \cdot (-s)/(\omega_c \cdot Q)$ which is also the output of the first integrator block. Equation (8-70) shows the flow of V_{OLP3} which corresponds to the third term of Equation (8-67).

$$V_{OLP3} = V_{OLP} \cdot \left(-\frac{s}{\omega_c \cdot Q} \right) \cdot (1) \cdot \left(-\frac{\omega_c}{Q \cdot s} \right) \cdot \left(-\frac{\omega_c \cdot Q}{s} \right)$$

$$= -\frac{V_{OLP} \cdot \dfrac{\omega_c}{Q} \cdot s}{s^2} \qquad (8\text{-}70)$$

Since VOLP is the sum of VOLP1, VOLP2, and VOLP3, the block diagram is the implementation of Equation (8-67). But why make the modification of the third term on the right side of Equation (8-67)? What if the block diagram output is taken from the output of the first integrator rather than the output of the second integrator? Figure 8-33 shows the block diagram with the output from the first integrator.

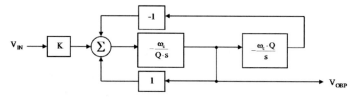

Figure 8-33 First Alternate Output

There are still three terms in the output equation due to the summer. The three expressions making up the output are shown in Equation (8-71).

$$V_{OBP} = V_{IN} \cdot (K) \cdot \left(-\frac{\omega_c}{Q \cdot s} \right)$$

$$+ V_{OBP} \cdot \left(-\frac{\omega_c \cdot Q}{s} \right) \cdot (-1) \cdot \left(-\frac{\omega_c}{Q \cdot s} \right)$$

$$+ V_{OBP} \cdot (1) \cdot \left(-\frac{\omega_c}{Q \cdot s} \right)$$

$$V_{OBP} = -\frac{V_{IN} \cdot K \cdot \omega_c}{Q \cdot s} - \frac{V_{OBP} \cdot \omega_c^2}{s^2} - \frac{V_{OBP} \cdot \omega_c}{Q \cdot s} \qquad (8\text{-}71)$$

Solving Equation (8-71) for V_{OBP}/V_{IN} yields Equation (8-72).

$$V_{OBP} = -\frac{V_{IN} \cdot K \cdot \omega_c}{Q \cdot s} - \frac{V_{OBP} \cdot \omega_c^2}{s^2} - \frac{V_{OBP} \cdot \omega_c}{Q \cdot s}$$

$$V_{OBP} \cdot \left(1 + \frac{\omega_c}{Q \cdot s} + \frac{\omega_c^2}{s^2} \right) = -\frac{V_{IN} \cdot K \cdot \omega_c}{Q \cdot s}$$

Multiplying both sides by s^2 yields:

$$V_{OBP} \cdot \left(s^2 + \frac{\omega_c}{Q} \cdot s + \omega_c^2 \right) = -\frac{V_{IN} \cdot K \cdot \omega_c \cdot s}{Q}$$

$$\frac{V_{OBP}}{V_{IN}} = -\frac{K \cdot \dfrac{\omega_c}{Q} \cdot s}{s^2 + \dfrac{\omega_c}{Q} \cdot s + \omega_c^2} \qquad (8\text{-}72)$$

Equation (8-72) is of the form of Equation (8-41), the equation for a band pass filter. Since, for a band pass filter, ω_0/Q is the bandwidth B, letting ω_c be ω_0 will yield Equation (8-41) but with a negative sign as was the case for the design of Section 8-5.

Next consider the block diagram output being from the summer as shown in Figure 8-34.

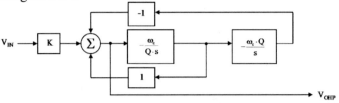

Figure 8-34 Second Alternate Output

There are still three terms in the output equation due to the summer. The three expressions making up the output are shown in Equation (8-73).

$$V_{OHP} = V_{IN} \cdot (K)$$

$$+ V_{OHP} \cdot \left(-\frac{\omega_c}{Q \cdot s} \right) \cdot \left(-\frac{\omega_c \cdot Q}{s} \right) \cdot (-1)$$

$$+ V_{OHP} \cdot \left(-\frac{\omega_c}{Q \cdot s} \right) \cdot (1)$$

$$V_{OHP} = V_{IN} \cdot K - \frac{V_{OHP} \cdot \omega_c^2}{s^2} - \frac{V_{OHP} \cdot \omega_c}{Q \cdot s} \qquad (8\text{-}73)$$

Solving Equation (8-73) for V_{OHP}/V_{IN} yields Equation (8-74).

$$V_{OHP} = V_{IN} \cdot K - \frac{V_{OHP} \cdot \omega_c^2}{s^2} - \frac{V_{OHP} \cdot \omega_c}{Q \cdot s}$$

$$V_{OHP} \cdot \left(1 + \frac{\omega_c}{Q \cdot s} + \frac{\omega_c^2}{s^2} \right) = V_{IN} \cdot K$$

Multiplying both sides by s^2 yields:

$$V_{OHP} \cdot \left(s^2 + \frac{\omega_c}{Q} \cdot s + \omega_c^2 \right) = V_{IN} \cdot K \cdot s^2$$

$$\frac{V_{OHP}}{V_{IN}} = -\frac{K \cdot s^2}{s^2 + \frac{\omega_c}{Q} \cdot s + \omega_c^2} \qquad (8\text{-}74)$$

Equation (8-74) is of the form of the 2 pole transfer function of Equation (8-30), the equation for a high pass filter. The versatility of the state variable filter is now apparent. The same filter has three separate outputs – one for a low pass realization, one for a high pass realization, and one for a band pass realization.

The next step is to convert the block diagram to an actual active circuit. The two integrator blocks (the ones with an s in the denominator) can be implemented with the circuit of Figure 8-35.

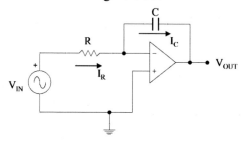

Figure 8-35 Integrator Circuit

Writing the nodal equation for the negative input node of the op-amp yields Equation (8-75).

$$I_R = I_C \tag{8-75}$$

Since the positive input of the op-amp is connected to ground, the negative op-amp input must be at aero potential for the output to be a finite value. Using this and converting Equation (8-75) to nodal voltages yields Equation (8-76).

$$\frac{V_{IN} - 0}{R} = \frac{0 - V_{OUT}}{\dfrac{1}{C \cdot s}} \Rightarrow \frac{V_{OUT}}{V_{IN}} = -\frac{1}{R \cdot C \cdot s} \tag{8-76}$$

The summer block and the three gain blocks (K, 1, and -1) can be implemented with the circuit of Figure 8-36.

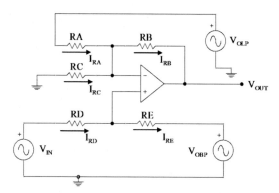

Figure 8-36 Summer Circuit

First, write the nodal equation for the op-amp positive input node as shown in Equation (8-77).

$$I_{RD} = I_{RE}$$

$$\frac{V_{IN} - V^+}{RD} = \frac{V^+ - V_{OBP}}{RE} \Rightarrow V^+ \cdot \left(\frac{1}{RD} + \frac{1}{RE}\right) = \frac{V_{IN}}{RD} + \frac{V_{OBP}}{RE}$$

$$V^+ = \frac{V_{IN} \cdot RE}{RD + RE} + \frac{V_{OBP} \cdot RD}{RD + RE} \tag{8-77}$$

Next, write the nodal equation for the op-amp negative input node as shown in Equation (8-78).

$$I_{RA} + I_{RC} = I_{RB}$$

$$\frac{V_{OLP} - V^-}{RA} + \frac{0 - V^-}{RC} = \frac{V^- - V_{OUT}}{RB}$$

$$\Rightarrow \frac{V_{OUT}}{RB} = V^- \cdot \left(\frac{1}{RA} + \frac{1}{RB} + \frac{1}{RC}\right) - \frac{V_{OLP}}{RA}$$

$$\Rightarrow V_{OUT} = V^- \cdot RB \cdot \left(\frac{1}{RA} + \frac{1}{RB} + \frac{1}{RC}\right) - \frac{V_{OLP} \cdot RB}{RA} \tag{8-78}$$

Since the op-amp positive and negative inputs must be equal for the output to be finite, V+ from Equation (8-77) can be substituted into Equation (8-78) for V- to yield Equation (8-79).

$$V_{OUT} = \frac{V_{IN} \cdot RE \cdot RB}{RD + RE} \cdot \left(\frac{1}{RA} + \frac{1}{RB} + \frac{1}{RC}\right)$$
$$+ \frac{V_{OBP} \cdot RD \cdot RB}{RD + RE} \left(\frac{1}{RA} + \frac{1}{RB} + \frac{1}{RC}\right) \tag{8-79}$$
$$- \frac{V_{OLP} \cdot RB}{RA}$$

In order to make Equation (8-79) agree with the block diagram, the assumptions of Equations (8-80) will be required.

$$RB = RA$$
$$RC = \frac{RA}{K-1} \quad \text{where } K \geq 1$$
$$RD = \frac{RA}{K} \tag{80}$$
$$RE = RA$$

Substituting Equations (8-80) into Equation (8-79) yields Equation (8-81) which agrees with the block diagram.

$$V_{OUT} = \frac{V_{IN} \cdot RA \cdot RA}{\dfrac{RA}{K} + RA} \cdot \left(\frac{1}{RA} + \frac{1}{RA} + \frac{K-1}{RA} \right)$$

$$+ \frac{V_{OBP} \cdot \dfrac{RA}{K} \cdot RA}{\dfrac{RA}{K} + RA} \left(\frac{1}{RA} + \frac{1}{RA} + \frac{K-1}{RA} \right)$$

$$- \frac{V_{OLP} \cdot RA}{RA}$$

$$V_{OUT} = V_{IN} \cdot K + V_{OBP} - V_{OLP} \tag{8-81}$$

The complete state variable circuit is shown in Figure 8-37.

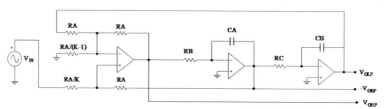

Figure 8-37 State Variable Filter Circuit

The design procedure starts by determining the required transfer function as was done for the previous active realizations with an even number of poles. Then select a standard value for CA and calculate RB from Equation (8-82).

$$-\frac{1}{RB \cdot CA \cdot s} = -\frac{\omega_c}{Q \cdot s} \Rightarrow RB = \frac{Q}{CA \cdot \omega_c} \tag{8-82}$$

Next, select a standard value for CB and calculate RC from Equation (8-83).

$$-\frac{1}{RC \cdot CB \cdot s} = -\frac{Q \cdot \omega_c}{s} \Rightarrow RC = \frac{1}{CB \cdot Q \cdot \omega_c} \tag{8-83}$$

Finally, select a standard value for RA and calculate RA/(K-1) and RA/K from Equations (8-84).

$$\text{Resistor RA/(K-1)} = \frac{RA}{K-1} \quad \text{Note: For } K = 1, \left(\frac{RA}{K-1} \right) \text{ is open}$$

$$\text{Resistor RA/K} = \frac{RA}{K} \tag{8-84}$$

If the resistor values are less than desirable, select other standard values and recalculate the resistor values. Repeat this process until acceptable resistor values are obtained. Since the filter calculations are rather lengthy, an EXCEL spreadsheet or math software can be used to speed the process and minimize errors.

The filter stages can then be cascaded to form the final filter design. In order to keep the individual filter stages from saturating, always place the circuit with the lowest value of K closest to the input and follow in the cascading operation in order of increasing K.

Example 8-7

Implement the low pass filter of Example 8-1 using the state variable filter realization.

From Example 8-1, the transfer function equation is:

$$G(s) = \cfrac{1}{\left(\left(\dfrac{s}{7,274}\right)^2 + \dfrac{1}{0.5412}\cdot\left(\dfrac{s}{7,274}\right)+1\right)} \cdot$$

$$\cfrac{1}{\left(\left(\dfrac{s}{7,274}\right)^2 + \dfrac{1}{1.3066}\cdot\left(\dfrac{s}{7,274}\right)+1\right)}$$

The filter can be implemented with two 2 pole stages using the filter circuit of Figure 8-37. The stage 1 requirements are:

$$K_1 = 1$$
$$\omega_{c1} = 7,274$$
$$Q_1 = 0.5412$$

The stage 2 requirements are:

$$K_2 = 1$$
$$\omega_{c2} = 7,274$$
$$Q_2 = 1.3066$$

For stage 1, select a standard value for CA_1 of 0.0068 microfarads and calculate RB_1 from Equation (8-82).

$$RB_1 = \frac{Q_1}{CA_1 \cdot \omega_{c1}} = 10.941 \text{ kohms}$$

Select a standard value of 11.0 kohms for RB_1. Next, select a standard value for CB_1 of 0.015 microfarads and calculate RC_1 from Equation (8-83).

$$RC_1 = \frac{1}{CB_1 \cdot Q_1 \cdot \omega_{c1}} = 16.935 \text{ kohms}$$

Select a standard value of 16.9 kohms for RC_1. Finally, select a standard value for RA_1 of 10.0 kohms and calculate $RA_1/(K_1-1)$ and RA_1/K_1 from Equations (8-84).

$$\text{Resistor } RA_1/(K_1-1) = \frac{RA_1}{K_1 - 1} \text{ which is an open since } K_1 = 1$$

$$\text{Resistor } RA_1/K_1 = \frac{RA_1}{K_1} = 10.0 \text{ kohms}$$

Select a standard value of 10.0 kohms for RA_1/K_1.

For stage 2, select a standard value for CA_2 of 0.012 microfarads and calculate RB_2 from Equation (8-82).

$$RB_2 = \frac{Q_2}{CA_2 \cdot \omega_{c2}} = 14.97 \text{ kohms}$$

Select a standard value of 15.0 kohms for RB_2. Next, select a standard value for CB_2 of 0.01 microfarads and calculate RC_2 from Equation (8-83).

$$RC_2 = \frac{1}{CB_2 \cdot Q_2 \cdot \omega_{c2}} = 10.522 \text{ kohms}$$

Select a standard value of 10.5 kohms for RC_2. Finally, select a standard value for RA_2 of 10.0 kohms and calculate $RA_2/(K_2-1)$ and RA_2/K_2 from Equations (8-84).

$$\text{Resistor } RA_2/(K_2-1) = \frac{RA_2}{K_2 - 1} \text{ which is an open since } K_2 = 1$$

$$\text{Resistor } RA_2/K_2 = \frac{RA_2}{K_2} = 10.0 \text{ kohms}$$

Select a standard value of 10.0 kohms for RA_2/K_2. The final circuit is shown in Figure 8-38.

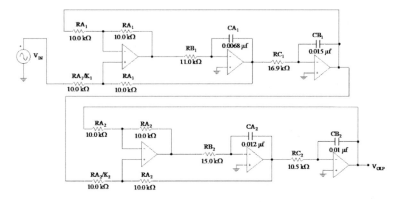

Figure 8-38 State Variable Filter for Example 8-7

Figure 8-39 is the frequency response of the circuit of Figure 8-38 using a computer software circuit analysis program. The specification is superimposed in red lines. Note that the filter response exceeds both the pass band and stop band specifications.

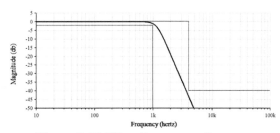

Figure 8-39 Filter Frequency Response

Note that the filter response is almost identical to the frequency response of Figure 8-11 of Example 8-1. However, the number of op-amps for Example 8-7 is three times the number of op-amps required for Example 8-1. Why then would a designer use the state variable circuit? Since the state variable filter has the versatility to be either a low pass, high pass, or band pass filter, manufacturers have placed the op-amps in a single integrated circuit which reduces the number of devices required to implement the filter.

Example 8-8

Implement the high pass filter of Example 8-4 using the state variable filter realization.

From Example 8-4, the transfer function equation is:

216

$$G(s) = \frac{5.723 \cdot s^2}{s^2 + \dfrac{61,085}{0.81} \cdot s + 61,085^2} \cdot$$

$$\frac{1.527 \cdot s^2}{s^2 + \dfrac{31,872}{3.7479} \cdot s + 31,872^2}$$

The filter can be implemented with two 2 pole stages using the filter circuit of Figure 8-37. The stage 1 requirements are:

$$K_1 = 5.723$$

$$\omega_{c1} = 61,085$$

$$Q_1 = 0.81$$

The stage 2 requirements are:

$$K_2 = 1.527$$

$$\omega_{c2} = 31,872$$

$$Q_2 = 3.7479$$

For stage 1, select a standard value for CA_1 of 0.001 microfarads and calculate RB_1 from Equation (8-82).

$$RB_1 = \frac{Q_1}{CA_1 \cdot \omega_{c1}} = 13.26 \text{ kohms}$$

Select a standard value of 13.3 kohms for RB_1. Next, select a standard value for CB_1 of 0.0012 microfarads and calculate RC_1 from Equation (8-83).

$$RC_1 = \frac{1}{CB_1 \cdot Q_1 \cdot \omega_{c1}} = 16.84 \text{ kohms}$$

Select a standard value of 16.9 kohms for RC_1. Finally, select a standard value for RA_1 of 60.4 kohms and calculate $RA_1/(K_1-1)$ and RA_1/K_1 from Equations (8-84).

$$\text{Resistor } RA_1/(K_1-1) = \frac{RA_1}{K_1 - 1} = 12.788 \text{ kohms}$$

$$\text{Resistor } RA_1/K_1 = \frac{RA_1}{K_1} = 10.55 \text{ kohms}$$

Select a standard values of 12.7 kohms for $RA_1/(K_1-1)$ and 10.5 kohms for RA_1/K_1

For stage 2, select a standard value for CA_2 of 0.01 microfarads and calculate RB_2 from Equation (8-82).

$$RB_2 = \frac{Q_2}{CA_2 \cdot \omega_{c2}} = 11.76 \text{ kohms}$$

Select a standard value of 11.8 kohms for RB_2. Next, select a standard value for CB_2 of 820 picofarads and calculate RC_2 from Equation (8-83).

$$RC_2 = \frac{1}{CB_2 \cdot Q_2 \cdot \omega_{c2}} = 10.21 \text{ kohms}$$

Select a standard value of 10.2 kohms for RC_2. Finally, select a standard value for RA_2 of 15.4 kohms and calculate $RA_2/(K_2-1)$ and RA_2/K_2 from Equations (8-84).

$$\text{Resistor } RA_2/(K_2-1) = \frac{RA_2}{K_2-1} = 29.22 \text{ kohms}$$

$$\text{Resistor } RA_2/K_2 = \frac{RA_2}{K_2} = 10.09 \text{ kohms}$$

Select a standard values of 29.4 kohms for $RA_1/(K_1-1)$ and 10.0 kohms for RA_1/K_1 The final circuit is shown in Figure 8-40.

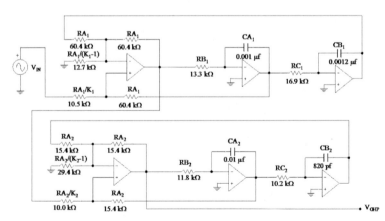

Figure 8-40 State Variable Filter for Example 8-8

218

Figure 8-41 is the frequency response of the circuit of Figure 8-40 using a computer software circuit analysis program. The specification is superimposed in red lines. Note that the filter response exceeds both the pass band and stop band specifications.

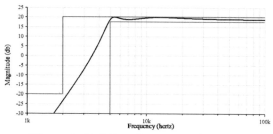

Figure 8-41 Filter Frequency Response

Example 8-9

Implement the band pass filter of Example 8-5 using the state variable filter realization.

From Example 8-5, the transfer function equation is:

$$G(s) = \frac{1382 \cdot s}{\left(s^2 + 1053.4 \cdot s + 4.61383 \cdot 10^7\right)} \cdot$$

$$\frac{1382 \cdot s}{\left(s^2 + 901.3 \cdot s + 3.378 \cdot 10^7\right)}$$

$$= \frac{1.3119 \cdot 1053.4 \cdot s}{\left(s^2 + 1053.4 \cdot s + 6793^2\right)} \cdot$$

$$\frac{1.5333 \cdot 901.3 \cdot s}{\left(s^2 + 901.3 \cdot s + 5812^2\right)}$$

The filter can be implemented with two 2 pole stages using the filter circuit of Figure 8-37. The stage 1 requirements are:

$$K_1 = 1.3119$$

$$\omega_{c1} = 6793$$

$$B_1 = 1053.4$$

$$Q_1 = \frac{\omega_{c1}}{B_1} = 6.449$$

The stage 2 requirements are:

219

$$K_2 = 1.5333$$

$$\omega_{c2} = 5812$$

$$B_2 = 901.3$$

$$Q_2 = \frac{\omega_{c2}}{B_2} = 6.449$$

For stage 1, select a standard value for CA_1 of 0.068 micro-farads and calculate RB_1 from Equation (8-82).

$$RB_1 = \frac{Q_1}{CA_1 \cdot \omega_{c1}} = 13.96 \text{ kohms}$$

Select a standard value of 14.0 kohms for RB_1. Next, select a standard value for CB_1 of 0.0018 microfarads and calculate RC_1 from Equation (8-83).

$$RC_1 = \frac{1}{CB_1 \cdot Q_1 \cdot \omega_{c1}} = 12.68 \text{ kohms}$$

Select a standard value of 12.7 kohms for RC_1. Finally, select a standard value for RA_1 of 14.7 kohms and calculate $RA_1/(K_1-1)$ and RA_1/K_1 from Equations (8-84).

$$\text{Resistor } RA_1/(K_1-1) = \frac{RA_1}{K_1 - 1} = 47.13 \text{ kohms}$$

$$\text{Resistor } RA_1/K_1 = \frac{RA_1}{K_1} = 11.21 \text{ kohms}$$

Select a standard values of 47.5 kohms for $RA_1/(K_1-1)$ and 11.3 kohms for RA_1/K_1

For stage 2, select a standard value for CA_2 of 0.068 micro-farads and calculate RB_2 from Equation (8-82).

$$RB_2 = \frac{Q_2}{CA_2 \cdot \omega_{c2}} = 16.32 \text{ kohms}$$

Select a standard value of 16.2 kohms for RB_2. Next, select a standard value for CB_2 of 0.0015 microfarads and calculate RC_2 from Equation (8-83).

$$RC_2 = \frac{1}{CB_2 \cdot Q_2 \cdot \omega_{c2}} = 17.79 \text{ kohms}$$

Select a standard value of 17.8 kohms for RC_2. Finally, select a standard value for RA_2 of 15.4 kohms and calculate $RA_2/(K_2-1)$ and RA_2/K_2 from Equations (8-84).

$$\text{Re sistor } RA_2/(K_2-1) = \frac{RA_2}{K_2-1} = 28.88 \text{ kohms}$$

$$\text{Re sistor } RA_2/K_2 = \frac{RA_2}{K_2} = 10.04 \text{ kohms}$$

Select a standard values of 28.7 kohms for $RA_1/(K_1-1)$ and 10.0 kohms for RA_1/K_1 The final circuit is shown in Figure 8-42.

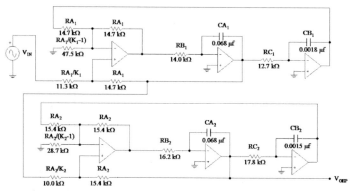

Figure 8-42 State Variable Filter for Example 8-9

Figure 8-43 is the frequency response of the circuit of Figure 8-42 using a computer software circuit analysis program. The specification is superimposed in red lines. Note that the filter response exceeds both the pass band and stop band specifications.

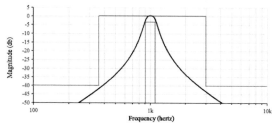

Figure 8-43 Filter Frequency Response

The state variable circuit doesn't have a notch filter output. However, the method used in Section 8-6 can also be applied to the state variable circuit. In Section 8-6 the notch transfer function was split into the sum of low pass and high pass transfer functions. The separate low pass and high pass transfer functions were designed and then applied to a summer circuit to form the notch filter.

The state variable realization has the advantage of having both a low pass and a high pass output from the same circuit that can then be applied to the summer circuit. Equation (8-61) is the notch transfer function for a single stage. Equation (8-61) can be split into high pass and low pass functions and related to the state variable transfer functions as shown in Equation (8-85).

$$G(s) = \frac{K \cdot \left(s^2 + \omega_0^2\right)}{s^2 + N \cdot s + \omega_b^2}$$

Let: $\omega_b = \omega_c$ and $N = \dfrac{\omega_c}{Q}$

$$G(s) = \left(\frac{K \cdot s^2}{s^2 + \dfrac{\omega_c}{Q} \cdot s + \omega_c^2}\right) + \left(\frac{K \cdot \left(\dfrac{\omega_0^2}{\omega_c^2}\right) \cdot \omega_c^2}{s^2 + \dfrac{\omega_c}{Q} \cdot s + \omega_c^2}\right) \tag{8-85}$$

The low pass and high pass K value for the state variable circuit will be 1 and the notch K value will be implemented in the summer circuit. The state variable notch filter will have a negated output just as the notch filter of Section 8-6. Figure 8-44 shows the state variable implementation of a notch filter stage.

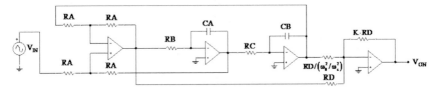

Figure 8-44 State Variable Notch Filter Circuit

The design procedure starts by determining the required transfer function as was done in Section 8-6 and then splitting each stage into high pass and low pass components in the form of Equation (8-85). The state variable method above can be used to implement the state variable stage as shown in Figure 8-44. Select a value for RD for the summing op-amp and calculate the other two resistors from Equations (8-86).

Resistor $(K \cdot RD) = K \cdot RD$

Resistor $\left(RD / \left(\omega_0^2 / \omega_c^2\right)\right) = \dfrac{RD}{\omega_0^2 / \omega_c^2}$ \hspace{1cm} (8-86)

The filter stages can then be cascaded to form the final filter design.

Example 8-10

Implement the notch filter of Example 8-6 using the state variable filter realization.

From Example 8-6, the transfer function equation is:

$$G(s) = \frac{s^2 + 3.9478 \cdot 10^7}{\left(s^2 + 8793 \cdot s + 1.0836 \cdot 10^8\right)} \cdot \frac{s^2 + 3.9478 \cdot 10^7}{\left(s^2 + 3203 \cdot s + 1.4382 \cdot 10^7\right)}$$

The filter can be implemented with two 2 pole stages using the filter circuit of Figure 8-44.

Splitting the first stage into high pass and low pass components and placing in the form of Equation (8-85) yields:

$$G_1(s) = \frac{s^2 + 3.9478 \cdot 10^7}{\left(s^2 + 8793 \cdot s + 1.0836 \cdot 10^8\right)}$$

$$= \frac{s^2 + 6283^2}{\left(s^2 + 8793 \cdot s + 10410^2\right)}$$

$$= \frac{s^2}{\left(s^2 + 8793 \cdot s + 10410^2\right)}$$

$$+ \frac{0.3643 \cdot 10410^2}{\left(s^2 + 8793 \cdot s + 10410^2\right)}$$

The stage 1 requirements are:

$$\omega_0 = 6283$$

$$K_{1N} = 1$$

$$K_{1HP} = 1$$

$$K_{1LP} = 0.3643$$

$$\omega_{c1} = 10410$$

$$\frac{\omega_{c1}}{Q_1} = 8793$$

$$Q_1 = 1.1839$$

For stage 1, select a standard value for CA_1 of 0.01 microfarads and calculate RB_1 from Equation (8-82).

$$RB_1 = \frac{Q_1}{CA_1 \cdot \omega_{c1}} = 11.37 \text{ kohms}$$

Select a standard value of 11.3 kohms for RB_1. Next, select a standard value for CB_1 of 0.0068 microfarads and calculate RC_1 from Equation (8-83).

$$RC_1 = \frac{1}{CB_1 \cdot Q_1 \cdot \omega_{c1}} = 11.93 \text{ kohms}$$

Select a standard value of 11.8 kohms for RC_1. Finally, select a standard value for RA_1 of 10.0 kohms.

Select a standard value of 10.0 kohms for RD_1 and use Equation (8-86) to calculate the remaining resistors.

Resistor $\left(K_{1N} \cdot RD_1 \right) = 10.0$ kohms

Resistor $\left(RD_1 / \left(\omega_0^2 / \omega_{c1}^2 \right) \right) = 27.45$ kohms

Select standard values of 10.0 kohms for $K_{1N} \cdot RD_1$ and 27.4 kohms for $RD_1 / \left(\omega_0^2 / \omega_{c1}^2 \right)$.

Splitting the second stage into high pass and low pass components and placing in the form of Equation (8-85) yields:

$$
\begin{aligned}
G_2(s) &= \frac{s^2 + 3.9478 \cdot 10^7}{\left(s^2 + 3203 \cdot s + 1.4382 \cdot 10^7 \right)} \\
&= \frac{s^2 + 6283^2}{\left(s^2 + 3203 \cdot s + 3792^2 \right)} \\
&= \frac{s^2}{\left(s^2 + 3203 \cdot s + 3792^2 \right)} \\
&\quad + \frac{2.745 \cdot 3792^2}{\left(s^2 + 3203 \cdot s + 3792^2 \right)}
\end{aligned}
$$

The stage 2 requirements are:

$$\omega_0 = 6283$$

$$K_{2N} = 1$$

$$K_{2HP} = 1$$

$$K_{2LP} = 2.745$$

$$\omega_{c2} = 3792$$

$$\frac{\omega_{c2}}{Q_2} = 3203$$

$$Q_2 = 1.1839$$

For stage 2, select a standard value for CA_2 of 0.018 microfarads and calculate RB_2 from Equation (8-82).

$$RB_2 = \frac{Q_2}{CA_2 \cdot \omega_{c2}} = 17.35 \text{ kohms}$$

Select a standard value of 17.4 kohms for RB_2. Next, select a standard value for CB_2 of 0.018 microfarads and calculate RC_2 from Equation (8-83).

$$RC_2 = \frac{1}{CB_2 \cdot Q_2 \cdot \omega_{c2}} = 12.37 \text{ kohms}$$

Select a standard value of 12.4 kohms for RC_2. Finally, select a standard value for RA_2 of 10.0 kohms.

Select a standard value of 30.1 kohms for RD_2 and use Equation (8-86) to calculate the remaining resistors.

Resistor $\left(K_{2N} \cdot RD_2 \right) = 30.1$ kohms

Resistor $\left(RD_2 / \left(\omega_0^2 / \omega_{c2}^2 \right) \right) = 10.96$ kohms

Select standard values of 30.1 kohms for $K_{2N} \cdot RD_2$ and 11.0 kohms for $RD_2 / \left(\omega_0^2 / \omega_{c2}^2 \right)$.

The final circuit is shown in Figure 8-45. Figure 8-46 is the frequency response of the circuit of Figure 8-45 using a computer software circuit analysis program. The specification is superimposed in red lines. Note that the filter response exceeds both the pass band and stop band specifications.

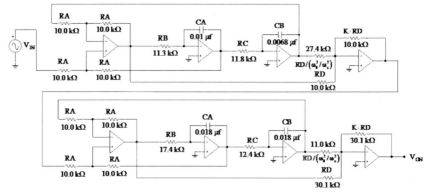

Figure 8-45 State Variable Filter for Example 8-10

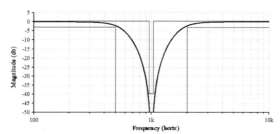

Figure 8-46 Filter Frequency Response

PROBLEMS

8-1. Design an active Butterworth low pass filter with a maximum pass band magnitude of 20 db, a minimum pass band magnitude of 18 db out to 2,000 hertz, and a maximum stop band magnitude of -20 db from 10,000 hertz on.

8-2. Design an active Chebyshev low pass filter with a maximum pass band magnitude of 0 db, a minimum pass band magnitude of -2 db out to 1000 hertz, and a maximum stop band magnitude of -40 db from 5000 hertz on.

8-3. Design an active Butterworth high pass filter with a maximum pass band magnitude of 20 db, a minimum pass band magnitude of 18 db from 10000 hertz on out, and a maximum stop band magnitude of -20 db below 2000 hertz.

226

8-4. Design an active Chebyshev high pass filter with a maximum pass band magnitude of 20 db, a minimum pass band magnitude of 18 db from 3000 hertz on out, and a maximum stop band magnitude of -20 db below 1000 hertz.

8-5. Design an active band pass Butterworth filter with a maximum pass band magnitude of 20 db at a center frequency of 1000 hertz and a 900 hertz bandwidth at the 16.9897 db points. There shall be a maximum stop band magnitude of -20 db below 50 hertz and above 20 khertz.

8-6. Design an active notch Butterworth filter with a maximum pass band magnitude of 20 db. The notch center frequency is 2000 hertz with a 2300 hertz notchwidth at the 16.9897 db points. There shall be a maximum stop band magnitude of -20 db above 1900 hertz and below 2100 hertz.

8-7. Implement the low pass filter of Problem 8-1 using the state variable filter realization.

8-8. Implement the high pass filter of Problem 8-3 using the state variable filter realization.

8-9. Implement the band pass filter of Problem 8-5 using the state variable filter realization.

8-10. Implement the notch filter of Problem 8-6 using the state variable filter realization.

8-11. Verify that the notch filter circuit of Figure 8-28 implements the transfer function of Equation (8-61).

APPENDIX A
NODAL VOLTAGE ANALYSIS

This analysis method can be used for basic steady state DC circuit analysis, basic steady state sinusoidal AC circuit analysis, and advanced transient and steady state circuit analysis. It can be applied to the DC circuit directly since the sources and impedances are real numbers. The sinusoidal AC circuit must be in impedance form with sources and impedances represented with complex numbers prior to application. The advanced circuit analysis circuit must be transformed into the Laplace domain with sources and impedances Laplace transformed prior to application.

A circuit consists of branches and nodes. A circuit branch is any circuit element (impedance or source) and a circuit node is the junction of two or more circuit branches. Figure A-1 is a part of a larger circuit but is representative of circuit nodes and branches. The circuit nodes are labeled N1 through N5 and the circuit branches are the impedances Z1 through Z4. Sources will be discussed later.

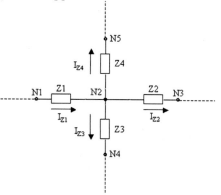

Figure A-1 Portion of a Larger Circuit

In order to derive the set of simultaneous equations for this portion of the circuit, it will be necessary to define the branch currents flowing through the impedances. The direction of the currents is arbitrary and any direction is acceptable but once a direction is defined it cannot be changed. Branch currents I_{Z1} through I_{Z4} have been defined in Figure A-1.

Kirchhoff's current law is used to write the equation for node N2. The sum of the currents into a node must equal the sum of the currents leaving the node as shown in Equation (A-1).

$$I_{Z1} = I_{Z2} + I_{Z3} + I_{Z4} \qquad \text{(A-1)}$$

These currents can be redefined in terms of node voltages (the voltage at the nodes with respect to a common reference node) using Ohm's law. A branch current can then be redefined as the node voltage at the tail of the current arrow minus the node voltage at the tip of the current arrow all divided by the branch impedance. Equation (A-1) is redefined in terms of node voltages in Equation (A-2).

$$\frac{V_{N1} - V_{N2}}{Z1} = \frac{V_{N2} - V_{N3}}{Z2} + \frac{V_{N2} - V_{N4}}{Z3} + \frac{V_{N2} - V_{N5}}{Z4} \tag{A-2}$$

Equation (A-2) can be rearranged by collecting each of the node voltage expressions on the left side of the equation as shown in Equation (A-3).

$$\left(\frac{1}{Z1}\right) \cdot V_{N1} - \left(\frac{1}{Z1} + \frac{1}{Z2} + \frac{1}{Z3} + \frac{1}{Z4}\right) \cdot V_{N2}$$

$$+ \left(\frac{1}{Z2}\right) \cdot V_{N3} + \left(\frac{1}{Z3}\right) \cdot V_{N4} + \left(\frac{1}{Z4}\right) \cdot V_{N5} = 0 \tag{A-3}$$

This process is repeated for each unknown node (known and unknown nodes will be discussed later) in the total circuit creating a set of simultaneous equations equal to the number of unknown nodes in the circuit.

What about sources? First, consider current sources. Figure A-2 shows Figure A-1 modified to include a current source. In this case the current flowing through the current source is the direction and magnitude of the current source rather than an arbitrarily defined direction. Again, using Kirchhoff's current law to write the equation for node N2 yields Equation (A-4).

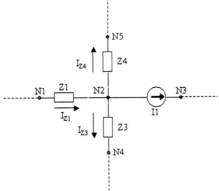

Figure A-2 Portion of a Larger Circuit with a Current Source

$$I_{Z1} = I1 + I_{Z3} + I_{Z4} \tag{A-4}$$

These currents will be redefined in terms of the node voltages as in Equation (A-2) with the exception of the current source which will remain I1 since it is a known value. Equation (A-4) is then redefined as Equation (A-5).

$$\frac{V_{N1} - V_{N2}}{Z1} = I1 + \frac{V_{N2} - V_{N4}}{Z3} + \frac{V_{N2} - V_{N5}}{Z4} \tag{A-5}$$

Equation (A-5) can then be rearranged by collecting each of the node voltage expressions on the left side of the equation and leaving the I1 expression on the right side of the equation as shown in Equation (A-6).

$$\left(\frac{1}{Z1}\right) \cdot V_{N1} - \left(\frac{1}{Z1} + \frac{1}{Z3} + \frac{1}{Z4}\right) \cdot V_{N2} + \left(\frac{1}{Z3}\right) \cdot V_{N4} + \left(\frac{1}{Z4}\right) \cdot V_{N5} = I1 \tag{A-6}$$

Again, this process is repeated for each unknown node in the total circuit creating a set of simultaneous equations equal to the number of unknown nodes in the circuit. The presence of the current source actually made the node equation simpler.

Next, consider voltage sources. Voltage sources are more difficult to deal with than current sources. The two circuit nodes that a voltage source is connected between determine how the voltage source is treated. If one end of the voltage source is connected to the reference node, it is called a grounded voltage source. The reference node is considered ground whether or not it is actually tied to ground because it is the zero voltage point in the circuit.

If neither end of the voltage source is connected to the reference node, it is called a floating voltage source. With current sources, grounded and floating current sources are treated the same. Grounded voltage sources will be discussed first. Figure A-3 shows a portion of a larger circuit with a grounded voltage source.

Note that the reference node is N0 and that node N1 has become V1 which is the value of the voltage source. Node V1 is a known node and no Kirchhoff's current law equation needs to be developed for known nodes. They only need to be developed for unknown

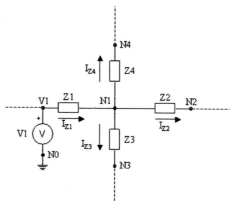

Figure A-3 Portion of a Larger Circuit with a Grounded Voltage Source

nodes. Also note that the reference node N0 is also a known node since its value is zero. The Kirchhoff's current law equation for node N1 of Figure A-3 is shown in Equation (A-7).

$$I_{Z1} = I_{Z2} + I_{Z3} + I_{Z4} \tag{A-7}$$

Redefining the currents of Equation (A-7) in terms of node voltages yields Equation (A-8).

$$\frac{V1 - V_{N1}}{Z1} = \frac{V_{N1} - V_{N2}}{Z2} + \frac{V_{N1} - V_{N3}}{Z3} + \frac{V_{N1} - V_{N4}}{Z4} \tag{A-8}$$

Equation (A-8) can then be rearranged by collecting each of the unknown node voltage expressions on the left side of the equation and moving the known V1 node voltage expression to the right side of the equation as shown in Equation (A-9).

$$-\left(\frac{1}{Z1} + \frac{1}{Z2} + \frac{1}{Z3} + \frac{1}{Z4}\right) \cdot V_{N1} + \left(\frac{1}{Z2}\right) \cdot V_{N2}$$

$$+\left(\frac{1}{Z3}\right) \cdot V_{N3} + \left(\frac{1}{Z4}\right) \cdot V_{N4} = -\frac{V1}{Z1} \tag{A-9}$$

Again, this process is repeated for each unknown node in the total circuit creating a set of simultaneous equations equal to the number of unknown nodes in the circuit. Note that the grounded voltage source has actually reduced the number of unknown nodes for this portion of the circuit by one.

Next, consider a floating voltage source. The floating voltage source is a voltage source between two nodes neither of which is the reference node. Figure A-4 shows a portion of a larger circuit that contains a floating voltage source.

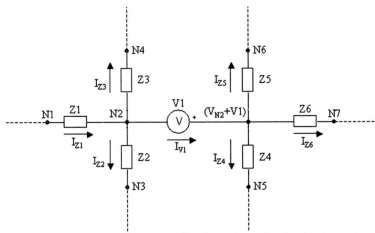

Figure A-4 Portion of a Larger Circuit with a Floating Voltage Source

The floating voltage source establishes a fixed voltage between two nodes. Therefore, only one of the nodes is unknown since the other is the value of the voltage source different from it. Either end of the floating voltage source can be the unknown node but the convention used in this book is that the negative side of the floating voltage source will be the unknown node and the positive side is the known node.

Figure A-4 shows that the negative side of source V1 is the unknown node N2 and the positive side is the known node (V_{N2} + V1). Additional branches and nodes were added to the circuit to permit developing the Kirchhoff's current law equation for node N2. It is also necessary to assign a current through the voltage source V1 which will flow from the negative end to the positive end. The Kirchhoff's current law equation for node N2 of Figure A-4 is shown in Equation (A-10).

$$I_{Z1} = I_{Z2} + I_{Z3} + I_{V1} \qquad\qquad (A\text{-}10)$$

The next step would normally be to redefine the currents in terms of node voltages but I_{V1} cannot be redefined that way since its value is not known and it is a source and not an impedance. The additional branches and nodes permit the development of a Kirchhoff's current law equation for node (V_{N2}+V1) to redefine I_{V1} as shown in Equation (A-11).

$$I_{V1} = I_{Z4} + I_{Z5} + I_{Z6} \qquad\qquad (A\text{-}11)$$

Equation (A-10) and Equation (A-11) can be combined to yield a Kirchhoff's current law equation that has all impedance currents as shown in Equation (A-12).

$$I_{Z1} = I_{Z2} + I_{Z3} + I_{Z4} + I_{Z5} + I_{Z6} \qquad\qquad (A\text{-}12)$$

Redefining the currents of Equation (A-12) in terms of node voltages yields Equation (A-13).

$$
\begin{aligned}
\frac{V_{N1} - V_{N2}}{Z1} &= \frac{V_{N2} - V_{N3}}{Z2} + \frac{V_{N2} - V_{N4}}{Z3} \\
&+ \frac{V_{N2} + V1 - V_{N5}}{Z4} + \frac{V_{N2} + V1 - V_{N6}}{Z5} + \frac{V_{N2} + V1 - V_{N7}}{Z6}
\end{aligned}
\qquad (A\text{-}13)
$$

Equation (A-13) can then be rearranged by collecting each of the unknown node voltage expressions on the left side of the equation and leaving the known V1 voltage expressions on the right side of the equation as shown in Equation (A-14).

$$\left(\frac{1}{Z1}\right)\cdot V_{N1} - \left(\frac{1}{Z1} + \frac{1}{Z2} + \frac{1}{Z3} + \frac{1}{Z4} + \frac{1}{Z5} + \frac{1}{Z6}\right)\cdot V_{N2}$$

$$+\left(\frac{1}{Z2}\right)\cdot V_{N3} + \left(\frac{1}{Z3}\right)\cdot V_{N4} + \left(\frac{1}{Z4}\right)\cdot V_{N5} + \left(\frac{1}{Z5}\right)\cdot V_{N6}$$

$$+\left(\frac{1}{Z6}\right)\cdot V_{N7} = \left(\frac{1}{Z4} + \frac{1}{Z5} + \frac{1}{Z6}\right)\cdot V1 \qquad \text{(A-14)}$$

Again, this process is repeated for each unknown node in the total circuit creating a set of simultaneous equations equal to the number of unknown nodes in the circuit. Note that the floating voltage source also reduced the number of unknown nodes for this portion of the circuit by one although the known node still required the development of a Kirchhoff's current law equation which was combined with the unknown node equation.

At this point all of the formal nodal voltage analysis concepts have been discussed and a procedure can be developed.

Step 1 – Determine the reference node. If the circuit has a ground node it must be the reference node. If the circuit does not have a ground node, select the node with the most voltage sources tied to it. If this still leaves more than one possibility, select the one of those with the most branches. If this still leaves more than one possibility, arbitrarily select one. Note that the reference node cannot be changed once it has been established.

Step 2 – Label all of the known grounded voltage source nodes V1, V2, etc.

Step 3 – Label all of the unknown nodes N1, N2, etc. Then label all of the known floating voltage source nodes $(V_{NX}+V1)$, $(V_{NX}+V2)$, etc.

Step 4 – Assign impedance branch currents and label them. Also label all floating voltage source currents.

Step 5 – Develop a set of simultaneous Kirchhoff's current law equations for each unknown node in the circuit. The total number of unknown nodes and equations is the total number of nodes less one for each voltage source and less one more for the reference node.

Step 6 – Redefine the set of simultaneous equations in terms of node voltages. Collect all unknown node voltage expressions on the left side of the equations and all known source expressions on the right side of the equations.

Step 7 – Solve the set of simultaneous equations for the unknown node voltages.

If the solution of step 7 is a basic steady state DC circuit analysis, this is the final result. If the solution is for a basic steady state sinusoidal AC circuit analysis, the sinusoids need to be added back to yield the final result. If the solution is for an advanced transient and steady state circuit analysis, the results need to be inverse Laplace transformed to yield the final result.

This seven step procedure along with the discussions for each type of circuit configuration will provide the solution of any basic steady state DC circuit analysis and any advanced transient and steady state circuit analysis. It will only provide the solution for a basic steady state sinusoidal AC circuit analysis with single frequency sources.

Basic steady state sinusoidal AC circuit analyses with multiple sources of different frequencies require the use of superposition. A sub circuit for each frequency involved is created. Each sub circuit is analyzed using the nodal voltage analysis method and the sinusoids are added back into each sub circuit result prior to combining the nodal voltage results from all of the sub circuits.

Step 7 requires the solution of simultaneous equations. The customary method for solving simultaneous equations is to use matrix algebra. The coefficients of the unknown node voltages as seen in the final nodal analysis equations are all forms of the reciprocal of impedance values. The reciprocal of impedance is called admittance and is customarily represented by the symbol Y. The coefficients of the simultaneous equations can be represented in matrix form as shown in Equation (A-15).

$$[Y] = \begin{bmatrix} y_{1,1} & y_{1,2} & \cdots & y_{1,m} \\ y_{2,1} & y_{2,2} & \cdots & y_{2,m} \\ \vdots & \vdots & \vdots & \vdots \\ y_{m,1} & y_{m,2} & \cdots & y_{m,m} \end{bmatrix} \tag{A-15}$$

Notice that the matrix in Equation (A-15) is square, the number of rows equals the number of columns. This is because the number of equations must equal the number of unknowns for a solution to exist. Also, not every unknown node voltage will exist in every equation in the set of simultaneous equations. When this situation exists, the coefficient for that unknown node voltage is zero and a zero must be placed at the appropriate location in the [Y] matrix.

The unknown node voltages of the simultaneous equations can be represented as a voltage matrix as shown in Equation (A-16). Notice that the number of rows of [V] and [Y] must be equal.

$$[V] = \begin{bmatrix} V_{N1} \\ V_{N2} \\ \vdots \\ V_{Nm} \end{bmatrix} \tag{A-16}$$

The portion of the set of equations on the left side of the equal sign can then be represented using matrix multiplication as shown in Equation (A-17).

$$\begin{bmatrix} y_{1,1} & y_{1,2} & \cdots & y_{1,m} \\ y_{2,1} & y_{2,2} & \cdots & y_{2,m} \\ \vdots & \vdots & \vdots & \vdots \\ y_{m,1} & y_{m,2} & \cdots & y_{m,m} \end{bmatrix} \cdot \begin{bmatrix} V_{N1} \\ V_{N2} \\ \vdots \\ V_{Nm} \end{bmatrix} \tag{A-17}$$

The known quantities on the right side of the equal sign in the final nodal analysis equations are all forms of currents, either a known current source value or a known voltage source value divided by an impedance. The right side of the system of simultaneous equations can then be represented by a current matrix as shown in Equation (A-18).

$$[I] = \begin{bmatrix} I_1 \\ I_2 \\ \vdots \\ I_m \end{bmatrix} \tag{A-18}$$

Notice that the number of rows of [I] must also equal the number of rows of [Y] and [V]. The final matrix representation of the simultaneous equations is shown in equation (A-19).

$$\begin{bmatrix} y_{1,1} & y_{1,2} & \cdots & y_{1,m} \\ y_{2,1} & y_{2,2} & \cdots & y_{2,m} \\ \vdots & \vdots & \vdots & \vdots \\ y_{m,1} & y_{m,2} & \cdots & y_{m,m} \end{bmatrix} \cdot \begin{bmatrix} V_{N1} \\ V_{N2} \\ \vdots \\ V_{Nm} \end{bmatrix} = \begin{bmatrix} I_1 \\ I_2 \\ \vdots \\ I_m \end{bmatrix} \tag{A-19}$$

Since the [Y] admittance matrix is square the matrix inverse operation can be used to obtain a solution. Since $[Y]^{-1} \cdot [Y] = [1]$, the identity matrix

(1's along the diagonal and 0's everywhere else), both sides of the equation can be multiplied by $[Y]^{-1}$ yielding Equation (A-20).

$$
\begin{bmatrix} V_{N1} \\ V_{N2} \\ \vdots \\ V_{Nm} \end{bmatrix} = \begin{bmatrix} y_{1,1} & y_{1,2} & \cdots & y_{1,m} \\ y_{2,1} & y_{2,2} & \cdots & y_{2,m} \\ \vdots & \vdots & \vdots & \vdots \\ y_{m,1} & y_{m,2} & \cdots & y_{m,m} \end{bmatrix}^{-1} \cdot \begin{bmatrix} I_1 \\ I_2 \\ \vdots \\ I_m \end{bmatrix}
\tag{A-20}
$$

Computer software and scientific calculators can be used to calculate the matrix inverse and the matrix multiplication required to obtain a solution for the unknown node voltages. If any of the matrix elements has algebraic expressions such as Laplace transforms, the computer software or scientific calculator must support symbolic math calculations.

It is interesting to note that the inverse of the admittance matrix is an impedance matrix and Equation (A-20) is actually a matrix version of Ohm's law, $[V] = [Z] \cdot [I]$.

APPENDIX B
LAPLACE TRANSFORM TABLES

The Laplace transform tables consist of three separate tables. The first is a table of some of the properties of the Laplace transform, the second is a table of Laplace transform pairs of typical circuit waveforms, and the third is a table of Laplace transform pairs of circuit responses from the typical circuit waveforms.

TABLE OF LAPLACE TRANSFORM PROPERTIES

Property	$f(t)$	$F(s)$
Laplace Transform	$f(t) \cdot u(t)$	$\int_0^\infty \left[f(t) \cdot e^{-s \cdot t} \right] dt$
Linearity	$K_1 \cdot f_1(t) \cdot u(t) \pm K_2 \cdot f_2(t) \cdot u(t) \pm \cdots$	$K_1 \cdot F_1(s) \pm K_2 \cdot F_2(s) \pm \cdots$
Real Translation	$f(t-a) \cdot u(t-a)$	$e^{-a \cdot s} \cdot F(s)$
Complex Translation	$e^{-a \cdot t} \cdot f(t) \cdot u(t)$	$F(s+a)$
Real Differentiation	$\left[\dfrac{d\,f(t)}{dt} \right] \cdot u(t)$	$s \cdot F(s) - f(0^-)$
Complex Differentiation	$-t \cdot f(t) \cdot u(t)$	$\dfrac{dF(s)}{ds}$
Real Integration	$\left[\int_0^t f(t)dt \right] \cdot u(t)$	$\dfrac{F(s)}{s}$
Complex Integration	$\left[\dfrac{f(t)}{t} \right] \cdot u(t)$	$\int_s^\infty F(s)ds$

TABLE OF LAPLACE TRANSFORM PAIRS OF CIRCUIT WAVEFORMS

No.	$f(t)$	$F(s)$
Impulse	$K \cdot \delta(t)$	K
Step	$K \cdot u(t)$	$\dfrac{K}{s}$
Ramp	$(K \cdot t) \cdot u(t)$	$\dfrac{K}{s^2}$
t^n	$\left(K \cdot t^n \right) \cdot u(t)$	$\dfrac{K \cdot (n!)}{s^{n+1}}$
Exponential	$\left(K \cdot e^{-\alpha \cdot t} \right) \cdot u(t)$	$\dfrac{K}{s+\alpha}$
Sine	$\left[K \cdot \sin(\omega \cdot t) \right] \cdot u(t)$	$\dfrac{K \cdot \omega}{s^2 + \omega^2}$
Cosine	$\left[K \cdot \cos(\omega \cdot t) \right] \cdot u(t)$	$\dfrac{K \cdot s}{s^2 + \omega^2}$
Phase Shifted Sine	$\left[K \cdot \sin(\omega \cdot t + \theta) \right] \cdot u(t)$	$\dfrac{K \cdot \sin(\theta) \cdot \left[s + \dfrac{\omega}{\tan(\theta)} \right]}{s^2 + \omega^2}$

The angle(x,y) function referenced in this table is a 4 quadrant arctangent function. It can be calculated using a scientific calculator by, first, calculating the values for x and y, then, inputting them as rectangular coordinates, and finally, using the rectangular to polar conversion feature where the angle argument is the evaluation of the angle(x,y) function.

TABLE OF LAPLACE TRANSFORM PAIRS OF CIRCUIT RESPONSES

No.	$F(s)$	$f(t)$
1	$\dfrac{K}{(s+a)}$	$\left(K\cdot e^{-at}\right)\cdot u(t)$
2	$\dfrac{K}{s\cdot(s+a)}$	$\left[\dfrac{K\cdot\left(1-e^{-at}\right)}{a}\right]\cdot u(t)$
3	$\dfrac{K}{s^2\cdot(s+a)}$	$\left[\dfrac{K\cdot\left(a\cdot t-1+e^{-at}\right)}{a^2}\right]\cdot u(t)$
4	$\dfrac{K}{\left(s^2+\omega^2\right)\cdot(s+a)}$	$\left[\dfrac{K\cdot e^{-at}}{a^2+\omega^2}+\dfrac{K\cdot\sin(\omega\cdot t+\theta)}{\omega\cdot\sqrt{a^2+\omega^2}}\right]\cdot u(t)$ where $\theta=\text{angle}(x,y)$ and $x=a$; $y=-\omega$
5	$\dfrac{K\cdot(s+\alpha)}{\left(s^2+\omega^2\right)\cdot(s+a)}$	$\left[\dfrac{K\cdot(\alpha-a)\cdot e^{-at}}{a^2+\omega^2}+\dfrac{K}{\omega}\cdot\sqrt{\dfrac{\alpha^2+\omega^2}{a^2+\omega^2}}\cdot\sin(\omega\cdot t+\theta)\right]\cdot u(t)$ where $\theta=\text{angle}(x,y)$ and $x=\omega^2+a\cdot\alpha$; $y=\omega\cdot(a-\alpha)$
6	$\dfrac{K\cdot s}{(s+a)}$	$K\cdot\delta(t)-\left(K\cdot a\cdot e^{-at}\right)\cdot u(t)$
7	$\dfrac{K\cdot s}{\left(s^2+\omega^2\right)\cdot(s+a)}$	$\left[\dfrac{K\cdot\sin(\omega\cdot t+\theta)}{\sqrt{a^2+\omega^2}}-\dfrac{K\cdot a\cdot e^{-at}}{a^2+\omega^2}\right]\cdot u(t)$ where $\theta=\text{angle}(x,y)$ and $x=\omega$; $y=a$
8	$\dfrac{K\cdot(s+\alpha)\cdot s}{\left(s^2+\omega^2\right)\cdot(s+a)}$	$\left[\dfrac{K\cdot\left(a^2-a\cdot\alpha\right)\cdot e^{-at}}{a^2+\omega^2}+K\cdot\sqrt{\dfrac{\alpha^2+\omega^2}{a^2+\omega^2}}\cdot\sin(\omega\cdot t+\theta)\right]\cdot u(t)$ where $\theta=\text{angle}(x,y)$ and $x=\omega\cdot(\alpha-a)$; $y=\omega^2+a\cdot\alpha$
9a	$\dfrac{K}{(s+a)\cdot(s+b)}$	$\left[\dfrac{K\cdot\left(e^{-at}-e^{-bt}\right)}{b-a}\right]\cdot u(t)$
9b	$\dfrac{K}{(s+a)^2}$	$\left(K\cdot t\cdot e^{-at}\right)\cdot u(t)$
10a	$\dfrac{K}{s\cdot(s+a)\cdot(s+b)}$	$\left[\dfrac{K}{a\cdot b}+\dfrac{K\cdot\left(b\cdot e^{-at}-a\cdot e^{-bt}\right)}{a\cdot b\cdot(a-b)}\right]\cdot u(t)$
10b	$\dfrac{K}{s\cdot(s+a)^2}$	$\left[\dfrac{K}{a^2}-\dfrac{K\cdot e^{-at}}{a^2}-\dfrac{K\cdot t\cdot e^{-at}}{a}\right]\cdot u(t)$

TABLE OF LAPLACE TRANSFORM PAIRS OF CIRCUIT RESPONSES

No.	F(s)	f(t)
11a	$\dfrac{K}{s^2\cdot(s+a)\cdot(s+b)}$	$\left[\dfrac{K\cdot\left(a^2\cdot e^{-bt}-b^2\cdot e^{-at}\right)}{a^2\cdot b^2\cdot(a-b)}+\dfrac{K\cdot t}{a\cdot b}-\dfrac{K\cdot(a+b)}{a^2\cdot b^2}\right]\cdot u(t)$
11b	$\dfrac{K}{s^2\cdot(s+a)^2}$	$\left[\dfrac{K\cdot t}{a^2}-\dfrac{2\cdot K}{a^3}+\dfrac{K\cdot t\cdot e^{-at}}{a^2}+\dfrac{2\cdot K\cdot e^{-at}}{a^3}\right]\cdot u(t)$
12a	$\dfrac{K}{\left(s^2+\omega^2\right)\cdot(s+a)\cdot(s+b)}$	$\left[\dfrac{K\cdot e^{-at}}{(b-a)\cdot\left(a^2+\omega^2\right)}+\dfrac{K\cdot e^{-bt}}{(a-b)\cdot\left(b^2+\omega^2\right)}\right.$ $\left.+\dfrac{K\cdot\sin(\omega\cdot t+\theta)}{\omega\cdot\sqrt{\left(a^2+\omega^2\right)\cdot\left(b^2+\omega^2\right)}}\right]\cdot u(t)$ where $\theta=$ angle (x,y) and $x=a\cdot b-\omega^2$; $y=-\omega\cdot(a+b)$
12b	$\dfrac{K}{\left(s^2+\omega^2\right)\cdot(s+a)^2}$	$\left[\dfrac{K\cdot t\cdot e^{-at}}{a^2+\omega^2}+\dfrac{2\cdot K\cdot a\cdot e^{-at}}{\left(a^2+\omega^2\right)^2}+\dfrac{K\cdot\sin(\omega\cdot t+\theta)}{\omega\cdot\left(a^2+\omega^2\right)}\right]\cdot u(t)$ where $\theta=$ angle (x,y) and $x=a^2-\omega^2$; $y=-2\cdot a\cdot\omega$
13a	$\dfrac{K\cdot(s+\alpha)}{\left(s^2+\omega^2\right)\cdot(s+a)\cdot(s+b)}$	$\left[\dfrac{K\cdot(\alpha-a)\cdot e^{-at}}{(b-a)\cdot\left(a^2+\omega^2\right)}+\dfrac{K\cdot(\alpha-b)\cdot e^{-bt}}{(a-b)\cdot\left(b^2+\omega^2\right)}\right.$ $\left.+\dfrac{K\cdot\sin(\omega\cdot t+\theta)}{\omega}\sqrt{\dfrac{\alpha^2+\omega^2}{\left(a^2+\omega^2\right)\cdot\left(b^2+\omega^2\right)}}\right]\cdot u(t)$ where $\theta=$ angle (x,y) and $x=a\cdot b\cdot\alpha+\omega^2\cdot(a+b-\alpha)$ $y=\omega\cdot\left(a\cdot b-a\cdot\alpha-b\cdot\alpha-\omega^2\right)$
13b	$\dfrac{K\cdot(s+\alpha)}{\left(s^2+\omega^2\right)\cdot(s+a)^2}$	$\left[\dfrac{K\cdot(\alpha-a)\cdot t\cdot e^{-at}}{a^2+\omega^2}+\dfrac{K\cdot\left(2\cdot\alpha\cdot a-a^2+\omega^2\right)\cdot e^{-at}}{\left(a^2+\omega^2\right)^2}\right.$ $\left.+\dfrac{K\cdot\sqrt{\alpha^2+\omega^2}\cdot\sin(\omega\cdot t+\theta)}{\omega\cdot\left(a^2+\omega^2\right)}\right]\cdot u(t)$ where $\theta=$ angle (x,y) and $x=a^2\alpha+\omega^2\cdot(2\cdot a-\alpha)$ $y=\omega\cdot\left(a^2-\omega^2-2\cdot a\cdot\alpha\right)$
14a	$\dfrac{K\cdot s}{(s+a)\cdot(s+b)}$	$\left[\dfrac{K\cdot\left(a\cdot e^{-at}-b\cdot e^{-bt}\right)}{a-b}\right]\cdot u(t)$
14b	$\dfrac{K\cdot s}{(s+a)^2}$	$\left[K\cdot(1-a\cdot t)\cdot e^{-at}\right]\cdot u(t)$

TABLE OF LAPLACE   TRANSFORM PAIRS OF CIRCUIT RESPONSES

No.	F(s)	f(t)
15a	$\dfrac{K \cdot s}{\left(s^2+\omega^2\right) \cdot (s+a) \cdot (s+b)}$	$\left[\dfrac{K \cdot a \cdot e^{-a \cdot t}}{(a-b) \cdot \left(a^2+\omega^2\right)} - \dfrac{K \cdot b \cdot e^{-b \cdot t}}{(a-b) \cdot \left(b^2+\omega^2\right)} \right.$ $\left. + \dfrac{K \cdot \sin(\omega \cdot t + \theta)}{\sqrt{\left(a^2+\omega^2\right) \cdot \left(b^2+\omega^2\right)}} \right] \cdot u(t)$ where $\theta = $ angle (x,y) and $x = \omega \cdot (a+b)$; $y = a \cdot b - \omega^2$
15b	$\dfrac{K \cdot s}{\left(s^2+\omega^2\right) \cdot (s+a)^2}$	$\left[\dfrac{K \cdot \left(\omega^2-a^2\right) \cdot e^{-a \cdot t}}{\left(a^2+\omega^2\right)^2} - \dfrac{K \cdot a \cdot t \cdot e^{-a \cdot t}}{a^2+\omega^2} + \dfrac{K \cdot \sin(\omega \cdot t + \theta)}{a^2+\omega^2}\right] \cdot u(t)$ where $\theta = $ angle (x,y) and $x = 2 \cdot a \cdot \omega$; $y = a^2 - \omega^2$
16a	$\dfrac{K \cdot (s+\alpha) \cdot s}{\left(s^2+\omega^2\right) \cdot (s+a) \cdot (s+b)}$	$\left[\dfrac{K \cdot \left(a^2-\alpha \cdot a\right) \cdot e^{-a \cdot t}}{(b-a) \cdot \left(a^2+\omega^2\right)} + \dfrac{K \cdot \left(b^2-\alpha \cdot b\right) \cdot e^{-b \cdot t}}{(a-b) \cdot \left(b^2+\omega^2\right)} \right.$ $\left. +K \cdot \sin(\omega \cdot t + \theta)\sqrt{\dfrac{\alpha^2+\omega^2}{\left(a^2+\omega^2\right) \cdot \left(b^2+\omega^2\right)}}\right] \cdot u(t)$ where $\theta = $ angle (x,y) and $x = \omega \cdot \left(\omega^2 + a \cdot \alpha + b \cdot \alpha - a \cdot b\right)$ $y = a \cdot b \cdot \alpha + \omega^2 \cdot (a+b-\alpha)$
16b	$\dfrac{K \cdot (s+\alpha) \cdot s}{\left(s^2+\omega^2\right) \cdot (s+a)^2}$	$\left[\dfrac{K \cdot \left(a^2-\alpha \cdot a\right) \cdot t \cdot e^{-a \cdot t}}{a^2+\omega^2} + \dfrac{K \cdot \left(\alpha \cdot \omega^2 - \alpha \cdot a^2 - 2 \cdot a \cdot \omega^2\right) \cdot e^{-a \cdot t}}{\left(a^2+\omega^2\right)^2} \right.$ $\left. + \dfrac{K \cdot \sqrt{\alpha^2+\omega^2} \cdot \sin(\omega \cdot t + \theta)}{a^2+\omega^2}\right] \cdot u(t)$ where $\theta = $ angle (x,y) and $x = \omega \cdot \left(\omega^2 + 2 \cdot a \cdot \alpha - a^2\right)$ $y = a^2 \cdot \alpha + \omega^2 \cdot (2 \cdot a - \alpha)$
17a	$\dfrac{K \cdot s^2}{(s+a) \cdot (s+b)}$	$K \cdot \delta(t) + \left[\dfrac{K \cdot \left(b^2 \cdot e^{-b \cdot t} - a^2 \cdot e^{-a \cdot t}\right)}{a-b}\right] \cdot u(t)$
17b	$\dfrac{K \cdot s^2}{(s+a)^2}$	$K \cdot \delta(t) + \left(K \cdot a^2 \cdot t \cdot e^{-a \cdot t} - 2 \cdot K \cdot a \cdot e^{-a \cdot t}\right) \cdot u(t)$

TABLE OF LAPLACE TRANSFORM PAIRS OF CIRCUIT RESPONSES

No.	F(s)	f(t)
18a	$\dfrac{K \cdot s^2}{\left(s^2 + \omega^2\right) \cdot (s+a) \cdot (s+b)}$	$\left[\dfrac{K \cdot a^2 \cdot e^{-a \cdot t}}{(b-a) \cdot \left(a^2 + \omega^2\right)} + \dfrac{K \cdot b^2 \cdot e^{-b \cdot t}}{(a-b) \cdot \left(b^2 + \omega^2\right)} \right.$ $\left. + \dfrac{K \cdot \omega \cdot \sin(\omega \cdot t + \theta)}{\sqrt{\left(a^2 + \omega^2\right) \cdot \left(b^2 + \omega^2\right)}} \right] \cdot u(t)$ where $\theta = \text{angle}\,(x,y)$ and $x = \omega^2 - a \cdot b$; $y = \omega \cdot (a+b)$
18b	$\dfrac{K \cdot s^2}{\left(s^2 + \omega^2\right) \cdot (s+a)^2}$	$\left[\dfrac{K \cdot a^2 \cdot t \cdot e^{-a \cdot t}}{a^2 + \omega^2} - \dfrac{2 \cdot K \cdot \omega^2 \cdot a \cdot e^{-a \cdot t}}{\left(a^2 + \omega^2\right)^2} + \dfrac{K \cdot \omega \cdot \sin(\omega \cdot t + \theta)}{a^2 + \omega^2} \right] \cdot u(t)$ where $\theta = \text{angle}\,(x,y)$ and $x = \omega^2 - a^2$; $y = 2 \cdot a \cdot \omega$
19a	$\dfrac{K \cdot (s+\alpha) \cdot s^2}{\left(s^2 + \omega^2\right) \cdot (s+a) \cdot (s+b)}$	$\left[\dfrac{K \cdot \left(\alpha \cdot a^2 - a^3\right) \cdot e^{-a \cdot t}}{(b-a) \cdot \left(a^2 + \omega^2\right)} + \dfrac{K \cdot \left(\alpha \cdot b^2 - b^3\right) \cdot e^{-b \cdot t}}{(a-b) \cdot \left(b^2 + \omega^2\right)} \right.$ $\left. + K \cdot \omega \cdot \sin(\omega \cdot t + \theta) \sqrt{\dfrac{\alpha^2 + \omega^2}{\left(a^2 + \omega^2\right) \cdot \left(b^2 + \omega^2\right)}} \right] \cdot u(t)$ where $\theta = \text{angle}\,(x,y)$ and $x = \omega^2 \cdot (\alpha - a - b) - a \cdot b \cdot \alpha$ $y = \omega \cdot \left(\omega^2 + a \cdot \alpha + b \cdot \alpha - a \cdot b\right)$
19b	$\dfrac{K \cdot (s+\alpha) \cdot s^2}{\left(s^2 + \omega^2\right) \cdot (s+a)^2}$	$\left[\dfrac{K \cdot a^2 \cdot (\alpha - a) \cdot t \cdot e^{-a \cdot t}}{a^2 + \omega^2} + \dfrac{K \cdot a \cdot \left(a^3 + 3 \cdot a \cdot \omega^2 - 2 \cdot \alpha \cdot \omega^2\right) \cdot e^{-a \cdot t}}{\left(a^2 + \omega^2\right)^2} \right.$ $\left. + \dfrac{K \cdot \omega \cdot \sqrt{\alpha^2 + \omega^2} \cdot \sin(\omega \cdot t + \theta)}{a^2 + \omega^2} \right] \cdot u(t)$ where $\theta = \text{angle}\,(x,y)$ and $x = \omega^2 \cdot (\alpha - 2 \cdot a) - a^2 \cdot \alpha$ $y = \omega \cdot \left(\omega^2 + 2 \cdot a \cdot \alpha - a^2\right)$
20	$\dfrac{K}{(s+a)^2 + b^2}$	$\left[\dfrac{K \cdot e^{-a \cdot t} \cdot \sin(b \cdot t)}{b}\right] \cdot u(t)$
21	$\dfrac{K}{s \cdot \left[(s+a)^2 + b^2\right]}$	$\left[\dfrac{K}{a^2 + b^2} - \dfrac{K \cdot e^{-a \cdot t} \cdot \sin(b \cdot t + \theta)}{b \cdot \sqrt{a^2 + b^2}}\right] \cdot u(t)$ where $\theta = \text{angle}\,(x,y)$ and $x = a$; $y = b$
22	$\dfrac{K}{s^2 \cdot \left[(s+a)^2 + b^2\right]}$	$\left[\dfrac{K \cdot t}{a^2 + b^2} - \dfrac{2 \cdot K \cdot a}{\left(a^2 + b^2\right)^2} + \dfrac{K \cdot e^{-a \cdot t} \cdot \sin(b \cdot t + \theta)}{b \cdot \left(a^2 + b^2\right)}\right] \cdot u(t)$ where $\theta = \text{angle}\,(x,y)$ and $x = a^2 - b^2$; $y = 2 \cdot a \cdot b$

TABLE OF LAPLACE TRANSFORM PAIRS OF CIRCUIT RESPONSES

No.	F(s)	f(t)
23	$$\dfrac{K}{\left(s^2+\omega^2\right)\cdot\left[(s+a)^2+b^2\right]}$$	$$\left[\dfrac{K\cdot\omega\cdot e^{-a\cdot t}\cdot\sin(b\cdot t+\theta)+K\cdot b\cdot\sin(\omega\cdot t+\phi)}{b\cdot\omega\cdot\sqrt{4\cdot a^2\cdot\omega^2+\left(a^2+b^2-\omega^2\right)^2}}\right]\cdot u(t)$$ where $\theta=$ angle (x_1,y_1) ; $\phi=$ angle (x_2,y_2) and $\quad x_1=a^2-b^2+\omega^2$; $y_1=2\cdot a\cdot b$ $\qquad x_2=a^2+b^2-\omega^2$; $y_2=-2\cdot a\cdot\omega$
24	$$\dfrac{K\cdot(s+\alpha)}{\left(s^2+\omega^2\right)\cdot\left[(s+a)^2+b^2\right]}$$	$$\left[\dfrac{K}{b}\sqrt{\dfrac{(\alpha-a)^2+b^2}{\left(a^2-b^2+\omega^2\right)^2+4\cdot b^2\cdot a^2}}\cdot e^{-a\cdot t}\cdot\sin(b\cdot t+\theta)\right.$$ $$\left.+\dfrac{K}{\omega}\cdot\sqrt{\dfrac{\alpha^2+\omega^2}{\left(a^2+b^2-\omega^2\right)^2+4\cdot\omega^2\cdot a^2}}\cdot\sin(\omega\cdot t+\phi)\right]\cdot u(t)$$ where $\theta=$ angle (x_1,y_1) ; $\phi=$ angle (x_2,y_2) and $\quad x_1=\alpha\cdot(a^2-b^2+\omega^2)-a\cdot(a^2+b^2+\omega^2)$ $\qquad y_1=b\cdot\left(\omega^2-a^2-b^2+2\cdot a\cdot\alpha\right)$ $\qquad x_2=\alpha\cdot(a^2+b^2-\omega^2)+2\cdot a\cdot\omega^2$ $\qquad y_2=\omega\cdot\left(a^2+b^2-\omega^2-2\cdot a\cdot\alpha\right)$
25	$$\dfrac{K\cdot s}{(s+a)^2+b^2}$$	$$\left[\dfrac{K\cdot\sqrt{a^2+b^2}\cdot e^{-a\cdot t}\cdot\sin(b\cdot t+\theta)}{b}\right]\cdot u(t)$$ where $\theta=$ angle (x,y) and $x=-a$; $y=b$
26	$$\dfrac{K\cdot s}{\left(s^2+\omega^2\right)\cdot\left[(s+a)^2+b^2\right]}$$	$$\left[\dfrac{K\cdot\sqrt{a^2+b^2}\cdot e^{-a\cdot t}\cdot\sin(b\cdot t+\theta)+K\cdot b\cdot\sin(\omega\cdot t+\phi)}{b\cdot\sqrt{4\cdot a^2\cdot\omega^2+\left(a^2+b^2-\omega^2\right)^2}}\right]\cdot u(t)$$ where $\theta=$ angle (x_1,y_1) ; $\phi=$ angle (x_2,y_2) and $\quad x_1=-a\cdot(a^2+b^2+\omega^2)$ $\qquad y_1=b\cdot\left(\omega^2-a^2-b^2\right)$ $\qquad x_2=2\cdot a\cdot\omega$ $\qquad y_2=a^2+b^2-\omega^2$

244

TABLE OF LAPLACE TRANSFORM PAIRS OF CIRCUIT RESPONSES

No.	F(s)	f(t)
27	$$\frac{K \cdot (s+\alpha) \cdot s}{\left(s^2+\omega^2\right) \cdot \left[(s+a)^2+b^2\right]}$$	$$\left[\frac{K}{b} \cdot \sqrt{\frac{\left(a^2+b^2\right) \cdot \left[(a-\alpha)^2+b^2\right]}{\left(a^2+b^2-\omega^2\right)^2+4 \cdot \omega^2 \cdot a^2}} \cdot e^{-a \cdot t} \cdot \sin(b \cdot t + \theta)\right.$$ $$\left. +K \cdot \sqrt{\frac{\alpha^2+\omega^2}{\left(a^2+b^2-\omega^2\right)^2+4 \cdot \omega^2 \cdot a^2}} \cdot \sin(\omega \cdot t + \phi)\right] \cdot u(t)$$ where $\theta = \text{angle}\left(x_1, y_1\right)$; $\phi = \text{angle}\left(x_2, y_2\right)$ and $x_1 = \left(a^2+b^2\right)^2+\omega^2 \cdot \left(a^2-b^2\right)-a \cdot \alpha \cdot \left(a^2+b^2+\omega^2\right)$ $y_1 = -b \cdot \left[\alpha \cdot \left(a^2+b^2-\omega^2\right)+2 \cdot a \cdot \omega^2\right]$ $x_2 = \omega \cdot \left(\omega^2-a^2-b^2+2 \cdot a \cdot \alpha\right)$ $y_2 = \omega^2 \cdot (2 \cdot a - \alpha) + \alpha \cdot \left(a^2+b^2\right)$
28	$$\frac{K \cdot s^2}{(s+a)^2+b^2}$$	$$K \cdot \delta(t) - \left[\frac{K \cdot \left(a^2+b^2\right) \cdot e^{-a \cdot t} \cdot \sin(b \cdot t + \theta)}{b}\right] \cdot u(t)$$ where $\theta = \text{angle}(x, y)$ and $x = b^2 - a^2$; $y = 2 \cdot a \cdot b$
29	$$\frac{K \cdot s^2}{\left(s^2+\omega^2\right) \cdot \left[(s+a)^2+b^2\right]}$$	$$\left[\frac{K \cdot \left(a^2+b^2\right) \cdot e^{-a \cdot t} \cdot \sin(b \cdot t + \theta) + K \cdot \omega \cdot b \cdot \sin(\omega \cdot t + \phi)}{b \cdot \sqrt{4 \cdot a^2 \cdot \omega^2 + \left(a^2+b^2-\omega^2\right)^2}}\right] \cdot u(t)$$ where $\theta = \text{angle}\left(x_1, y_1\right)$; $\phi = \text{angle}\left(x_2, y_2\right)$ and $x_1 = \left(a^2+b^2\right)^2+\omega^2 \cdot \left(a^2-b^2\right)$ $y_1 = -2 \cdot a \cdot b \cdot \omega^2$ $x_2 = \omega^2 - a^2 - b^2$ $y_2 = 2 \cdot a \cdot \omega$
30	$$\frac{K \cdot (s+\alpha) \cdot s^2}{\left(s^2+\omega^2\right) \cdot \left[(s+a)^2+b^2\right]}$$	$$\left[\frac{K \cdot \left(a^2+b^2\right)}{b} \cdot \sqrt{\frac{(a-\alpha)^2+b^2}{\left(a^2+b^2-\omega^2\right)^2+4 \cdot \omega^2 \cdot a^2}} \cdot e^{-a \cdot t} \cdot \sin(b \cdot t + \theta)\right.$$ $$\left. +K \cdot \omega \cdot \sqrt{\frac{\omega^2+\alpha^2}{\left(a^2+b^2-\omega^2\right)^2+4 \cdot \omega^2 \cdot a^2}} \cdot \sin(\omega \cdot t + \phi)\right] \cdot u(t)$$ where $\theta = \text{angle}\left(x_1, y_1\right)$; $\phi = \text{angle}\left(x_2, y_2\right)$ and $x_1 = \alpha \cdot \left[\left(a^2+b^2\right)^2+\omega^2 \cdot \left(a^2-b^2\right)\right]$ $-a \cdot \left[\left(a^2+b^2\right)^2+\omega^2 \cdot \left(a^2-3 \cdot b^2\right)\right]$ $y_1 = b \cdot \left[\left(a^2+b^2\right)^2+\omega^2 \cdot \left(3 \cdot a^2-b^2-2 \cdot a \cdot \alpha\right)\right]$ $x_2 = \alpha \cdot \left(\omega^2-a^2-b^2\right)-2 \cdot a \cdot \omega^2$ $y_2 = \omega \cdot \left(\omega^2-a^2-b^2+2 \cdot a \cdot \alpha\right)$

APPENDIX C
REVIEW OF COMPLEX NUMBERS

A complex number consists of a real part and an imaginary part. The real part is just a number such as the value of a resistor. The imaginary part is a real number multiplied by $\sqrt{-1}$. The total complex number is represented by Equation (C-1).

$$n = x + \sqrt{-1} \cdot y \qquad\qquad (C-1)$$

Rather than write $\sqrt{-1}$ all the time, it is customary to use the letter j for circuit analysis. Mathematicians usually use the letter i but this represents current in circuit analysis so the letter j is used instead.

Complex numbers are usually visualized on a two dimensional Cartesian coordinate system where the horizontal axis is the real axis and the vertical axis is the imaginary axis as shown in Figure C-1. The number $x + j \cdot y$ is represented in what is referred to as rectangular form. The same number can also be represented in polar form

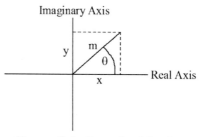

Figure C-1 Complex Number Visualization

as $m\angle\theta$ where m is the magnitude and θ is the angle between the number line and the positive real axis.

This form of a complex number is also referred to as a phasor when used in conjunction with sinusoidal analysis. The values of m and θ can be determined from x and y using Equation (C-2).

$$m = \sqrt{x^2 + y^2} \quad \text{and} \quad \theta = \arctan\left(\frac{y}{x}\right) \qquad\qquad (C-2)$$

Similarly the values of x and y can be determined from m and θ using Equation (C-3).

$$x = m \cdot \cos(\theta) \quad \text{and} \quad y = m \cdot \sin(\theta) \qquad\qquad (C-3)$$

The j operator can be described as a rotation of 90 degrees. If a positive real number is multiplied by j, it is rotated 90 degrees counter clockwise forming an imaginary number. If a positive real number is multiplied by j^2, it is rotated 180 degrees counter clockwise forming a negative real number. If a positive real number is multiplied by j^3, it is rotated 270 degrees counter clockwise forming a negative imaginary number. If a positive real number is multiplied by j^4, it is rotated 360 degrees counter clockwise forming a positive real number. Accordingly $j = \sqrt{-1}$, $j^2 = -1$, $j^3 = -\sqrt{-1}$, and $j^4 = 1$. The j operator can also be used to account for the 90 degree phase shift between voltage and current in capacitors and inductors.

During the course of circuit analysis it will be necessary to add, subtract, multiply, and divide complex numbers. Addition and subtraction are best performed using the rectangular form. The real parts and the imaginary parts are added or subtracted independently with the resultant new real and imaginary parts being the rectangular form of the resultant complex number as shown in Equation (C-4).

$$\left(x_1 + jy_1\right) \pm \left(x_2 + jy_2\right) = \left(x_1 \pm x_2\right) + j\left(y_1 \pm y_2\right) \tag{C-4}$$

Multiplication and division are best performed using the polar form. For multiplication the magnitudes are multiplied and the angles are added to form the resultant product also in polar form as shown in Equation (C-5).

$$\left(m_1 \angle \theta_1\right) \cdot \left(m_2 \angle \theta_2\right) = \left(m_1 \cdot m_2\right) \angle \left(\theta_1 + \theta_2\right) \tag{C-5}$$

For division the magnitudes are divided and the angles are subtracted to form the resultant quotient also in polar form as shown in Equation (C-6).

$$\frac{m_1 \angle \theta_1}{m_2 \angle \theta_2} = \left(\frac{m_1}{m_2}\right) \angle \left(\theta_1 - \theta_2\right) \tag{C-6}$$

The need to use complex numbers in circuit analysis calculations complicates the calculations required. Fortunately, most scientific calculators and computer math software handle complex numbers. All that is required is to enter the numbers in either rectangular or polar format along with the desired operation and the calculator or math software displays the result. Real and complex numbers can even be mixed and the results can be stored for reuse. This greatly simplifies the complications created by complex numbers.

Complex numbers can also be used to represent a sinusoidal source. In the analysis of steady state AC circuits the sinusoidal portion of the source is ignored and only the magnitude and phase of the sinusoid are used. The circuit cannot change the frequency of the sinusoid, only the magnitude and phase. Since a complex number has both a magnitude and an angle, it can be used to represent a sinusoidal signal with the magnitude being the signal magnitude and the angle being the signal phase. Once the analysis is complete the sinusoid can be added back. This process can only be used for sinusoids of the same frequency.

Complex number representation of sinusoids can also be used to combine sinusoids of the same frequency but different magnitudes and phases. The sinusoid magnitudes are the complex number magnitudes and the sinusoid phases are the complex number angles. The complex numbers are then combined to yield an equivalent single magnitude and an equivalent single angle. This magnitude is then the magnitude of the combined sinusoid and the angle is the combined sinusoid phase. Again, this process can only be used for sinusoids of the same frequency.

APPENDIX D
PARTIAL FRACTION EXPANSION

Given a function consisting of a numerator polynomial and a denominator polynomial, with the denominator polynomial in factored form, partial fraction expansion can be used to convert the function to a sum of simpler functions with each one having a denominator equal to one of the original function denominator factors. Equation (D-1) is an example of such a function.

$$F(s) = \frac{5s + 17}{(s+3)\cdot(s+4)} \tag{D-1}$$

Partial fraction expansion is a mathematical process that can convert Equation (D-1) to the form of Equation (D-2).

$$F(s) = \frac{2}{s+3} + \frac{3}{s+4} \tag{D-2}$$

Partial fraction expansion requires that the order (highest power of s) of the numerator must be at least one less than the order of the denominator. If not, it is an improper fraction and simply divide the numerator by the denominator and apply partial fraction expansion to the remainder divided by the divisor. The quotient will also be part of the simpler functions.

Partial fraction expansion involves isolating the factors of the denominator of the function. These factors will be of two types - real or complex. Real factors will be of the form $(s+a)^r$ and complex factors will be of the form $\left((s+\alpha)^2 + \omega^2\right)^r$. For real factors $-a$ is a real root of the denominator polynomial and for complex factors $(-\alpha \pm j\cdot\omega)$ are the complex roots of the denominator polynomial. The partial fraction expansion for real factors of the first order is shown in Equation (D-3).

$$F(s) = \frac{N(s)}{(s+a_1)\cdot(s+a_2)\cdots(s+a_m)}$$

$$= \frac{A_1}{(s+a_1)} + \frac{A_2}{(s+a_2)} + \cdots + \frac{A_m}{(s+a_m)} \tag{D-3}$$

251

The partial fraction expansion for a real factor of multiple order is shown in Equation (D-4).

$$F(s) = \frac{N(s)}{(s+a)^r} = \frac{A_1}{(s+a)} + \frac{A_2}{(s+a)^2} + \cdots + \frac{A_r}{(s+a)^r} \tag{D-4}$$

The partial fraction expansion for complex factors of the first order is shown in Equation (D-5).

$$F(s) = \frac{N(s)}{\left((s+\alpha_1)^2 + \omega_1^2\right) \cdot \left((s+\alpha_2)^2 + \omega_2^2\right) \cdots \left((s+\alpha_m)^2 + \omega_m^2\right)}$$

$$= \frac{A_1 \cdot s + B_1}{\left((s+\alpha_1)^2 + \omega_1^2\right)} + \frac{A_2 \cdot s + B_2}{\left((s+\alpha_2)^2 + \omega_2^2\right)} + \cdots + \frac{A_m \cdot s + B_m}{\left((s+\alpha_m)^2 + \omega_m^2\right)} \tag{D-5}$$

The partial fraction expansion for a complex factor of multiple order is shown in Equation (D-6).

$$F(s) = \frac{N(s)}{\left((s+\alpha)^2 + \omega^2\right)^r}$$

$$= \frac{A_1 \cdot s + B_1}{\left((s+\alpha)^2 + \omega^2\right)} + \frac{A_2 \cdot s + B_2}{\left((s+\alpha)^2 + \omega^2\right)^2} + \cdots + \frac{A_r \cdot s + B_r}{\left((s+\alpha)^2 + \omega^2\right)^r} \tag{D-6}$$

Equations (D-3) through (D-6) are used to define the partial fraction expansion of the function based on the denominator factors that exist. The next step is to multiply both sides of the equation by the factored denominator. This eliminates any denominators on both sides of the equation.

At this point evaluate the equation with s equal to each of the real roots (-a values) of the original denominator. This will yield values for some of the A coefficients which should be entered into the equation. The last step is to expand the right side of the equation collecting the remaining A and B coefficients with the respective powers of s and create a set of simultaneous equations by equating like coefficients on each side of the equation to solve for the rest of the A and B coefficients.

As an example consider the function of Equation (D-1).

$$F(s) = \frac{5s + 17}{(s + 3) \cdot (s + 4)}$$

Expand the function using Equation (D-3):

$$\frac{5s + 17}{(s + 3) \cdot (s + 4)} = \frac{A_1}{(s + 3)} + \frac{A_2}{(s + 4)}$$

Multiply both sides of the equation by the denominator:

$$5s + 17 = A_1 \cdot (s + 4) + A_2 \cdot (s + 3)$$

Evaluate the equation at $s = -3$ and $s = -4$:

$$\left(5s + 17 = A_1 \cdot (s + 4) + A_2 \cdot (s + 3)\right)\big|_{s=-3}$$

$$2 = A_1$$

$$\left(5s + 17 = A_1 \cdot (s + 4) + A_2 \cdot (s + 3)\right)\big|_{s=-4}$$

$$-3 = -A_2 \Rightarrow 3 = A_2$$

The final result is:

$$\frac{5s + 17}{(s + 3) \cdot (s + 4)} = \frac{2}{(s + 3)} + \frac{3}{(s + 4)}$$

This agrees with Equation (D-2). An example with multiple order real factors is shown in Equation (D-7).

$$F(s) = \frac{s+2}{(s+4)^3} \tag{D-7}$$

Expand the function using Equation (D-4):

$$\frac{s+2}{(s+4)^3} = \frac{A_1}{(s+4)} + \frac{A_2}{(s+4)^2} + \frac{A_3}{(s+4)^3}$$

Multiply both sides of the equation by the denominator:

$$s+2 = A_1 \cdot (s+4)^2 + A_2 \cdot (s+4) + A_3$$

Evaluate the equation at $s = -4$:

$$\left(s+2 = A_1 \cdot (s+4)^2 + A_2 \cdot (s+4) + A_3\right)\big|_{s=-4}$$

$$-2 = A_3$$

Expand the right side of the equation:

$$s+2 = A_1 \cdot s^2 + \left(8 \cdot A_1 + A_2\right) \cdot s + \left(16 \cdot A_1 + 4 \cdot A_2 - 2\right)$$

Equate left side coefficients with right side coefficients:

$$0 = A_1$$

$$1 = 8 \cdot A_1 + A_2 \Rightarrow 1 = A_2$$

$$2 = 16 \cdot A_1 + 4 \cdot A_2 - 2 \Rightarrow 1 = A_2$$

The final result is:

$$\frac{s+2}{(s+4)^3} = \frac{1}{(s+4)^2} - \frac{2}{(s+4)^3}$$

Next, consider the example with a few special cases shown in Equation (D-8).

$$F(s) = \frac{3 \cdot s^3 + 4 \cdot s^2 + 14 \cdot s + 16}{s \cdot (s^2 + 4)} \tag{D-8}$$

For this equation the numerator and denominator are of the same order. Partial fraction expansion requires the numerator order to be less than the denominator. To correct this issue, expand the denominator and divide the numerator by the denominator:

$$\begin{array}{r} 3 \\ (s^3 + 4 \cdot s) \overline{\smash{\big)} (3 \cdot s^3 + 4 \cdot s^2 + 14 \cdot s + 16)} \\ \underline{-3 \cdot s^3 \qquad\qquad -12 \cdot s} \\ 4 \cdot s^2 + 2 \cdot s + 16 \end{array}$$

The new function then becomes the quotient plus the remainder divided by the divisor:

$$\frac{3 \cdot s^3 + 4 \cdot s^2 + 14 \cdot s + 16}{s \cdot (s^2 + 4)} = 3 + \frac{4 \cdot s^2 + 2 \cdot s + 16}{s \cdot (s^2 + 4)}$$

Now partial fraction expansion can be performed on the fraction portion using Equations (D-3) and (D-5) since the s factor is $(s + a)$ with a equal to zero and the $(s^2 + 4)$ factor is $((s + \alpha)^2 + \omega^2)$ with α equal to zero and ω equal to 2.

$$\frac{4 \cdot s^2 + 2 \cdot s + 16}{s \cdot (s^2 + 4)} = \frac{A_1}{s} + \frac{A_2 \cdot s + B_2}{s^2 + 4}$$

Multiply both sides of the equation by the denominator:

$$4 \cdot s^2 + 2 \cdot s + 16 = A_1 \cdot (s^2 + 4) + A_2 \cdot s^2 + B_2 \cdot s$$

Evaluate the equation at s equal to 0:

$$16 = 4 \cdot A_1 \Rightarrow 4 = A_1$$

Expand the right side of the equation with A_1 equal to 4:

$$4 \cdot s^2 + 2 \cdot s + 16 = (4 + A_2) \cdot s^2 + B_2 \cdot s + 16$$

Equate left side coefficients with right side coefficients:

$$4 = 4 + A_2 \Rightarrow 0 = A_2$$
$$2 = B_2$$
$$16 = 16$$

The final result is:

$$\frac{3 \cdot s^3 + 4 \cdot s^2 + 14 \cdot s + 16}{s \cdot (s^2 + 4)} = 3 + \frac{4}{s} + \frac{2}{s^2 + 4}$$

Finally, consider a typical example containing both real and complex factors shown in Equation (D-9).

$$F(s) = \frac{s^4 + 48 \cdot s^3 + 278 \cdot s^2 + 1006 \cdot s + 975}{(s+1) \cdot \left((s+2)^2 + 9\right) \cdot \left((s+3)^2 + 16\right)} \qquad \text{(D-9)}$$

Expand the function using Equations (D-3) and (D-5):

$$\frac{s^4 + 48 \cdot s^3 + 278 \cdot s^2 + 1006 \cdot s + 975}{(s+1) \cdot \left((s+2)^2 + 9\right) \cdot \left((s+3)^2 + 16\right)} =$$

$$\frac{A_1}{s+1} + \frac{A_2 \cdot s + B_2}{(s+2)^2 + 9} + \frac{A_3 \cdot s + B_3}{(s+3)^2 + 16}$$

Multiply both sides of the equation by the denominator:

$$s^4 + 48 \cdot s^3 + 278 \cdot s^2 + 1006 \cdot s + 975 =$$

$$A_1 \cdot \left((s+2)^2 + 9\right) \cdot \left((s+3)^2 + 16\right) +$$

$$\left(A_2 \cdot s + B_2\right) \cdot (s+1) \cdot \left((s+3)^2 + 16\right) +$$

$$\left(A_3 \cdot s + B_3\right) \cdot (s+1) \cdot \left((s+2)^2 + 9\right)$$

Evaluate the equation at s equal to -1:

$$1 - 48 + 278 - 1006 + 975 = A_1 \cdot (10) \cdot (20)$$

$$200 = A_1 \cdot 200 \Rightarrow 1 = A_1$$

Expand the right side of the equation with A_1 equal to 1:

$$s^4 + 48 \cdot s^3 + 278 \cdot s^2 + 1006 \cdot s + 975 = \left(1 + A_2 + A_3\right) \cdot s^4 +$$

$$\left(10 + 7 \cdot A_2 + B_2 + 5 \cdot A_3 + B_3\right) \cdot s^3 +$$

$$\left(62 + 31 \cdot A_2 + 7 \cdot B_2 + 17 \cdot A_3 + 5 \cdot B_3\right) \cdot s^2 +$$

$$\left(178 + 25 \cdot A_2 + 31 \cdot B_2 + 13 \cdot A_3 + 17 \cdot B_3\right) \cdot s +$$

$$325 + 25 \cdot B_2 + 13 \cdot B_3$$

Equate left side coefficients with right side coefficients:

$$1 = 1 + A_2 + A_3 \Rightarrow 0 = A_2 + A_3$$

$$48 = 10 + 7 \cdot A_2 + B_2 + 5 \cdot A_3 + B_3$$

$$\Rightarrow 38 = 7 \cdot A_2 + B_2 + 5 \cdot A_3 + B_3$$

$$278 = 62 + 31 \cdot A_2 + 7 \cdot B_2 + 17 \cdot A_3 + 5 \cdot B_3$$

$$\Rightarrow 216 = 31 \cdot A_2 + 7 \cdot B_2 + 17 \cdot A_3 + 5 \cdot B_3$$

$$1006 = 178 + 25 \cdot A_2 + 31 \cdot B_2 + 13 \cdot A_3 + 17 \cdot B_3$$

$$\Rightarrow 828 = 25 \cdot A_2 + 31 \cdot B_2 + 13 \cdot A_3 + 17 \cdot B_3$$

$$975 = 325 + 25 \cdot B_2 + 13 \cdot B_3 \Rightarrow 650 = 25 \cdot B_2 + 13 \cdot B_3$$

Put the first 4 equations in matrix format and solve for the remaining A and B coefficients:

$$\begin{bmatrix} 1 & 1 & 0 & 0 \\ 7 & 5 & 1 & 1 \\ 31 & 17 & 7 & 5 \\ 25 & 13 & 31 & 17 \end{bmatrix} \cdot \begin{bmatrix} A_2 \\ A_3 \\ B_2 \\ B_3 \end{bmatrix} = \begin{bmatrix} 0 \\ 38 \\ 216 \\ 828 \end{bmatrix}$$

$$\begin{bmatrix} A_2 \\ A_3 \\ B_2 \\ B_3 \end{bmatrix} = \begin{bmatrix} 1 & 1 & 0 & 0 \\ 7 & 5 & 1 & 1 \\ 31 & 17 & 7 & 5 \\ 25 & 13 & 31 & 17 \end{bmatrix}^{-1} \cdot \begin{bmatrix} 0 \\ 38 \\ 216 \\ 828 \end{bmatrix} = \begin{bmatrix} 0 \\ 0 \\ 13 \\ 25 \end{bmatrix}$$

The final result is:

$$\frac{s^4 + 48 \cdot s^3 + 278 \cdot s^2 + 1006 \cdot s + 975}{(s+1) \cdot \left((s+2)^2 + 9\right) \cdot \left((s+3)^2 + 16\right)} =$$

$$\frac{1}{s+1} + \frac{13}{(s+2)^2 + 9} + \frac{25}{(s+3)^2 + 16}$$

Computer software is also available that will convert functions to partial fraction expansion.

PROBLEM SOLUTIONS

Problem Solutions Chapter 2

2-1. Determine the mathematical representation for $v(t)$ that is a DC voltage of 12 volts switched on at t equal to 0.

$$v(t) = 12 \cdot u(t)$$

2-2. A voltage described by the equation $2 \cdot t + 3 \cdot t^2$ is applied to a circuit at t equal to 0. Determine the mathematical representation for the applied voltage $v(t)$.

$$v(t) = \left(2 \cdot t + 3 \cdot t^2\right) \cdot u(t)$$

2-3. A ramp voltage with a slope of 4 volts per second is applied to a circuit at t equal to 0. Determine the mathematical representation for the applied voltage $v(t)$.

$$v(t) = 4 \cdot t \cdot u(t)$$

2-4. Plot the voltage $v(t) = -0.5 \cdot t \cdot u(t)$.

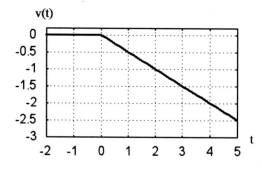

2-5. Determine the damping factor and time constant for:

$$v(t) = 12 \cdot e^{-(10 \cdot t)} \cdot u(t)$$

Damping Factor $= 10$

Time Constant $= \dfrac{1}{10} = 0.1$ seconds

2-6. A voltage $v(t)$ applied at t equal to 0 starts at 8 volts and decays exponentially with a time constant of 0.5 seconds. Determine the mathematical representation for the applied voltage.

Damping Factor $= \dfrac{1}{\text{Time Constant}} = \dfrac{1}{0.5} = 2$

$$v(t) = 8 \cdot e^{-2t} \cdot u(t)$$

2-7. Plot the voltage $v(t) = 10 \cdot \sin(1000 \cdot t + 30^\circ) \cdot u(t)$ and determine the amplitude, period, frequency, and radian frequency.

Amplitude $= 10$

Radian Frequency $= 1000$ radians/second

Frequency $= \dfrac{1000}{2 \cdot \pi} = 159.155$ hertz

Period $= \dfrac{1}{\text{Frequency}} = 0.0062832$ seconds

$10 \cdot \sin(30^\circ) = 5$ which is the starting voltage at t = 0

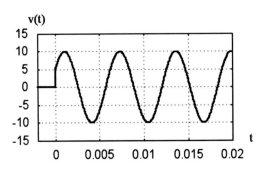

2-8. Plot the voltage $v(t) = 10 \cdot e^{-2t} \cdot \sin(20 \cdot \pi \cdot t) \cdot u(t)$.

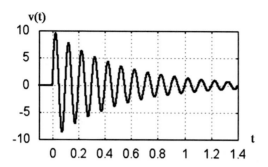

2-9. Determine the mathematical representation for $v(t)$ that is a DC voltage of 5 volts switched on at t equal to 3 seconds.

$$v(t) = 5 \cdot u(t-3)$$

2-10. A voltage described by the equation $3 \cdot t + 4$ is applied to a circuit at t equal to 2 seconds. Determine the mathematical representation for the applied voltage $v(t)$.

$$v(t) = 3 \cdot (t-2) \cdot u(t-2) + 4 \cdot u(t-2)$$
$$\text{or } v(t) = (3 \cdot (t-2) + 4) \cdot u(t-2)$$

2-11. A ramp voltage with a slope of 5 volts per second is applied to a circuit at t equal to 2 seconds. Determine the mathematical representation for the applied voltage $v(t)$.

$$v(t) = 5 \cdot (t-2) \cdot u(t-2)$$

2-12. Determine the mathematical representation for the voltage shown below:

The slope of the ramp is 3 volts/second and the ramp is time shifted by 2 seconds.

$$v(t) = 3 \cdot (t-2) \cdot u(t-2)$$

2-13. Determine the mathematical representation for the voltage shown below:

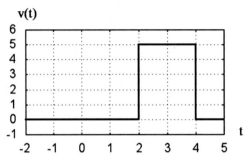

This function can be created using 2 step functions. The first has an amplitude of 5 volts and is time shifted by 2 seconds. The second also has an amplitude of 5 volts but is time shifted by 4 seconds and is subtracted from the first.

$$v(t) = 5 \cdot u(t-2) - 5 \cdot u(t-4)$$

2-14. Determine the mathematical representation for the voltage shown below:

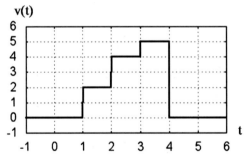

This function can be created using 4 step functions. The first has an amplitude of 2 volts and is time shifted by 1 second. The second also has an amplitude of 2 volts but is time shifted by 2 seconds and is added to the first. The third has an amplitude of 1 volt and is time shifted by 3 seconds and is added to the first two. The fourth has an amplitude of 5 volts and is time shifted by 4 seconds and is subtracted from the first three.

$$v(t) = 2 \cdot u(t-1) + 2 \cdot u(t-2) + u(t-3) - 5 \cdot u(t-4)$$

2-15. A sinusoidal source with an amplitude of 2 and a frequency of 5 is turned on at t equal to 0.2 seconds. Determine its mathematical representation.

$$\omega = 2 \cdot \pi \cdot f = 2 \cdot \pi \cdot 5 = 10 \cdot \pi$$

$$v(t) = 2 \cdot \sin(10 \cdot \pi \cdot (t - 0.2)) \cdot u(t - 0.2)$$

2-16. Plot the voltage $v(t) = 5 \cdot (t-2) \cdot u(t-2)$.

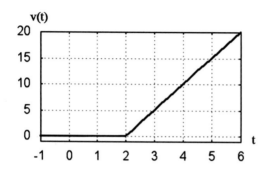

2-17. Plot the voltage $v(t) = 3 \cdot u(t-2) - 3 \cdot u(t-4)$.

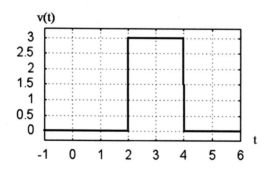

2-18. Plot the voltage:

$$v(t) = 5 \cdot t \cdot u(t) - 5 \cdot (t-1) \cdot u(t-1) + 5 \cdot (t-2) \cdot u(t-2)$$
$$-5 \cdot (t-3) \cdot u(t-3)$$

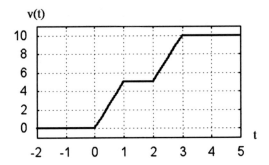

2-19. Determine the mathematical representation for an impulse function with an area of 3 that starts at t equal to 2.

$$v(t) = 3 \cdot \delta(t-2)$$

2-20. Determine the mathematical representation in summation format for the repeating voltage that starts at t equal to 0 shown below:

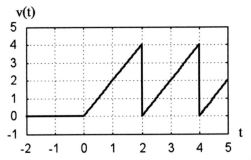

This function can be created with a ramp function with a slope of 2 volts per second and a series of 4 volt time shifted step functions (at 2 second intervals) subtracted from the ramp function. When the ramp function reaches 2 seconds (4 volts), the first step function jumps the function back to zero but the ramp function continues to rise until it reaches 4 seconds where the second step function jumps the function back to zero and the ramp function continues to rise etc.

$$v(t) = 2 \cdot t \cdot u(t) - 4 \cdot u(t-2) - 4 \cdot u(t-4) - 4 \cdot u(t-6) \cdots$$

Putting this in summation format yields:

$$v(t) = 2 \cdot t \cdot u(t) - 4 \cdot \sum_{n=1}^{n=\infty} u(t-2 \cdot n)$$

2-21. Determine the mathematical representation in summation format for the repeating half-wave rectified sine wave voltage that starts at t equal to 0 shown below:

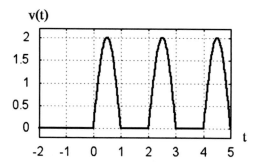

The amplitude of the sine wave is 2 volts and the period is 2 seconds (from the figure half the period is 1 second). The frequency is 1/2 hertz. The radian frequency is then $\omega = 2 \cdot \pi \cdot f = 2 \cdot \pi \cdot 0.5 = \pi$. The sinusoid representation is then $2 \cdot \sin(\pi \cdot t) \cdot u(t)$. This, however, is a repeating sine wave not just the first half cycle. To cancel out all but the first half cycle, a second time shifted sinusoid of the same amplitude and frequency starting at t = 1 must be added to the first sinusoid. The resulting equation is:

$$2 \cdot \sin(\pi \cdot t) \cdot u(t) + 2 \cdot \sin(\pi \cdot (t-1)) \cdot u(t-1)$$

This is only the first half cycle of the repeating half-wave rectified sine wave. To make this a repeating function this must be repeated every 2 seconds:

$$v(t) = \left(2 \cdot \sin(\pi \cdot t) \cdot u(t) + 2 \cdot \sin(\pi \cdot (t-1)) \cdot u(t-1)\right)$$
$$+ \left(2 \cdot \sin(\pi \cdot (t-2)) \cdot u(t-2)\right)$$
$$+ 2 \cdot \sin(\pi \cdot (t-3)) \cdot u(t-3)\right) \cdots$$

Putting this in summation format yields:

$$v(t) = 2 \cdot \sum_{n=0}^{n=\infty} \left(\sin(\pi \cdot (t-n)) \cdot u(t-n)\right)$$

Problem Solutions Chapter 3

3-1. What is the expression for the current through a resistor given the voltage across the resistor?

$$i(t) = \frac{v(t)}{R}$$

3-2. What is the expression for the capacitor current given the voltage across the capacitor?

$$i(t) = C \cdot \frac{d(v(t))}{dt}$$

3-3. What is the expression for the current through an inductor given the voltage across the inductor?

$$i(t) = \frac{1}{L} \cdot \int_0^t (v(t))\, dt + I_0$$

3-4. What is the expression for the voltage across a resistor given the current through the resistor?

$$v(t) = R \cdot i(t)$$

3-5. What is the expression for the voltage across a capacitor given the capacitor current?

$$v(t) = \frac{1}{C} \cdot \int_0^t (i(t))\, dt + V_0$$

3-6. What is the expression for the voltage across an inductor given the current through the inductor?

$$v(t) = L \cdot \frac{d(i(t))}{dt}$$

3-7. Use nodal voltage analysis to determine the integral-differential equation for the following circuit:

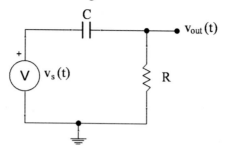

The reference node for the above circuit is the ground node. The node connecting $v_s(t)$ and C is a grounded voltage source node and is equal to $v_s(t)$. The node connecting C and R is the output node, $v_{out}(t)$, which is the only unknown node. Branch currents $i_C(t)$, and $i_R(t)$ have been assigned based on expected current direction. The figure below is the updated schematic:

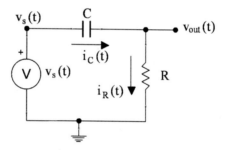

Applying Kirchhoff's current law to node $v_{out}(t)$ yields:

$$i_C(t) = i_R(t)$$

Substituting Equation (3-6) for $i_C(t)$ and Equation (3-2) for $i_R(t)$ yields:

$$C \cdot \frac{d(v_s(t) - v_{out}(t))}{dt} = \frac{v_{out}(t)}{R}$$

This is the integral-differential equation for the circuit.

3-8. Use nodal voltage analysis to determine the integral-differential equation for the following circuit:

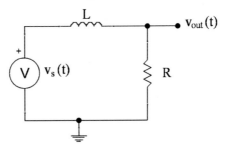

The reference node for the above circuit is the ground node. The node connecting $v_s(t)$ and L is a grounded voltage source node and is equal to $v_s(t)$. The node connecting L and R is the output node, $v_{out}(t)$, which is the only unknown node. Branch currents $i_L(t)$, and $i_R(t)$ have been assigned based on expected current direction. The figure below is the updated schematic:

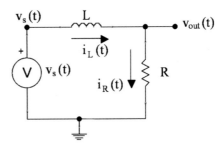

Applying Kirchhoff's current law to node $v_{out}(t)$ yields:

$$i_L(t) = i_R(t)$$

Substituting Equation (3-12) for $i_L(t)$ and Equation (3-2) for $i_R(t)$ yields:

$$\frac{1}{L} \cdot \int_0^t \left(v_s(t) - v_{out}(t) \right) dt + I_{L0} = \frac{v_{out}(t)}{R}$$

This is the integral-differential equation for the circuit and I_{L0} is the current flowing in the inductor at t equal to zero.

3-9. Use nodal voltage analysis to determine the integral-differential equation for the following circuit:

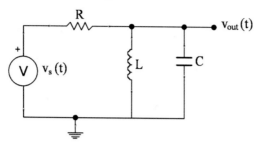

The reference node for the above circuit is the ground node. The node connecting $v_s(t)$ and R is a grounded voltage source node and is equal to $v_s(t)$. The node connecting L, C, and R is the output node, $v_{out}(t)$, which is the only unknown node. Branch currents $i_R(t)$, $i_L(t)$, and $i_C(t)$ have been assigned based on expected current direction. The figure below is the updated schematic:

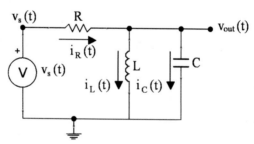

Applying Kirchhoff's current law to node $v_{out}(t)$ yields:

$$i_R(t) = i_L(t) + i_C(t)$$

Substituting Equation (3-2) for $i_R(t)$, Equation (3-12) for $i_L(t)$ and Equation (3-6) for $i_C(t)$ yields:

$$\frac{v_S(t) - v_{out}(t)}{R} = \frac{1}{L} \cdot \int_0^t \left(v_{out}(t)\right) dt + I_{L0} + C \cdot \frac{d\left(v_{out}(t)\right)}{dt}$$

This is the integral-differential equation for the circuit and I_{L0} is the current flowing in the inductor at t equal to zero.

3-10. Use nodal voltage analysis to determine the integral-differential equation for the following circuit:

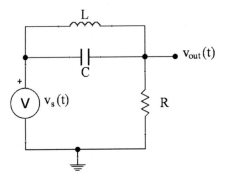

The reference node for the above circuit is the ground node. The node connecting $v_S(t)$, L, and C is a grounded voltage source node and is equal to $v_S(t)$. The node connecting L, C, and R is the output node, $v_{out}(t)$, which is the only unknown node. Branch currents $i_C(t)$, $i_L(t)$, and $i_R(t)$ have been assigned based on expected current direction. The figure below is the updated schematic:

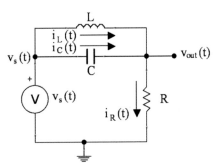

Applying Kirchhoff's current law to node $v_{out}(t)$ yields:

$$i_L(t) + i_C(t) = i_R(t)$$

Substituting, Equation (3-12) for $i_L(t)$, Equation (3-6) for $i_C(t)$, and Equation (3-2) for $i_R(t)$ yields:

$$\frac{1}{L} \cdot \int_0^t \left(v_S(t) - v_{out}(t) \right) dt + I_{L0} + C \cdot \frac{d\left(v_S(t) - v_{out}(t) \right)}{dt}$$

$$= \frac{v_{out}(t)}{R}$$

This is the integral-differential equation for the circuit and I_{L0} is the current flowing in the inductor at t equal to zero.

Problem Solutions Chapter 4

4-1. What are the two classifications of current and voltage waveforms?

Continuous time (analog) and discrete time (digital).

4-2. What method is used to transform periodic waveforms from the time domain to the frequency domain?

The Fourier series.

4-3. What two methods are used to transform non-periodic waveforms from the time domain to the frequency domain?

The Fourier transform and the Laplace transform.

4-4. What are the differences between the Fourier and Laplace transforms?

The Laplace transform assumes that all waveforms are zero for t less than 0 while the Fourier transform assumes time to be continuous from negative to positive infinity. Also, the Laplace transform has an additional decaying real exponential making it useable with more functions than the Fourier transform.

4-5. What is the linearity property of the Laplace transform?

$$\mathscr{L}\left(K_1 \cdot f_1(t) \pm K_2 \cdot f_2(t) \pm \cdots\right) = K_1 \cdot F_1(s) \pm K_2 \cdot F_2(s) \pm \cdots$$

4-6. What is the Laplace transform of a step function of magnitude 5?

$$\frac{5}{s}$$

4-7. What is the Laplace transform of a ramp function of slope 4?

$$\frac{4}{s^2}$$

4-8. What is the Laplace transform of an impulse function of area 3?

3

4-9. What is the Laplace transform of $6 \cdot e^{-7t} \cdot u(t)$?

$$\frac{6}{s+7}$$

4-10. What is the complex translation property of the Laplace transform?

$$\mathscr{L}\left(e^{-\alpha t} \cdot f(t)\right) = F(s+\alpha)$$

4-11. What is the Laplace transform of $3 \cdot \sin(5 \cdot t) \cdot u(t)$?

$$\frac{15}{s^2 + 25}$$

4-12. What is the Laplace transform of $4 \cdot e^{-2t} \cdot \sin(3 \cdot t) \cdot u(t)$?

$$\mathcal{L}\left(4 \cdot e^{-2t} \cdot \sin(3 \cdot t) \cdot u(t)\right) = \mathcal{L}\left(4 \cdot \sin(3 \cdot t) \cdot u(t)\right)_{s=s+2}$$

$$= \frac{12}{(s+2)^2 + 9}$$

4-13. What is the Laplace transform of $6 \cdot \sin(3 \cdot t + 60°) \cdot u(t)$?

From the second table of Appendix B:

$$\mathcal{L}\left(K \cdot \sin(\omega \cdot t + \theta) \cdot u(t)\right) = \frac{K \cdot \sin(\theta)\left(s + \dfrac{\omega}{\tan(\theta)}\right)}{s^2 + \omega^2}$$

where: $K = 6 \quad \omega = 3 \quad \theta = 60°$

$$\mathcal{L}\left(6 \cdot \sin(3 \cdot t + 60°) \cdot u(t)\right) = \frac{5.196 \cdot (s + 1.732)}{s^2 + 9}$$

4-14. What is the Laplace transform of $6 \cdot e^{-2t} \cdot \sin(3 \cdot t + 60°) \cdot u(t)$?

Using the complex translation property: $\mathcal{L}\left(e^{-\alpha t} \cdot f(t)\right) = F(s + \alpha)$
From the second table of Appendix B:

$$\mathcal{L}\left(K \cdot \sin(\omega \cdot t + \theta) \cdot u(t)\right) = \frac{K \cdot \sin(\theta)\left(s + \dfrac{\omega}{\tan(\theta)}\right)}{s^2 + \omega^2}$$

where: $K = 6 \quad \omega = 3 \quad \theta = 60°$

$$\mathcal{L}\left(6 \cdot \sin(3 \cdot t + 60°) \cdot u(t)\right) = \frac{5.196 \cdot (s + 1.732)}{s^2 + 9}$$

$$\mathcal{L}\left(6 \cdot e^{-2t} \cdot \sin(3 \cdot t + 60°) \cdot u(t)\right) = \left(\frac{5.196 \cdot (s + 1.732)}{s^2 + 9}\right)\Bigg|_{s=s+2}$$

$$\mathcal{L}\left(6 \cdot e^{-2t} \cdot \sin(3 \cdot t + 60°) \cdot u(t)\right) = \frac{5.196 \cdot (s + 3.732)}{(s+2)^2 + 9}$$

4-15. What is the real translation property of the Laplace transform?

$$\mathcal{L}(f(t-a)\cdot u(t-a)) = e^{-as}\cdot F(s)$$

4-16. What is the Laplace transform of $5\cdot(t-2)\cdot u(t-2)$?

$$\frac{5\cdot e^{-2s}}{s^2}$$

4-17. What is the Laplace transform of a rectangular pulse with an amplitude of 3 volts and a width of 0.5 seconds that starts at t equal to 0?

$$\frac{3}{s} - \frac{3\cdot e^{-0.5s}}{s}$$

4-18. What is the real differentiation property of the Laplace transform?

$$\mathcal{L}\left(\frac{df(t)}{dt}\right) = s\cdot\mathcal{L}(f(t)) - f(0)$$

4-19. What is the complex differentiation property of the Laplace transform?

$$\mathcal{L}(-t\cdot f(t)) = \frac{dF(s)}{ds}$$

4-20. What is the real integration property of the Laplace transform?

$$\mathcal{L}\left(\int_0^t f(t)\,dt\right) = \frac{\mathcal{L}(f(t))}{s}$$

4-21. What is the complex integration property of the Laplace transform?

$$\mathcal{L}\left(\frac{f(t)}{t}\right) = \int_s^\infty F(s)\,ds$$

4-22. The integral-differential equation for the circuit of Chapter 3 problem 7 is (see problem solutions chapter 3 problem 7 for details):

$$C\cdot\frac{d(v_s(t) - v_{out}(t))}{dt} = \frac{v_{out}(t)}{R}$$

Assume zero initial values, convert the equation to the Laplace domain, and solve for $V_{out}(s)$.

$$s \cdot C \cdot \left(V_s(s) - V_{out}(s)\right) = \frac{V_{out}(s)}{R}$$

$$V_{out}(s) \cdot \left(\frac{1}{R} + s \cdot C\right) = s \cdot C \cdot V_s(s)$$

$$V_{out}(s) = V_s(s) \cdot \frac{s}{s + \dfrac{1}{R \cdot C}}$$

4-23. The integral-differential equation for the circuit of Chapter 3 problem 8 is (see problem solutions chapter 3 problem 8 for details):

$$\frac{1}{L} \cdot \int_0^t \left(v_s(t) - v_{out}(t)\right) dt + I_{L0} = \frac{v_{out}(t)}{R}$$

Assume zero initial values, convert the equation to the Laplace domain, and solve for $V_{out}(s)$.

$$\frac{1}{s \cdot L}\left(V_s(s) - V_{out}(s)\right) = \frac{V_{out}(s)}{R}$$

$$V_{out}(s) \cdot \left(\frac{1}{R} + \frac{1}{s \cdot L}\right) = \frac{V_s(s)}{s \cdot L}$$

$$V_{out}(s) = V_s(s) \cdot \frac{\dfrac{R}{L}}{s + \dfrac{R}{L}}$$

4-24. The integral-differential equation for the circuit of Chapter 3 problem 9 is (see problem solutions chapter 3 problem 9 for details):

$$\frac{v_s(t) - v_{out}(t)}{R} = \frac{1}{L} \cdot \int_0^t \left(v_{out}(t)\right) dt + I_{L0} + C \cdot \frac{d\left(v_{out}(t)\right)}{dt}$$

Assume zero initial values, convert the equation to the Laplace domain, and solve for $V_{out}(s)$.

$$\frac{V_s(s) - V_{out}(s)}{R} = \frac{V_{out}(s)}{s \cdot L} + s \cdot C \cdot V_{out}(s)$$

$$V_{out}(s) \cdot \left(\frac{1}{s \cdot L} + s \cdot C + \frac{1}{R}\right) = \frac{V_s(s)}{R}$$

$$V_{out}(s) = V_s(s) \cdot \frac{\dfrac{s}{R \cdot C}}{s^2 \cdot R \cdot L \cdot C + \dfrac{s}{R \cdot C} + \dfrac{1}{L \cdot C}}$$

4-25. The integral-differential equation for the circuit of Chapter 3 problem 10 is (see problem solutions chapter 3 problem 10 for details):

$$\frac{1}{L} \cdot \int_0^t \left(v_s(t) - v_{out}(t) \right) dt + I_{L0} + C \cdot \frac{d\left(v_s(t) - v_{out}(t) \right)}{dt} = \frac{v_{out}(t)}{R}$$

Assume zero initial values, convert the equation to the Laplace domain, and solve for $V_{out}(s)$.

$$\frac{1}{s \cdot L} \cdot \left(V_s(s) - V_{out}(s) \right) + s \cdot C \cdot \left(V_s(s) - V_{out}(s) \right) = \frac{V_{out}(s)}{R}$$

$$V_{out}(s) \cdot \left(\frac{1}{R} + \frac{1}{s \cdot L} + s \cdot C \right) = V_s(s) \cdot \left(s \cdot C + \frac{1}{s \cdot L} \right)$$

$$V_{out}(s) \cdot \left(\frac{s \cdot L + R + s^2 \cdot R \cdot L \cdot C}{s \cdot L \cdot R} \right) = V_s(s) \cdot \left(\frac{s^2 \cdot L \cdot C + 1}{s \cdot L} \right)$$

$$V_{out}(s) = V_s(s) \cdot \frac{s^2 + \dfrac{1}{L \cdot C}}{s^2 + \dfrac{s}{R \cdot C} + \dfrac{1}{L \cdot C}}$$

4-26. What is the inverse Laplace transform of $\dfrac{8}{s+3}$?

This is of the form of transform pair number 1 in Appendix B.

$$\mathcal{L}^{-1}\left(\frac{K}{s+a} \right) = \left(K \cdot e^{-at} \right) \cdot u(t)$$

where: $K = 8 \quad a = 3$

$$\mathcal{L}^{-1}\left(\frac{8}{s+3} \right) = \left(8 \cdot e^{-3t} \right) \cdot u(t)$$

4-27. What is the inverse Laplace transform of $\dfrac{32}{s^2 + 64}$?

This is of the form of the Sine transform pair in Appendix B.

$$\mathcal{L}^{-1}\left(\frac{K \cdot \omega}{s^2 + \omega^2}\right) = (K \cdot \sin(\omega \cdot t)) \cdot u(t)$$

where: $K = 4$ $\omega = 8$

$$\mathcal{L}^{-1}\left(\frac{32}{s^2 + 64}\right) = (4 \cdot \sin(8 \cdot t)) \cdot u(t)$$

4-28. What is the inverse Laplace transform of $\dfrac{6 \cdot s + 15}{s^2 + 9}$?

This is of the form of the Shifted Sine transform pair in Appendix B.

$$\mathcal{L}^{-1}\left(\frac{K \cdot \sin(\theta) \cdot \left(s + \dfrac{\omega}{\tan(\theta)}\right)}{s^2 + \omega^2}\right) = (K \cdot \sin(\omega \cdot t + \theta)) \cdot u(t)$$

where: $K \cdot \sin(\theta) = 6$ $\omega = 3$ $\dfrac{\omega}{\tan(\theta)} = 2.5$

$\tan(\theta) = 1.2$ $\theta = 50.2°$ $K = 7.81$

$$\mathcal{L}^{-1}\left(\frac{6 \cdot s + 15}{s^2 + 9}\right) = (7.81 \cdot \sin(3 \cdot t + 50.2°)) \cdot u(t)$$

4-29. What is the inverse Laplace transform of $\dfrac{4 \cdot s - 22}{s^2 + 4 \cdot s + 29}$?

This can be rearranged to:

$$\frac{4 \cdot ((s+2) - 7.5)}{(s+2)^2 + 5^2}$$

The complex translation property states that:

$$\mathcal{L}(e^{-\alpha t} \cdot f(t) \cdot u(t)) = \mathcal{L}(f(t))_{s=s+\alpha}$$

Thus:

$$\mathcal{L}^{-1}\left(\frac{4 \cdot ((s+2) - 7.5)}{(s+2)^2 + 5^2}\right) = e^{-2t} \cdot \mathcal{L}^{-1}\left(\frac{4 \cdot (s - 7.5)}{s^2 + 5^2}\right)$$

This is of the form of the Shifted Sine transform pair in Appendix B.

$$\mathcal{L}^{-1}\left(\frac{K \cdot \sin(\theta) \cdot \left(s + \frac{\omega}{\tan(\theta)}\right)}{s^2 + \omega^2}\right) = (K \cdot \sin(\omega \cdot t + \theta)) \cdot u(t)$$

where: $K \cdot \sin(\theta) = 4 \quad \omega = 5 \quad \dfrac{\omega}{\tan(\theta)} = -7.5$

$$\tan(\theta) = \frac{5}{-7.5} \quad \theta = 146.31° \quad K = 7.211$$

$$\mathcal{L}^{-1}\left(\frac{6 \cdot s + 15}{s^2 + 9}\right) = (7.211 \cdot \sin(5 \cdot t + 146.31°)) \cdot u(t)$$

$$\mathcal{L}^{-1}\left(\frac{4 \cdot s - 22}{s^2 + 4 \cdot s + 29}\right) = (7.211 \cdot e^{-2t} \cdot \sin(5 \cdot t + 146.31°)) \cdot u(t)$$

4-30. What is the inverse Laplace transform of $\dfrac{3 \cdot s + 10}{s + 2}$?

This can be rearranged to:

$$\frac{3 \cdot (s + 2)}{s + 2} + \frac{4}{s + 2} = 3 + \frac{4}{s + 2}$$

The inverse Laplace transform of the first term is an impulse function with an area of 3.

The second term is of the form of transform pair number 1 in Appendix B.

$$\mathcal{L}^{-1}\left(\frac{K}{s + a}\right) = (K \cdot e^{-a \cdot t}) \cdot u(t)$$

where: $K = 4 \quad a = 2$

$$\mathcal{L}^{-1}\left(\frac{4}{s + 2}\right) = (4 \cdot e^{-2t}) \cdot u(t)$$

$$\mathcal{L}^{-1}\left(\frac{3 \cdot s + 10}{s + 2}\right) = 3 \cdot \delta(t) + (4 \cdot e^{-2t}) \cdot u(t)$$

4-31. Determine the voltage across R for the circuit below:

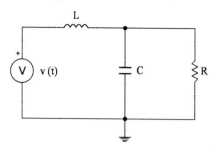

The source voltage $v(t)$ is a step function of 3 volts at t equal to zero. Also, L is 10 millihenry, C is 0.1 microfarad, R is 1000 ohms, and the capacitor initial voltage and inductor initial current are zero. Also plot the time domain solution.

First transform the circuit to the Laplace domain as shown below:

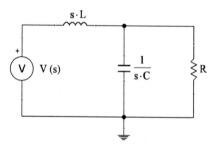

The voltage divider rule will be used to determine the voltage across R. First, the resistor and capacitor must be combined in parallel to form $Z(s)$ as shown below:

$$Z(s) = \frac{R \cdot \dfrac{1}{s \cdot C}}{R + \dfrac{1}{s \cdot C}} = \frac{R}{s \cdot R \cdot C + 1}$$

Since the voltage across R is the same as the voltage across $Z(s)$, $V_R(s)$ can be determined using the voltage divider rule as shown below:

$$V_R(s) = V(s) \cdot \dfrac{\dfrac{R}{s \cdot R \cdot C + 1}}{s \cdot L + \dfrac{R}{s \cdot R \cdot C + 1}}$$

$$V_R(s) = V(s) \cdot \dfrac{R}{s^2 \cdot R \cdot L \cdot C + s \cdot L + R}$$

$$= V(s) \cdot \dfrac{\dfrac{1}{L \cdot C}}{s^2 + s \cdot \dfrac{1}{R \cdot C} + \dfrac{1}{L \cdot C}}$$

Substituting the values for R, L, and C along with the Laplace transform of $v(t)$ into the above equation yields:

$$V_R(s) = \left(\dfrac{3}{s}\right) \cdot \left(\dfrac{1 \cdot 10^9}{s^2 + 1 \cdot 10^4 \cdot s + 1 \cdot 10^9}\right)$$

$$= \dfrac{3 \cdot 10^9}{s \cdot \left((s + 5000)^2 + 31225^2\right)}$$

The above equation is of the form of Laplace transform pair number 21 in Appendix B:

$$v_R(t) = \mathcal{L}^{-1}\left(\dfrac{K}{s \cdot \left((s+a)^2 + b^2\right)}\right)$$

$$= \left(\dfrac{K}{a^2 + b^2} - \dfrac{K \cdot e^{-at} \cdot \sin(b \cdot t + \theta)}{b \cdot \sqrt{a^2 + b^2}}\right) \cdot u(t)$$

where $\theta = \text{angle}(x, y)$

and $x = a$; $y = b$

Comparing the above two equations yields:

$$K = 3 \cdot 10^9 \quad a = 5000 \quad b = 31225$$

$$\theta = \tan^{-1}\left(\dfrac{31225}{5000}\right) = 80.9°$$

Substituting in these values yields the time domain solution for the voltage across R:

$$v_R(t) = \left(3 - 3.038 \cdot e^{-5000t} \cdot \sin(31225 \cdot t + 80.9°)\right) \cdot u(t)$$

Below is a plot of the voltage across R:

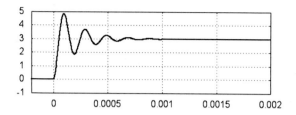

Of special note here is that no sinusoidal waveform was applied to the circuit, yet there is a transient decaying sinusoidal response.

4-32. Determine the voltage across R for the circuit below:

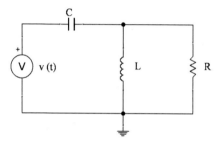

The source voltage $v(t)$ is a pulse at t equal to zero with an amplitude of 2 volts and a width of 1 millisecond. Also, L is 10 millihenry, C is 0.1 microfarad, R is 1000 ohms, and the capacitor initial voltage and inductor initial current are zero. Also plot the time domain solution.

First transform the circuit to the Laplace domain as shown below:

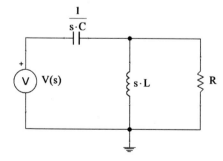

The voltage divider rule will be used to determine the voltage across R. First, the resistor and inductor must be combined in parallel to form $Z(s)$ as shown below:

$$Z(s) = \frac{s \cdot L \cdot R}{s \cdot L + R}$$

Since the voltage across R is the same as the voltage across $Z(s)$, $V_R(s)$ can be determined using the voltage divider rule as shown below:

$$V_R(s) = V(s) \cdot \frac{\dfrac{s \cdot L \cdot R}{s \cdot L + R}}{\dfrac{1}{s \cdot C} + \dfrac{s \cdot L \cdot R}{s \cdot L + R}}$$

$$= V(s) \cdot \frac{R \cdot L \cdot C \cdot s^2}{R \cdot L \cdot C \cdot s^2 + L \cdot s + R}$$

$$= V(s) \cdot \frac{s^2}{s^2 + s \cdot \dfrac{1}{R \cdot C} + \dfrac{1}{L \cdot C}}$$

The waveform $v(t)$ is a 2 volt step function minus a 2 volt step function delayed by 1 millisecond:

$$v(t) = 2 \cdot u(t) - 2 \cdot u(t - 0.001)$$

The Laplace transform of $v(t)$ is then:

$$\mathscr{L}(v(t)) = \frac{2}{s} - e^{-0.001 \cdot s} \cdot \frac{2}{s} = \frac{2}{s} \cdot \left(1 - e^{-0.001 \cdot s}\right)$$

Substituting the values for R, L, and C along with the Laplace transform of $v(t)$ into the above equation yields:

$$V_R(s) = \left(\frac{2}{s}\right) \cdot \left(\frac{s^2}{s^2 + 1 \cdot 10^4 \cdot s + 1 \cdot 10^9}\right) \cdot \left(1 - e^{-0.001 \cdot s}\right)$$

$$= \frac{2 \cdot s}{(s + 5000)^2 + 31225^2} \cdot \left(1 - e^{-0.001 \cdot s}\right)$$

The first part of the above equation is of the form of Laplace transform pair number 25 in Appendix B:

280

$$\mathcal{L}^{-1}\left(\frac{K \cdot s}{(s+a)^2 + b^2}\right) = \frac{K \cdot \sqrt{a^2 + b^2} \cdot e^{-a \cdot t} \cdot \sin(b \cdot t + \theta)}{b} \cdot u(t)$$

where $\theta = $ angle (x, y)

and $x = -a \; ; \; y = b$

The solution for $v_R(t)$ is the inverse Laplace transform above minus the same inverse Laplace transform but delayed in time by 1 millisecond.

Comparing the above two equations yields:

$$K = 2 \quad a = 5000 \quad b = 31225$$

$$\theta = \tan^{-1}\left(\frac{31225}{-5000}\right) = 99.1°$$

Substituting in these values yields the time domain solution for inverse Laplace transform:

$$v(t) = \left(2.025 \cdot e^{-5000 t} \cdot \sin(31225 \cdot t + 99.1°)\right) \cdot u(t)$$

The solution for $v_R(t)$ then is:

$$v_R(t) = \left(2.025 \cdot e^{-5000 t} \cdot \sin(31225 \cdot t + 99.1°)\right) \cdot u(t)$$
$$-\left(2.025 \cdot e^{-5000 \cdot (t - 0.001)} \cdot \sin(31225 \cdot (t - 0.001) + 99.1°)\right)$$
$$\cdot u(t - 0.001)$$

Below is a plot of the voltage across R:

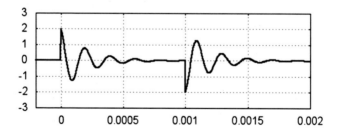

Note that the steady state solution is zero and only the transient solution resulting from each edge of the pulse exists and is a decaying sinusoid from each edge of the pulse.

4-33. Determine the voltage across R for the circuit of Problem 4-31 with the source voltage $v(t)$ a pulse at t equal to zero with an amplitude of 3 volts and a width of 200 microseconds. Also, L is 100 millihenry, C is 0.01 microfarad, R is 1000 ohms, and the capacitor initial voltage and inductor initial current are zero. Also plot the time domain solution.

The circuit transform to the Laplace domain is the same as for Problem 4-31 as shown below:

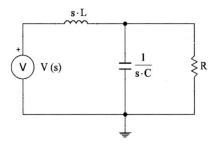

The voltage divider rule will be used to determine the voltage across R. First, the resistor and capacitor must be combined in parallel to form $Z(s)$ as shown below:

$$Z(s) = \frac{R \cdot \dfrac{1}{s \cdot C}}{R + \dfrac{1}{s \cdot C}} = \frac{R}{s \cdot R \cdot C + 1}$$

Since the voltage across R is the same as the voltage across $Z(s)$, $V_R(s)$ can be determined using the voltage divider rule as shown below:

$$V_R(s) = V(s) \cdot \frac{\dfrac{R}{s \cdot R \cdot C + 1}}{s \cdot L + \dfrac{R}{s \cdot R \cdot C + 1}}$$

$$V_R(s) = V(s) \cdot \frac{R}{s^2 \cdot R \cdot L \cdot C + s \cdot L + R}$$

$$= V(s) \cdot \frac{\dfrac{1}{L \cdot C}}{s^2 + s \cdot \dfrac{1}{R \cdot C} + \dfrac{1}{L \cdot C}}$$

The waveform $v(t)$ is a 3 volt step function minus a 3 volt step function delayed by 200 microseconds:

$$v(t) = 3 \cdot u(t) - 3 \cdot u(t - 0.0002)$$

The Laplace transform of $v(t)$ is then:

$$\mathcal{L}(v(t)) = \frac{3}{s} - e^{-0.0002 \cdot s} \cdot \frac{3}{s} = \frac{3}{s} \cdot (1 - e^{-0.0002 \cdot s})$$

Substituting the values for R, L, and C along with the Laplace transform of $v(t)$ into the above equation yields:

$$V_R(s) = \left(\frac{3}{s}\right) \cdot \left(\frac{1 \cdot 10^9}{s^2 + 1 \cdot 10^5 \cdot s + 1 \cdot 10^9}\right) \cdot (1 - e^{-0.0002 \cdot s})$$

$$= \left(\frac{3 \cdot 10^9}{s \cdot (s + 11270) \cdot (s + 88730)}\right) \cdot (1 - e^{-0.0002 \cdot s})$$

The first part of the above equation is of the form of Laplace transform pair number 10a in Appendix B:

$$\mathcal{L}^{-1}\left(\frac{K}{s \cdot (s+a) \cdot (s+b)}\right) = \left(\frac{K}{a \cdot b} - \frac{K \cdot (b \cdot e^{-a \cdot t} - a \cdot e^{-b \cdot t})}{a \cdot b \cdot (a - b)}\right) \cdot u(t)$$

The solution for $v_R(t)$ is the inverse Laplace transform above minus the same inverse Laplace transform but delayed in time by 200 microseconds.

Comparing the above two equations yields:

$$K = 3 \cdot 10^9 \quad a = 11270 \quad b = 88730$$

Substituting in these values yields the time domain solution for inverse Laplace transform:

$$v(t) = (3 - 3.4365 \cdot e^{-11270 \cdot t} + 0.4365 \cdot e^{-88730 \cdot t}) \cdot u(t)$$

The solution for $v_R(t)$ then is:

$$v_R(t) = (3 - 3.4365 \cdot e^{-11270 \cdot t} + 0.4365 \cdot e^{-88730 \cdot t}) \cdot u(t)$$
$$- (3 - 3.4365 \cdot e^{-11270 \cdot (t - 0.0002)} + 0.4365 \cdot e^{-88730 \cdot (t - 0.0002)})$$
$$\cdot u(t - 0.0002)$$

Below is a plot of the voltage across R:

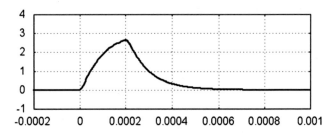

Of special note here is that the pulse width is less than 5 time constants of the longest exponential and the voltage across R never reaches steady state from the pulse first edge before the pulse second edge occurs. Also the component values for this problem did not produce a decaying sinusoid like Problems 4-31 and 32.

Problem Solutions Chapter 5

5-1.   Construct the pole zero plot for the system transfer function below and discuss the system stability.

$$G(s) = \frac{s+1}{s \cdot (s+2) \cdot \left((s+3)^2 + 16 \right)}$$

The gain is 1 and there is 1 real zero at -1, 2 real poles, one at 0 and the other at -2, and 1 complex pole at $-3 \pm j \cdot 4$. Below is the pole zero plot for the poles and zeros:

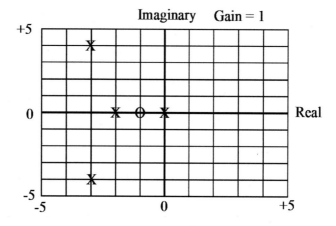

There are no poles in the right half plane and there is 1 pole on the imaginary axis at 0 making the system marginally stable.

5-2.   Construct the pole zero plot for the system transfer function below and discuss the system stability.

$$G(s) = \frac{s^2 \cdot (s-4)}{(s+3) \cdot (s+2)^2}$$

The gain is 1 and there is a second order real zero at the origin, a real zero at +4, a second order real pole at -2, and a real pole at -3. Below is the pole zero plot for the poles and zeros:

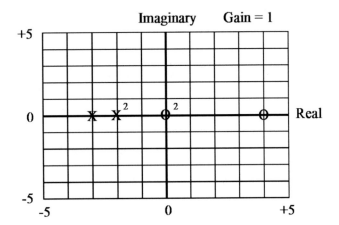

All of the poles are in the left half plane so the system is stable.

5-3. Construct the pole zero plot for the system transfer function below and discuss the system stability.

$$G(s) = \frac{(s-3)^2 + 4}{(s+3) \cdot (s-4) \cdot (s+1)^2}$$

The gain is 1 and there is a complex zero at $+3 \pm j \cdot 2$, a real pole at -3, a real pole at +4, and a second order real pole at -1. Below is the pole zero plot for the poles and zeros:

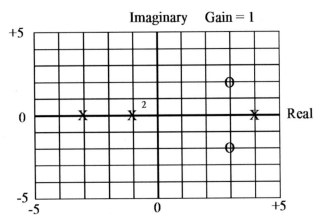

There is a pole at +4 in the right half plane making the system unstable.

5-4. Construct the pole zero plot for the system transfer function below and discuss the system stability.

$$G(s) = \frac{(s+2)\cdot(s-4)}{(s+1)\cdot(s^2+9)}$$

The gain is 1 and there is a real zero at -2, a real zero at +4, a real pole at -1, and a complex pole at $\pm j \cdot 3$. Below is the pole zero plot for the poles and zeros:

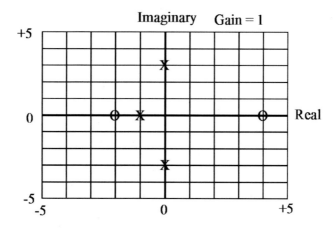

There are no poles in the right half plane and there is 1 complex pole on the imaginary axis at $\pm j \cdot 3$ making the system marginally stable.

5-5. Construct the pole zero plot for the system transfer function below and discuss the system stability.

$$G(s) = \frac{(s^2+4)\cdot\left((s-1)^2+9\right)}{s\cdot(s+5)\cdot(s+4)}$$

The gain is 1 and there is a complex zero at $\pm j \cdot 2$, a complex zero at $+1 \pm j \cdot 3$, a real pole at 0, a real pole at -4, and a real pole at -5. Below is the pole zero plot for the poles and zeros:

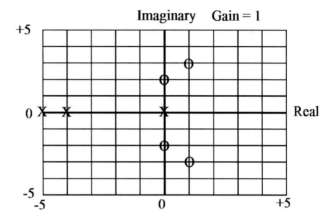

There are no poles in the right half plane and there is a real pole at 0 making the system marginally stable.

5-6. Construct the bode straight line approximation plot for the system transfer function below and compare it with a computer generated bode plot.

$$G(s) = \frac{10}{s+10}$$

All of the poles are in the left half plane so the system is stable. Converting the transfer function to break frequency form yields:

$$G(s) = \frac{10}{s+10} = \frac{1}{\dfrac{s}{10}+1}$$

The bode straight line approximation plots are created using Chapter 5 Figure 5-6 and are shown below:

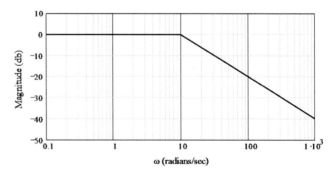

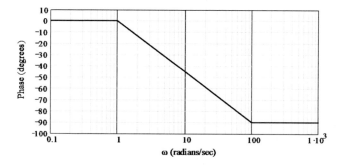

These plots (red dashed lines) are superimposed on the computer generated plots shown below:

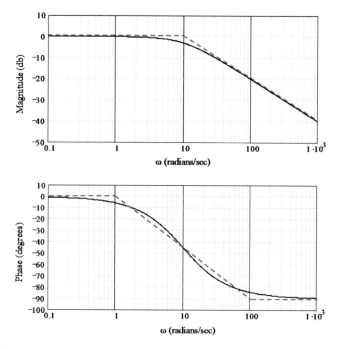

The bode straight line approximations compare quite well with the computer generated bode plots.

5-7.   Construct the bode straight line approximation plot for the system transfer function below and compare it with a computer generated bode plot.

$$G(s) = \frac{10 \cdot (s+10)}{s+100}$$

All of the poles are in the left half plane so the system is stable. Converting the transfer function to break frequency form yields:

$$G(s) = \frac{10 \cdot (s+10)}{s+100} = \frac{\frac{s}{10}+1}{\frac{s}{100}+1} = \left(\frac{s}{10}+1\right) \cdot \left(\frac{1}{\frac{s}{100}+1}\right)$$

The pole and zero have been color coded to match the plots. The bode magnitude straight line approximation plots are created using Chapter 5 Figure 5-9a for the zero and Figure 5-6a for the pole and are shown below:

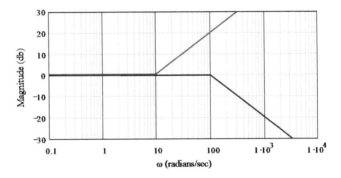

Both the pole and zero are 0 db out to $\omega = 10$. At $\omega = 10$, the zero (red line) rises at +20 db/decade. The combined plot follows the zero out to $\omega = 100$. At $\omega = 100$, the pole (blue line) falls at -20 db/decade canceling out the zero slope and creating a flat line for the rest of the plot. The combined plot is shown below:

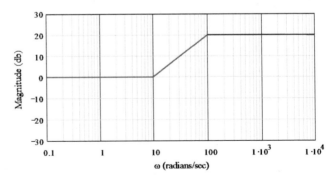

The above plot is superimposed (dashed red line) on the computer generated bode magnitude plot in the figure below:

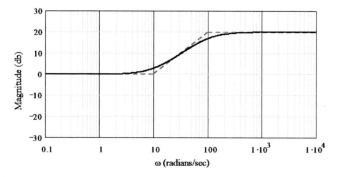

The bode magnitude straight line approximation compares quite well with the computer generated bode magnitude plot.

The bode phase straight line approximation plots are created using Chapter 5 Figure 5-9b for the zero and Figure 5-6b for the pole and are shown below:

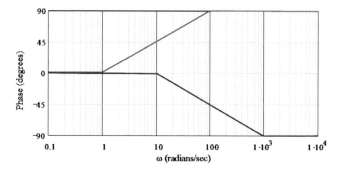

Both the pole and zero are 0 degrees out to $\omega = 1$. At $\omega = 1$, the zero (red line)rises at +45 deg/decade and the combined plot follows the zero out to $\omega = 10$. At $\omega = 10$, the pole (blue line) falls at -45 deg/decade canceling out the slope of the zero and creating a flat line out to $\omega = 100$, At $\omega = 100$, the zero slope stops and becomes flat at +90 degrees. The pole slope continues causing a falling slope of -45 deg/decade out to $\omega = 1000$. At $\omega = 1000$, the pole slope stops and becomes flat causing the combined plot slope to also become flat at 0 degrees for the rest of the plot. The combined plot is shown below:

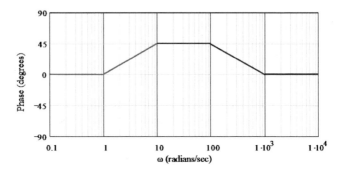

The above plot is superimposed (dashed red line) on the computer generated bode phase plot in the figure below:

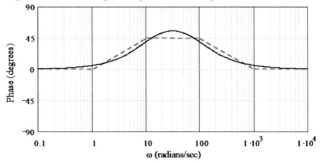

The bode phase straight line approximation compares quite well with the computer generated bode phase plot.

5-8. Construct the bode straight line approximation plot for the system transfer function below and compare it with a computer generated bode plot.

$$G(s) = \frac{10,000}{(s+50)^2 + 7500}$$

All of the poles are in the left half plane so the system is stable. Converting the transfer function to break frequency form yields:

$$G(s) = \frac{10,000}{(s+50)^2 + 7500} = \frac{1}{\left(\dfrac{s}{100}\right)^2 + \dfrac{1}{1} \cdot \left(\dfrac{s}{100}\right) + 1}$$

The bode straight line approximation plots are created using Chapter 5 Figure 5-14 and are shown below:

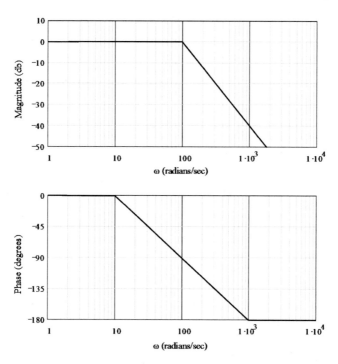

These plots (red dashed lines) are superimposed on the computer generated plots shown below:

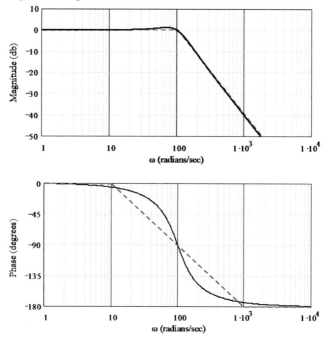

The bode straight line approximation magnitude plot is very close to the bode computer generated magnitude plot. The bode straight line approximation phase plot is not as good as the computer generated bode phase plot. A higher slope for the phase approximation would give a better comparison.

5-9. Construct the bode straight line approximation plot for the system transfer function below and compare it with a computer generated bode plot.

$$G(s) = \frac{1000 \cdot s}{(s+10) \cdot (s+100)}$$

All of the poles are in the left half plane so the system is stable. Converting the transfer function to break frequency form yields:

$$G(s) = \frac{1000 \cdot s}{(s+10) \cdot (s+100)} = \frac{10 \cdot \dfrac{s}{10}}{\left(\dfrac{s}{10}+1\right) \cdot \left(\dfrac{s}{100}+1\right)}$$

$$= 10 \cdot \left(\frac{s}{10}\right) \cdot \left(\frac{1}{\dfrac{s}{10}+1}\right) \cdot \left(\frac{1}{\dfrac{s}{100}+1}\right)$$

The break frequency for the zero at the origin is chosen to be the first break frequency which is the pole at $\omega = 10$. The poles and zero have been color coded to match the plots. The bode magnitude straight line approximation plots are created using Chapter 5 Figure 5-11a for the zero and Figure 5-6a for the poles and are shown below:

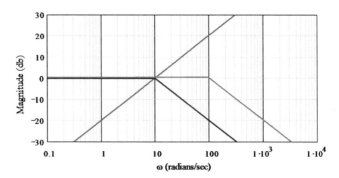

The gain factor of 10 will be dealt with later. Both the poles are 0 db out to $\omega = 10$. The combined plot will then follow the

+20 db/decade slope of the zero (red line) out to $\omega = 10$. At $\omega = 10$, the first pole (blue line) falls at -20 db/decade canceling out the slope of the zero and creating a flat line out to $\omega = 100$. At $\omega = 100$, the second pole (magenta line) falls at -20 db/decade for the rest of the plot. The combined plot is shown below:

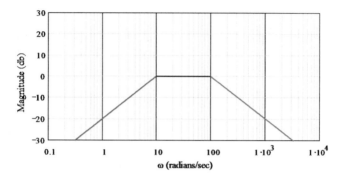

Convert the gain term of 10 to db yielding +20 db. This will then shift the above plot vertically by +20 db. The above plot (shifted) is superimposed (dashed red line) on the computer generated bode magnitude plot in the figure below:

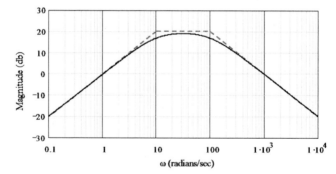

The bode magnitude straight line approximation compares quite well with the computer generated bode magnitude plot.

The bode phase straight line approximation plots are created using Chapter 5 Figure 5-11b for the zero and Figure 5-6b for the poles and are shown below:

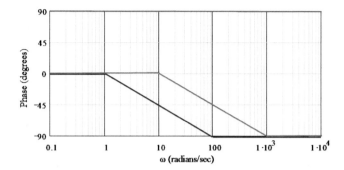

Both the poles are 0 degrees out to $\omega = 1$ so the combined plot will be +90 degrees out to $\omega = 1$ (red line) due to the phase of the zero. At $\omega = 1$, the first pole (blue line) falls at -45 deg/decade out to $\omega = 10$. At $\omega = 10$, the second pole (magenta line) falls at an additional -45 deg/decade creating a slope of -90 deg/decade out to $\omega = 100$, At $\omega = 100$, the first pole slope stops and becomes flat at -90 degrees. The second pole slope continues causing a falling slope of -45 deg/decade out to $\omega = 1000$. At $\omega = 1000$, the second pole slope stops and becomes flat causing the combined plot slope to also become flat at -90 degrees for the rest of the plot. The combined plot is shown below:

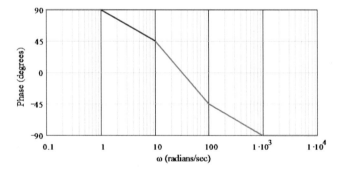

The above plot is superimposed (dashed red line) on the computer generated bode phase plot in the figure below:

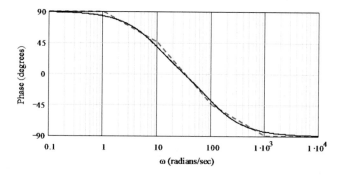

The bode phase straight line approximation compares quite well with the computer generated bode phase plot.

5-10. Construct the bode straight line approximation plot for the system transfer function below and compare it with a computer generated bode plot.

$$G(s) = \frac{10 \cdot \left((s+7)^2 + 51 \right)}{(s+100) \cdot (s+1)}$$

All of the poles are in the left half plane so the system is stable. Converting the transfer function to break frequency form yields:

$$G(s) = \frac{10 \cdot \left((s+7)^2 + 51 \right)}{(s+100) \cdot (s+1)} = \frac{10 \cdot \left(\left(\dfrac{s}{10} \right)^2 + \dfrac{1}{0.714} \cdot \left(\dfrac{s}{10} \right) + 1 \right)}{\left(\dfrac{s}{100} + 1 \right) \cdot \left(\dfrac{s}{1} + 1 \right)}$$

$$= 10 \cdot \left(\frac{1}{\dfrac{s}{1}+1} \right) \cdot \left(\left(\frac{s}{10} \right)^2 + \frac{1}{0.714} \cdot \left(\frac{s}{10} \right) + 1 \right) \cdot \left(\frac{1}{\dfrac{s}{100}+1} \right)$$

The poles and zero have been color coded to match the plots. The bode magnitude straight line approximation plots are created using Chapter 5 Figure 5-6a for the poles and Figure 5-17a for the zero and are shown below:

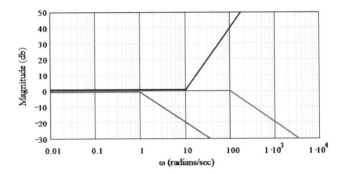

The gain factor of 10 will be dealt with later. Both the poles and the zero are 0 db out to $\omega = 1$. At $\omega = 1$, the first pole slopes down at -20 db/decade. The combined plot will then follow the first pole (red line) out to $\omega = 10$. At $\omega = 10$, the zero (blue line) rises at +40 db/decade canceling out the slope of the first pole and creating a +20 db/decade slope out to $\omega = 100$. At $\omega = 100$, the second pole (magenta line) falls at -20 db/decade canceling out the +20 db/decade slope and creating a flat line at 0 db for the rest of the plot. The combined plot is shown below:

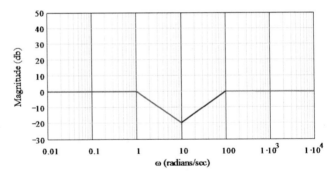

Convert the gain term of 10 to db yielding +20 db. This will then shift the above plot vertically by +20 db. The above plot (shifted) is superimposed (dashed red line) on the computer generated bode magnitude plot in the figure below:

298

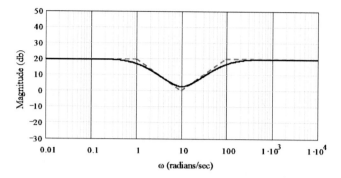

The bode magnitude straight line approximation compares quite well with the computer generated bode magnitude plot.

The bode phase straight line approximation plots are created using Chapter 5 Figure 5-6b for the poles and Figure 5-17b for the zero and are shown below:

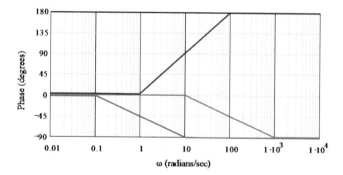

Both the poles and the zero are 0 degrees out to $\omega = 0.1$ so the combined plot will be 0 degrees out to $\omega = 0.1$ (red line). At $\omega = 0.1$, the first pole (red line) falls at -45 deg/decade out to $\omega = 1$. At $\omega = 1$, the zero (blue line) rises at +90 deg/decade canceling out the slope of the first pole and creating a slope of +45 deg/decade out to $\omega = 10$, At $\omega = 10$, the first pole slope stops and becomes flat at -90 degrees but the second pole starts falling at -45 deg/decade keeping the positive slope of +45 deg/decade out to $\omega = 100$. At $\omega = 100$, The zero slope stops leaving the second pole slope of -45 deg/decade out to $\omega = 1000$. At $\omega = 1000$, the second pole slope stops and becomes flat causing the combined plot slope to also become flat at 0 degrees for the rest of the plot. The combined plot is shown below:

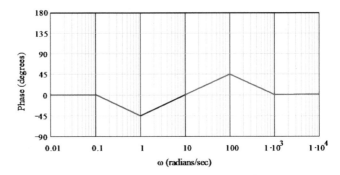

The above plot is superimposed (dashed red line) on the computer generated bode phase plot in the figure below:

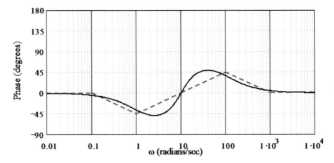

The approximation matches quite well except in the region $1 \leq \omega \leq 100$. This is due to the phase of the complex second order zero and a steeper slope for that approximation would have given a closer match.

Problem Solutions Chapter 6

6-1. Determine the transfer function of a Butterworth low pass filter with a maximum pass band magnitude of 0 db, a minimum pass band magnitude of -2 db out to 1000 hertz, and a maximum stop band magnitude of -30 db from 5000 hertz on. Create a specification graph and a computer generated bode plot of the transfer function and compare it with the filter specification.

First create the filter specification graph shown below:

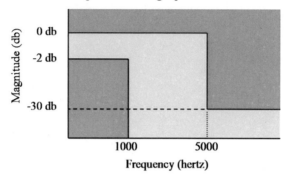

The frequencies must be converted from hertz to radians/sec to determine the filter transfer function:

$$\text{pass band frequency} = \omega_p = 2 \cdot \pi \cdot 1000 = 6,283$$
$$\text{stop band frequency} = \omega_s = 2 \cdot \pi \cdot 5000 = 31,416$$

The magnitude axis is already normalized to pass max equal to 0 db. The pass min value (A_p) is 2 db and the stop max value (A_s) is 30 db. Use Equation (6-14) to determine the number of poles:

$$n = \frac{\log_{10}\left(\dfrac{10^{(0.1 \cdot A_s)} - 1}{10^{(0.1 \cdot A_p)} - 1}\right)}{2 \cdot \log_{10}\left(\dfrac{\omega_s}{\omega_p}\right)} = \frac{\log_{10}\left(\dfrac{10^3 - 1}{10^{0.2} - 1}\right)}{2 \cdot \log_{10}\left(\dfrac{31,416}{6,283}\right)} = 2.312$$

Rounding n up to the next highest integer means that 3 poles will be required. Next use Equation (6-15) to determine the filter cutoff frequency ω_c:

$$\omega_c = \frac{\omega_p}{\left(10^{(0.1 \cdot A_p)} - 1\right)^{1/2 \cdot n}} = \frac{6,283}{\left(10^{0.2} - 1\right)^{1/6}} = 6,871$$

Using the Table 6-1 for the 3 pole entry and ω_c from above determine the filter transfer function:

$$G(s) = \frac{1}{\left(\dfrac{s}{6,871}+1\right)} \cdot \frac{1}{\left(\left(\dfrac{s}{6,871}\right)^2+\left(\dfrac{s}{6,871}\right)+1\right)}$$

Below is a computer generated bode magnitude plot of the filter transfer function with the specification superimposed in red lines:

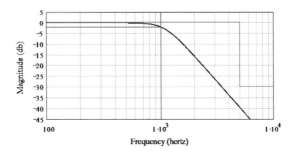

Note that the filter design meets the pass band specification and exceeds the stop band specification. Also, the bode plot was done in hertz rather than radians/sec so the specification could be easily compared.

6-2. Determine the transfer function of a Butterworth low pass filter with a maximum pass band magnitude of 20 db, a minimum pass band magnitude of 19 db out to 2000 hertz, and a maximum stop band magnitude of -20 db from 10000 hertz on. Create a specification graph and a computer generated bode plot of the transfer function and compare it with the filter specification.

First create the filter specification graph shown below:

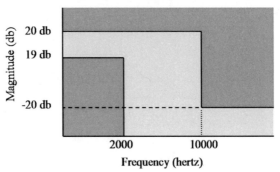

The frequencies must be converted from hertz to radians/sec to determine the filter transfer function:

pass band frequency $= \omega_p = 2 \cdot \pi \cdot 2{,}000 = 12{,}566$

stop band frequency $= \omega_s = 2 \cdot \pi \cdot 10{,}000 = 62{,}832$

The magnitude axis is needs to be normalized to pass max equal to 0 db.

Pass Max $= 20 - 20 = 0$ db

Pass Min $= 19 - 20 = -1$ db

Stop Max $= -20 - 20 = -40$ db

The pass min value (A_p) is 1 db and the stop max value (A_s) is 40 db. Use Equation (6-14) to determine the number of poles:

$$n = \frac{\log_{10}\left(\dfrac{10^{(0.1 \cdot A_s)} - 1}{10^{(0.1 \cdot A_p)} - 1}\right)}{2 \cdot \log_{10}\left(\dfrac{\omega_s}{\omega_p}\right)} = \frac{\log_{10}\left(\dfrac{10^4 - 1}{10^{0.1} - 1}\right)}{2 \cdot \log_{10}\left(\dfrac{62{,}832}{12{,}566}\right)} = 3.281$$

Rounding n up to the next highest integer means that 4 poles will be required. Next use Equation (6-15) to determine the filter cutoff frequency ω_c:

$$\omega_c = \frac{\omega_p}{\left(10^{(0.1 \cdot A_p)} - 1\right)^{1/2 \cdot n}} = \frac{12{,}566}{\left(10^{0.1} - 1\right)^{1/8}} = 14{,}878$$

Using the Table 6-1 for the 4 pole entry and ω_c from above determine the filter transfer function. The transfer function magnitudes must also be denormalized by adding a multiplying factor of 10 $(20 \cdot \log_{10} 10 = 20$ db$)$.

$$G(s) = 10 \cdot \frac{1}{\left(\left(\dfrac{s}{14{,}878}\right)^2 + \dfrac{1}{1.3066}\cdot\left(\dfrac{s}{14{,}878}\right) + 1\right)} \cdot \frac{1}{\left(\left(\dfrac{s}{14{,}878}\right)^2 + \dfrac{1}{0.5412}\cdot\left(\dfrac{s}{14{,}878}\right) + 1\right)}$$

Below is a computer generated bode magnitude plot of the filter transfer function with the specification superimposed in red lines:

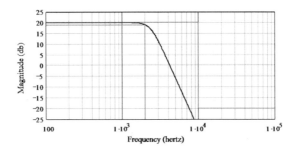

Note that the filter design meets the pass band specification and exceeds the stop band specification. Also, the bode plot was done in hertz rather than radians/sec so the specification could be easily compared.

6-3. Determine the transfer function of a Chebyshev low pass filter with a maximum pass band magnitude of 0 db, a minimum pass band magnitude of -2 db out to 1000 hertz, and a maximum stop band magnitude of -40 db from 5000 hertz on. Create a specification graph and a computer generated bode plot of the transfer function and compare it with the filter specification.

First create the filter specification graph shown below:

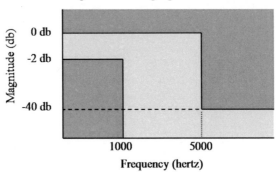

The frequencies must be converted from hertz to radians/sec to determine the filter transfer function:

$$\text{pass band frequency} = \omega_p = 2 \cdot \pi \cdot 1000 = 6,283$$

$$\text{stop band frequency} = \omega_s = 2 \cdot \pi \cdot 5000 = 31,416$$

The magnitude axis is already normalized to pass max equal to 0 db. The pass min value (A_p) is 2 db and the stop max value

(A_s) is 40 db. Use Equation (6-28) to determine the number of poles:

$$n = \frac{\cosh^{-1}\left(\sqrt{\dfrac{10^{(0.1 A_s)}-1}{10^{(0.1 A_p)}-1}}\right)}{\cosh^{-1}\left(\dfrac{\omega_s}{\omega_p}\right)} = \frac{\cosh^{-1}\left(\sqrt{\dfrac{10^4-1}{10^{0.2}-1}}\right)}{\cosh^{-1}\left(\dfrac{31,416}{6,283}\right)} = 2.428$$

Rounding n up to the next highest integer means that 3 poles will be required. Using Equation (6-19), determine ε:

$$\varepsilon = \sqrt{10^{(0.1 A_p)}-1} = \sqrt{10^{0.2}-1} = 0.7648$$

Using the Table 6-1 for the 4 pole entry, determine the Butterworth pole values:

$\alpha_1 = 0.5 \quad \omega_1 = 0.866$
$\alpha_2 = 1.0$

The Chebyshev transformation factors are:

$$K_\alpha = \sinh\left(\frac{1}{n}\cdot\sinh^{-1}\left(\frac{1}{\varepsilon}\right)\right)$$

$$= \sinh\left(\frac{1}{3}\cdot\sinh^{-1}\left(\frac{1}{0.7648}\right)\right) = 0.3689$$

$$K_\omega = \cosh\left(\frac{1}{n}\cdot\sinh^{-1}\left(\frac{1}{\varepsilon}\right)\right)$$

$$= \cosh\left(\frac{1}{3}\cdot\sinh^{-1}\left(\frac{1}{0.7648}\right)\right) = 1.0659$$

The Chebyshev pole values are then:

$\alpha_1 = 0.5\cdot 0.3689 = 0.1845$
$\omega_1 = 0.866\cdot 1.0659 = 0.9231$
$\alpha_2 = 1.0\cdot 0.3689 = 0.3689$

The normalized Chebyshev poles are then:

$$(s+0.3689)\cdot\left((s+0.1845)^2+0.9231^2\right)$$

Converting the poles to break frequency format yields:

$$\left(\frac{s}{0.3689}+1\right)\cdot\left(\left(\frac{s}{0.9413}\right)^2+\frac{1}{2.5516}\cdot\left(\frac{s}{0.9413}\right)+1\right)$$

Translate the poles to the pass band frequency by setting $s = s/\omega_p$:

$$\left(\frac{s}{2318}+1\right)\cdot\left(\left(\frac{s}{5914}\right)^2+\frac{1}{2.5516}\cdot\left(\frac{s}{5914}\right)+1\right)$$

Since n is odd the numerator of the transfer function must be 1 (the pass max value of 0 db). The Chebyshev low pass filter transfer function then is:

$$G(s) = \frac{1}{\left(\frac{s}{2318}+1\right)\cdot\left(\left(\frac{s}{5914}\right)^2+\frac{1}{2.5516}\cdot\left(\frac{s}{5914}\right)+1\right)}$$

Below is a computer generated bode magnitude plot of the filter transfer function with the specification superimposed in red lines.

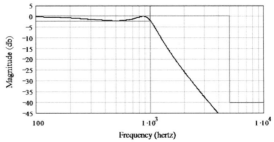

Note that the filter design meets the pass band specification and exceeds the stop band specification. Also, the bode plot was done in hertz rather than radians/sec so the specification could be easily compared.

6-4. Determine the transfer function of a Chebyshev low pass filter with a maximum pass band magnitude of 20 db, a minimum pass band magnitude of 19 db out to 2000 hertz, and a maximum stop band magnitude of -20 db from 5000 hertz on. Create a specification graph and a computer generated bode plot of the transfer function and compare it with the filter specification.

First create the filter specification graph shown below:

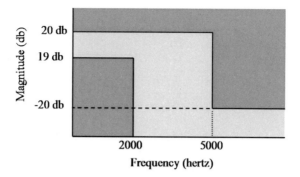

The frequencies must be converted from hertz to radians/sec to determine the filter transfer function:

pass band frequency $= \omega_p = 2 \cdot \pi \cdot 2000 = 12{,}566$

stop band frequency $= \omega_s = 2 \cdot \pi \cdot 5000 = 31{,}416$

The magnitude axis is needs to be normalized to pass max equal to 0 db.

Pass Max $= 20 - 20 = 0$ db

Pass Min $= 19 - 20 = -1$ db

Stop Max $= -20 - 20 = -40$ db

The pass min value (A_p) is 1 db and the stop max value (A_s) is 40 db. Use Equation (6-28) to determine the number of poles:

$$n = \dfrac{\cosh^{-1}\left(\sqrt{\dfrac{10^{(0.1 \cdot A_s)} - 1}{10^{(0.1 \cdot A_p)} - 1}}\right)}{\cosh^{-1}\left(\dfrac{\omega_s}{\omega_p}\right)} = \dfrac{\cosh^{-1}\left(\sqrt{\dfrac{10^4 - 1}{10^{0.1} - 1}}\right)}{\cosh^{-1}\left(\dfrac{31{,}416}{12{,}566}\right)} = 3.813$$

Rounding n up to the next highest integer means that 4 poles will be required. Using Equation (6-19), determine ε:

$$\varepsilon = \sqrt{10^{(0.1 \cdot A_p)} - 1} = \sqrt{10^{0.1} - 1} = 0.5088$$

Using the Table 6-1 for the 4 pole entry, determine the Butterworth pole values:

$\alpha_1 = 0.3827 \quad \omega_1 = 0.9239$

$\alpha_2 = 0.9239 \quad \omega_2 = 0.3827$

The Chebyshev transformation factors are:

$$K_\alpha = \sinh\left(\frac{1}{n}\cdot\sinh^{-1}\left(\frac{1}{\varepsilon}\right)\right)$$

$$= \sinh\left(\frac{1}{4}\cdot\sinh^{-1}\left(\frac{1}{0.5088}\right)\right) = 0.3646$$

$$K_\omega = \cosh\left(\frac{1}{n}\cdot\sinh^{-1}\left(\frac{1}{\varepsilon}\right)\right)$$

$$= \cosh\left(\frac{1}{4}\cdot\sinh^{-1}\left(\frac{1}{0.5088}\right)\right) = 1.0644$$

The Chebyshev pole values are then:

$$\alpha_1 = 0.3827\cdot0.3646 = 0.1395$$
$$\omega_1 = 0.9239\cdot1.0644 = 0.9834$$
$$\alpha_2 = 0.9239\cdot0.3646 = 0.3369$$
$$\omega_2 = 0.3827\cdot1.0644 = 0.4073$$

The normalized Chebyshev poles are then:

$$\left((s+0.1395)^2 + 0.9834^2\right)\cdot\left((s+0.3369)^2 + 0.4073^2\right)$$

Converting the poles to break frequency format yields:

$$\left(\left(\frac{s}{0.9932}\right)^2 + \frac{1}{3.559}\left(\frac{s}{0.9932}\right)+1\right)\cdot$$

$$\left(\left(\frac{s}{0.5286}\right)^2 + \frac{1}{0.7845}\left(\frac{s}{0.5286}\right)+1\right)$$

Translate the poles to the pass band frequency by setting $s = s/\omega_p$:

$$\left(\left(\frac{s}{12,481}\right)^2 + \frac{1}{3.559}\left(\frac{s}{12,481}\right)+1\right)\cdot$$

$$\left(\left(\frac{s}{6,642}\right)^2 + \frac{1}{0.7845}\left(\frac{s}{6,642}\right)+1\right)$$

Since n is even the numerator of the transfer function must be the pass min value converted using Equation (6-6):

$$\text{numerator} = \frac{1}{10^{(A_p/20)}} = \frac{1}{10^{0.05}} = 0.8913$$

The transfer function magnitudes must also be denormalized by adding a multiplying factor of 10 $(20 \cdot \log_{10} 10 = 20 \text{ db})$. The Chebyshev low pass filter transfer function then is:

$$G(s) = \frac{8.913}{\left(\left(\frac{s}{12{,}481}\right)^2 + \frac{1}{3.559} \cdot \left(\frac{s}{12{,}481}\right) + 1\right)} \cdot \frac{1}{\left(\left(\frac{s}{6{,}642}\right)^2 + \frac{1}{0.7845} \cdot \left(\frac{s}{6{,}642}\right) + 1\right)}$$

Below is a computer generated bode magnitude plot of the filter transfer function with the specification superimposed in red lines:

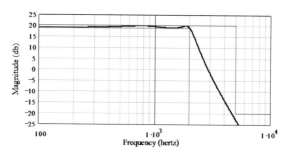

Note that the filter design meets the pass band specification and exceeds the stop band specification. Also, the bode plot was done in hertz rather than radians/sec so the specification could be easily compared.

6-5. Determine the transfer function of a Butterworth high pass filter with a maximum pass band magnitude of 0 db, a minimum pass band magnitude of -2 db from 5000 hertz on out, and a maximum stop band magnitude of -30 db below 1000 hertz. Create a specification graph and a computer generated bode plot of the transfer function and compare it with the filter specification.

First create the filter specification graph shown below:

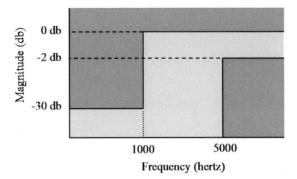

The frequencies must be converted from hertz to radians/sec to determine the filter transfer function:

$$\text{pass band frequency} = \omega_p = 2 \cdot \pi \cdot 5000 = 31,416$$

$$\text{stop band frequency} = \omega_s = 2 \cdot \pi \cdot 1000 = 6,283$$

The magnitude axis is already normalized to pass max equal to 0 db. The pass min value (A_p) is 2 db and the stop max value (A_s) is 30 db. Use Equation (6-31) to determine the number of poles:

$$n = \frac{\log_{10}\left(\dfrac{10^{(0.1 \cdot A_s)} - 1}{10^{(0.1 \cdot A_p)} - 1}\right)}{2 \cdot \log_{10}\left(\dfrac{\omega_p}{\omega_s}\right)} = \frac{\log_{10}\left(\dfrac{10^3 - 1}{10^{0.2} - 1}\right)}{2 \cdot \log_{10}\left(\dfrac{31,416}{6,283}\right)} = 2.312$$

Rounding n up to the next highest integer means that 3 poles will be required. Next use Equation (6-32) to determine the filter cutoff frequency ω_c:

$$\omega_c = \omega_p \cdot \left(10^{(0.1 \cdot A_p)} - 1\right)^{1/2 \cdot n} = 31,416 \cdot \left(10^{(0.2)} - 1\right)^{1/6} = 28,730$$

Using the Table 6-1 for the 3 pole entry, Equations (6-33) and (6-34), and ω_c from above determine the filter transfer function:

$$G(s) = \frac{s^3}{(s + 28,730) \cdot \left(s^2 + 28,730 \cdot s + 28,730^2\right)}$$

Below is a computer generated bode magnitude plot of the filter transfer function with the specification superimposed in red lines.

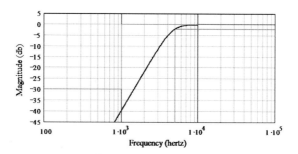

Note that the filter design meets the pass band specification and exceeds the stop band specification.

6-6. Determine the transfer function of a Butterworth high pass filter with a maximum pass band magnitude of 20 db, a minimum pass band magnitude of 19 db from 10000 hertz on out, and a maximum stop band magnitude of -20 db below 2000 hertz. Create a specification graph and a computer generated bode plot of the transfer function and compare it with the filter specification.

First create the filter specification graph shown below:

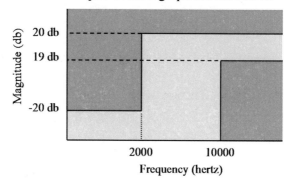

The frequencies must be converted from hertz to radians/sec to determine the filter transfer function:

$$\text{pass band frequency} = \omega_p = 2 \cdot \pi \cdot 10,000 = 62,832$$

$$\text{stop band frequency} = \omega_s = 2 \cdot \pi \cdot 2,000 = 12,566$$

The magnitude axis is needs to be normalized to pass max equal to 0 db.

Pass Max $= 20 - 20 = 0$ db

Pass Min $= 19 - 20 = -1$ db

Stop Max $= -20 - 20 = -40$ db

The pass min value (A_p) is 1 db and the stop max value (A_s) is 40 db. Use Equation (6-31) to determine the number of poles:

$$n = \frac{\log_{10}\left(\dfrac{10^{(0.1 \cdot A_s)} - 1}{10^{(0.1 \cdot A_p)} - 1}\right)}{2 \cdot \log_{10}\left(\dfrac{\omega_p}{\omega_s}\right)} = \frac{\log_{10}\left(\dfrac{10^4 - 1}{10^{0.1} - 1}\right)}{2 \cdot \log_{10}\left(\dfrac{62{,}832}{12{,}566}\right)} = 3.281$$

Rounding n up to the next highest integer means that 4 poles will be required. Next use Equation (6-32) to determine the filter cutoff frequency ω_c:

$$\omega_c = \omega_p \cdot \left(10^{(0.1 \cdot A_p)} - 1\right)^{1/2 \cdot n} = 62{,}832 \cdot \left(10^{(0.1)} - 1\right)^{1/8} = 53{,}067$$

Using the Table 6-1 for the 4 pole entry, Equation (6-34), and ω_c from above determine the filter transfer function. The transfer function magnitudes must also be denormalized by adding a multiplying factor of 10 $(20 \cdot \log_{10} 10 = 20\ db)$.

$$G(s) = \frac{10 \cdot s^4}{\left(s^2 + \dfrac{53{,}067}{1.3066} \cdot s + 53{,}067^2\right) \cdot \left(s^2 + \dfrac{53{,}067}{0.5412} \cdot s + 53{,}067^2\right)}$$

Below is a computer generated bode magnitude plot of the filter transfer function with the specification superimposed in red lines.

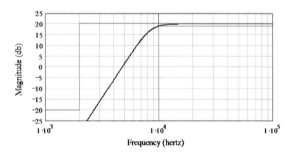

Note that the filter design meets the pass band specification and exceeds the stop band specification.

6-7. Determine the transfer function of a Chebyshev high pass filter with a maximum pass band magnitude of 0 db, a minimum pass band magnitude of -2 db from 5000 hertz on out, and a maximum stop band magnitude of -40 db below 1000 hertz. Create a specification graph and a computer generated bode plot of the transfer function and compare it with the filter specification.

First create the filter specification graph shown below:

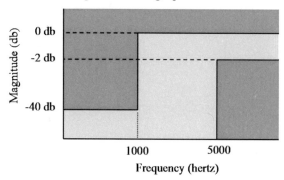

The frequencies must be converted from hertz to radians/sec to determine the filter transfer function:

$$\text{pass band frequency} = \omega_p = 2 \cdot \pi \cdot 5000 = 31,416$$

$$\text{stop band frequency} = \omega_s = 2 \cdot \pi \cdot 1000 = 6,283$$

The magnitude axis is already normalized to pass max equal to 0 db. The pass min value (A_p) is 2 db and the stop max value (A_s) is 40 db. Use Equation (6-39) to determine the number of poles:

$$n = \frac{\cosh^{-1}\left(\sqrt{\dfrac{10^{(0.1 \cdot A_s)} - 1}{10^{(0.1 \cdot A_p)} - 1}}\right)}{\cosh^{-1}\left(\dfrac{\omega_p}{\omega_s}\right)} = \frac{\cosh^{-1}\left(\sqrt{\dfrac{10^4 - 1}{10^{0.2} - 1}}\right)}{\cosh^{-1}\left(\dfrac{31,416}{6,283}\right)} = 2.428$$

Rounding n up to the next highest integer means that 3 poles will be required. Using Equation (6-19), determine ε:

$$\varepsilon = \sqrt{10^{(0.1 \cdot A_p)} - 1} = \sqrt{10^{0.2} - 1} = 0.7648$$

Using the Table 6-1 for the 4 pole entry, determine the Butterworth pole values:

$$\alpha_1 = 0.5 \quad \omega_1 = 0.866$$

$$\alpha_2 = 1.0$$

The Chebyshev transformation factors are:

$$K_{\alpha} = \sinh\left(\frac{1}{n} \cdot \sinh^{-1}\left(\frac{1}{\varepsilon}\right)\right)$$

$$= \sinh\left(\frac{1}{3} \cdot \sinh^{-1}\left(\frac{1}{0.7648}\right)\right) = 0.3689$$

$$K_{\omega} = \cosh\left(\frac{1}{n} \cdot \sinh^{-1}\left(\frac{1}{\varepsilon}\right)\right)$$

$$= \cosh\left(\frac{1}{3} \cdot \sinh^{-1}\left(\frac{1}{0.7648}\right)\right) = 1.0659$$

The Chebyshev pole values are then:

$$\alpha_1 = 0.5 \cdot 0.3689 = 0.1845$$
$$\omega_1 = 0.866 \cdot 1.0659 = 0.9231$$
$$\alpha_2 = 1.0 \cdot 0.3689 = 0.3689$$

The normalized Chebyshev poles are then:

$$(s + 0.3689) \cdot \left((s + 0.1845)^2 + 0.9231^2\right)$$

Converting the poles to break frequency format yields:

$$\left(\frac{s}{0.3689} + 1\right) \cdot \left(\left(\frac{s}{0.9413}\right)^2 + \frac{1}{2.5516} \cdot \left(\frac{s}{0.9413}\right) + 1\right)$$

Using Equations (6-36) and (6-38), convert these poles to the high pass form translated to ω_p.

$$s + \frac{31,416}{0.3689} = s + 85,161$$

$$s^2 + \frac{1}{2.5516} \cdot \left(\frac{31,416}{0.9413}\right) \cdot s + \left(\frac{31,416}{0.9413}\right)^2 = s^2 + \frac{33,375}{2.5516} \cdot s + 33,375^2$$

Since n is odd the numerator constant of the transfer function must be 1 (the pass max value of 0 db). The Chebyshev high pass transfer function then becomes:

$$G(s) = \frac{s^3}{(s + 85,161) \cdot \left(s^2 + \frac{33,375}{2.5516} \cdot s + 33,375^2\right)}$$

314

Below is a computer generated bode magnitude plot of the filter transfer function with the specification superimposed in red lines.

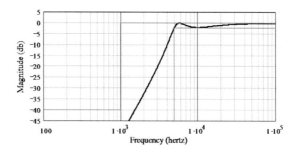

Note that the filter design meets the pass band specification and exceeds the stop band specification.

6-8. Determine the transfer function of a Chebyshev high pass filter with a maximum pass band magnitude of 20 db, a minimum pass band magnitude of 19 db from 5000 hertz on out, and a maximum stop band magnitude of -20 db below 2000 hertz. Create a specification graph and a computer generated bode plot of the transfer function and compare it with the filter specification.

First create the filter specification graph shown below:

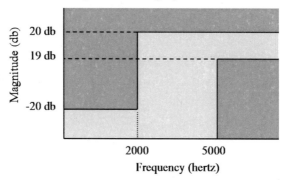

The frequencies must be converted from hertz to radians/sec to determine the filter transfer function:

$$\text{pass band frequency} = \omega_p = 2 \cdot \pi \cdot 5000 = 31,416$$

$$\text{stop band frequency} = \omega_s = 2 \cdot \pi \cdot 2000 = 12,566$$

The magnitude axis is needs to be normalized to pass max equal to 0 db.

Pass Max $= 20 - 20 = 0$ db

Pass Min $= 19 - 20 = -1$ db

Stop Max $= -20 - 20 = -40$ db

The pass min value (A_p) is 1 db and the stop max value (A_s) is 40 db. Use Equation (6-39) to determine the number of poles:

$$n = \frac{\cosh^{-1}\left(\sqrt{\dfrac{10^{(0.1 \cdot A_s)} - 1}{10^{(0.1 \cdot A_p)} - 1}}\right)}{\cosh^{-1}\left(\dfrac{\omega_p}{\omega_s}\right)} = \frac{\cosh^{-1}\left(\sqrt{\dfrac{10^4 - 1}{10^{0.1} - 1}}\right)}{\cosh^{-1}\left(\dfrac{31,416}{12,566}\right)} = 3.813$$

Rounding n up to the next highest integer means that 4 poles will be required. Using Equation (6-19), determine ε:

$$\varepsilon = \sqrt{10^{(0.1 \cdot A_p)} - 1} = \sqrt{10^{0.1} - 1} = 0.5088$$

Using the Table 6-1 for the 4 pole entry, determine the Butterworth pole values:

$$\alpha_1 = 0.3827 \quad \omega_1 = 0.9239$$
$$\alpha_2 = 0.9239 \quad \omega_2 = 0.3827$$

The Chebyshev transformation factors are:

$$K_\alpha = \sinh\left(\frac{1}{n} \cdot \sinh^{-1}\left(\frac{1}{\varepsilon}\right)\right)$$

$$= \sinh\left(\frac{1}{4} \cdot \sinh^{-1}\left(\frac{1}{0.5088}\right)\right) = 0.3646$$

$$K_\omega = \cosh\left(\frac{1}{n} \cdot \sinh^{-1}\left(\frac{1}{\varepsilon}\right)\right)$$

$$= \cosh\left(\frac{1}{4} \cdot \sinh^{-1}\left(\frac{1}{0.5088}\right)\right) = 1.0644$$

The Chebyshev pole values are then:

$$\alpha_1 = 0.3827 \cdot 0.3646 = 0.1395$$
$$\omega_1 = 0.9239 \cdot 1.0644 = 0.9834$$
$$\alpha_2 = 0.9239 \cdot 0.3646 = 0.3369$$
$$\omega_2 = 0.3827 \cdot 1.0644 = 0.4073$$

The normalized Chebyshev poles are then:

$$\left((s+0.1395)^2 +0.9834^2\right)\cdot\left((s+0.3369)^2 +0.4073^2\right)$$

Converting the poles to break frequency format yields:

$$\left(\left(\frac{s}{0.9932}\right)^2 +\frac{1}{3.559}\cdot\left(\frac{s}{0.9932}\right)+1\right)\cdot$$

$$\left(\left(\frac{s}{0.5286}\right)^2 +\frac{1}{0.7845}\cdot\left(\frac{s}{0.5286}\right)+1\right)$$

Using Equation (6-38), convert these poles to the high pass form translated to ω_p.

$$s^2 +\frac{1}{3.559}\cdot\left(\frac{31,416}{0.9932}\right)\cdot s+\left(\frac{31,416}{0.9932}\right)^2 =s^2 +\frac{31,631}{3.559}\cdot s+31,631^2$$

$$s^2 +\frac{1}{0.7845}\cdot\left(\frac{31,416}{0.5286}\right)\cdot s+\left(\frac{31,416}{0.5286}\right)^2 =s^2 +\frac{59,432}{0.7845}\cdot s+59,432^2$$

Since n is even the numerator constant of the transfer function must be the pass min value converted using Equation (6-6):

$$numerator =\frac{1}{10^{(A_p /20)}} =\frac{1}{10^{0.05}} =0.8913$$

The transfer function magnitudes must also be denormalized by adding a multiplying factor of 10 $(20\cdot \log_{10} 10 = 20\, db)$.

The Chebyshev high pass transfer function then becomes:

$$G(s) =\frac{8.913\cdot s^4}{\left(s^2 +\dfrac{31,631}{3.559}\cdot s+31,631^2\right)\cdot\left(s^2 +\dfrac{59,432}{0.7845}\cdot s+59,432^2\right)}$$

Below is a computer generated bode magnitude plot of the filter transfer function with the specification superimposed in red lines.

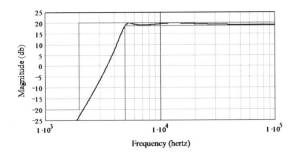

Frequency (hertz)

Note that the filter design meets the pass band specification and exceeds the stop band specification.

6-9. Determine the transfer function of a band pass filter with a maximum pass band magnitude of 0 db at a center frequency of 1000 hertz and a 250 hertz bandwidth at the -3.0103 db points. There shall be a maximum stop band magnitude of -40 db below 300 hertz and above 4,000 hertz. Create a specification graph and a computer generated bode plot of the transfer function and compare it with the filter specification.

First determine the filter Q using Equation (6-41):

$$Q = \frac{\omega_0}{B} = \frac{2 \cdot \pi \cdot 1000}{2 \cdot \pi \cdot 250} = 4$$

Since Q is greater than 0.5, use the narrow band approach.

Next, create the filter specification graph. Since the center frequency and bandwidth are given, they must be normalized and converted to the upper and lower pass frequencies. The normalized center frequency and bandwidth are:

$$\omega_0 = \frac{f_0}{1000} = 1 \quad B = \frac{250}{1000} = 0.25$$

The upper and lower pass frequencies can be calculated from Equation (6-40):

$$f_0 = 1000 = \sqrt{f_{ph} \cdot f_{pl}} \Rightarrow f_{pl} = \frac{1000^2}{f_{ph}}$$

$$B_f = 250 = f_{ph} - f_{pl} = f_{ph} - \frac{1000^2}{f_{ph}}$$

$$f_{ph}^2 - 250 \cdot f_{ph} - 1000^2 = 0 \Rightarrow f_{ph} = 1133 \quad \text{and} \quad f_{pl} = 883$$

318

The specification graph is shown below:

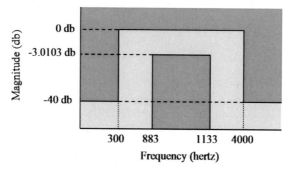

The next step is to normalize the specifications. The magnitude axis is already normalized since the pass max value is 0 db. To normalize the frequency axis divide all of the specifications by f_0 or 1000. Since this is a frequency divided by a frequency, the normalized axis can be considered either hertz or radians per second. The normalized frequency values are:

$$\omega_{pl} = \frac{883}{1000} = 0.883 \quad \omega_{ph} = \frac{1133}{1000} = 1.133$$

$$\text{lower stop} = \frac{300}{1000} = 0.3 \quad \text{upper stop} = \frac{4,000}{1000} = 4$$

The next step is to determine the low pass equivalent. To determine the low pass equivalent number of poles, determine the pass bandwidth B and the geometric stop bandwidth:

$$B = \frac{B_f}{f_0} = \frac{250}{1000} = 0.25$$

or $B = \omega_{ph} - \omega_{pl} = 1.133 - 0.883 = 0.25$

$1/(\text{lower stop}) = 1/0.3 = 3.333$

which is less than the upper stop

Therefore: $\omega_{sl} = 0.3$ and $\omega_{sh} = \frac{1}{0.3} = 3.333$

geometric stop width $= \omega_{sh} - \omega_{sl} = 3.333 - 0.3 = 3.033$

Determine the number of low pass equivalent poles using Equation (6-47):

$$n = \frac{\log_{10}\left(\dfrac{10^{(0.1 \cdot A_s)} - 1}{10^{(0.1 \cdot A_p)} - 1}\right)}{2 \cdot \log_{10}\left(\dfrac{\text{geometric stop width}}{\text{bandwidth}}\right)}$$

$$n = \frac{\log_{10}\left(\dfrac{10^{(4)} - 1}{10^{(0.30103)} - 1}\right)}{2 \cdot \log_{10}\left(\dfrac{3.033}{0.25}\right)} = 1.845$$

Rounding up to the next highest integer yields n equal to 2. Using Table 6-1 for the 2 pole entry, the normalized low pass equivalent transfer function is:

$$G(s)_{LP} = \frac{1}{\left(s^2 + \dfrac{s}{0.7071} + 1\right)}$$

Translating this to the pass band frequency of B yields:

$$G(s)_{LP} = \frac{1}{\left(\left(\dfrac{s}{0.25}\right)^2 + \dfrac{1}{0.7071} \cdot \left(\dfrac{s}{0.25}\right) + 1\right)}$$

Transforming this to band pass using Equation (6-50) yields:

$$G(s) = \frac{0.0625 \cdot s^2}{\left(s^4 + 0.3536 \cdot s^3 + 2.0625 \cdot s^2 + 0.3536 \cdot s + 1\right)}$$

Factoring the 4th order denominator yields:

$$G(s) = \frac{0.25 \cdot s}{\left(s^2 + 0.1612 \cdot s + 0.8376\right)} \cdot \frac{0.25 \cdot s}{\left(s^2 + 0.1924 \cdot s + 1.1939\right)}$$

Translating this out to the specification center frequency of 1000 hertz which is $2000 \cdot \pi$ radians per second yields:

$$G(s) = \frac{0.25 \cdot \left(\dfrac{s}{2000 \cdot \pi}\right)}{\left(\left(\dfrac{s}{2000 \cdot \pi}\right)^2 + 0.1612 \cdot \left(\dfrac{s}{2000 \cdot \pi}\right) + 0.8376\right)} \cdot$$

$$\frac{0.25 \cdot \left(\dfrac{s}{2000 \cdot \pi}\right)}{\left(\left(\dfrac{s}{2000 \cdot \pi}\right)^2 + 0.1924 \cdot \left(\dfrac{s}{2000 \cdot \pi}\right) + 1.1939\right)}$$

$$G(s) = \frac{1571 \cdot s}{\left(s^2 + 1013 \cdot s + 3.3067 \cdot 10^7\right)} \cdot \frac{1571 \cdot s}{\left(s^2 + 1209 \cdot s + 4.7133 \cdot 10^7\right)}$$

Below is a computer generated bode magnitude plot of the filter transfer function with the specification superimposed in red lines:

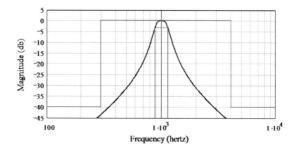

Note that the filter design meets the pass band specification and exceeds the stop band specification.

6-10. Determine the transfer function of a band pass filter with a maximum pass band magnitude of 0 db at a center frequency of 3000 hertz and an 8000 hertz bandwidth at the -3.0103 db points. There shall be a maximum stop band magnitude of -40 db below 100 hertz and above 90,000 hertz. Create a specification graph and a computer generated bode plot of the transfer function and compare it with the filter specification.

First determine the filter Q using Equation (6-41):

$$Q = \frac{\omega_0}{B} = \frac{2 \cdot \pi \cdot 3000}{2 \cdot \pi \cdot 8000} = 0.375$$

Since Q is less than 0.5, use the wide band approach.

Next, create the filter specification graph. Since the center frequency and bandwidth are given, they must be normalized and converted to the upper and lower pass frequencies. The normalized center frequency and bandwidth are:

$$\omega_0 = \frac{f_0}{3000} = 1 \quad B = \frac{8000}{3000} = 2.6667$$

The upper and lower pass frequencies can be calculated from Equation (6-40):

$$f_0 = 3000 = \sqrt{f_{ph} \cdot f_{pl}} \Rightarrow f_{pl} = \frac{3000^2}{f_{ph}}$$

$$B_f = 8000 = f_{ph} - f_{pl} = f_{ph} - \frac{3000^2}{f_{ph}}$$

$$f_{ph}^2 - 8000 \cdot f_{ph} - 3000^2 = 0 \Rightarrow f_{ph} = 9000 \quad \text{and} \quad f_{pl} = 1000$$

The specification graph is shown below:

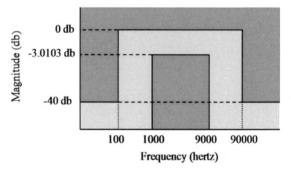

Since this is a wide band approach, it will be accomplished as separate low pass and high pass implementations. The band pass specification graph must be converted to two specification graphs, one for the low pass implementation and one for the high pass implementation. The low pass implementation will be considered first. Below is the low pass specification graph:

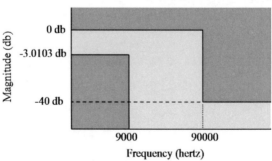

The frequencies must be converted from hertz to radians/sec to determine the filter transfer function:

pass band frequency $= \omega_p = 2 \cdot \pi \cdot 9{,}000 = 56{,}549$

stop band frequency $= \omega_s = 2 \cdot \pi \cdot 90{,}000 = 565{,}487$

The magnitude axis is already normalized to pass max equal to 0 db. The pass min value (A_p) is 3.0103 db and the stop max value (A_s) is 40 db. Use Equation (6-14) to determine the number of poles:

$$n = \frac{\log_{10}\left(\dfrac{10^{(0.1 \cdot A_s)} - 1}{10^{(0.1 \cdot A_p)} - 1}\right)}{2 \cdot \log_{10}\left(\dfrac{\omega_s}{\omega_p}\right)} = \frac{\log_{10}\left(\dfrac{10^4 - 1}{10^{0.30103} - 1}\right)}{2 \cdot \log_{10}\left(\dfrac{565{,}487}{56{,}549}\right)} = 2.0$$

Since n is already an integer, 2 poles will be required. Since the pass magnitude is 3.0103 db the cutoff frequency will be ω_p or 56,549.

Using the Table 6-1 for the 2 pole entry and ω_c equal to 56,549, determine the low pass transfer function:

$$G_{LP}(s) = \frac{1}{\left(\left(\dfrac{s}{56{,}549}\right)^2 + \dfrac{1}{0.7071} \cdot \left(\dfrac{s}{56{,}549}\right) + 1\right)}$$

Next consider the high pass implementation. Below is the high pass specification graph:

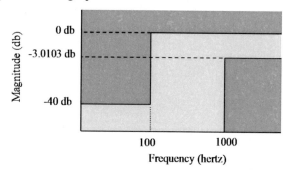

The frequencies must be converted from hertz to radians/sec to determine the filter transfer function:

pass band frequency $= \omega_p = 2 \cdot \pi \cdot 1000 = 6,283$

stop band frequency $= \omega_s = 2 \cdot \pi \cdot 100 = 628$

The magnitude axis is already normalized to pass max equal to 0 db. The pass min value (A_p) is 3.0103 db and the stop max value (A_s) is 40 db. Use Equation (6-31) to determine the number of poles:

$$n = \frac{\log_{10}\left(\dfrac{10^{(0.1 \cdot A_s)} - 1}{10^{(0.1 \cdot A_p)} - 1}\right)}{2 \cdot \log_{10}\left(\dfrac{\omega_p}{\omega_s}\right)} = \frac{\log_{10}\left(\dfrac{10^4 - 1}{10^{0.30103} - 1}\right)}{2 \cdot \log_{10}\left(\dfrac{6,283}{628}\right)} = 2.0$$

Since n is already an integer, 2 poles will be required. Since the pass magnitude is 3.0103 db the cutoff frequency will be ω_p or 6,283.

Using the Table 6-1 for the 2 pole entry and ω_c equal to 6,283, determine the high pass transfer function:

$$G_{HP}(s) = \frac{s^2}{\left(s^2 + \dfrac{6,283}{0.7071} \cdot s + 6,283^2\right)}$$

The band pass transfer function is then the product of the low pass and high pass transfer functions:

$$G(s) = G_{LP}(s) \cdot G_{HP}(s)$$

$$= \frac{1}{\left(\left(\dfrac{s}{56,549}\right)^2 + \dfrac{1}{0.7071} \cdot \left(\dfrac{s}{56,549}\right) + 1\right)} \cdot$$

$$\frac{s^2}{\left(s^2 + \dfrac{6,283}{0.7071} \cdot s + 6,283^2\right)}$$

Below is a computer generated bode magnitude plot of the band pass filter transfer function with the specification superimposed in red lines. Note that both the pass band and stop band specifications just meet. This is because the calculation for the number of poles came out to be an integer and no rounding up was required. In other problems the number of poles was rounded up causing the stop band specification to be exceeded.

324

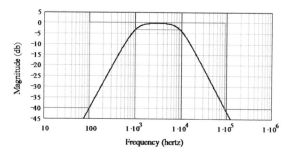

6-11. Determine the transfer function of a notch filter with a maximum pass band magnitude of 0 db. The notch center frequency is 2000 hertz with a 2100 hertz notchwidth at the -3.0103 db points. There shall be a maximum stop band magnitude of -40 db above 1900 hertz and below 2100 hertz. Create a specification graph and a computer generated bode plot of the transfer function and compare it with the filter specification.

First determine the filter Q using Equation (6-53):

$$Q = \frac{\omega_0}{N} = \frac{2 \cdot \pi \cdot 2000}{2 \cdot \pi \cdot 2100} = 0.9524$$

Since Q is greater than 0.5, use the narrow band approach.

Next, create the filter specification graph. Since the center frequency and notchwidth are given, they must be normalized and converted to the upper and lower pass frequencies. The normalized center frequency and notchwidth are:

$$\omega_0 = \frac{f_0}{2000} = 1 \quad N = \frac{2100}{2000} = 1.05$$

The upper and lower pass frequencies can be calculated from Equation (6-52):

$$f_0 = 2000 = \sqrt{f_{ph} \cdot f_{pl}} \Rightarrow f_{pl} = \frac{2000^2}{f_{ph}}$$

$$N_f = 2100 = f_{ph} - f_{pl} = f_{ph} - \frac{2000^2}{f_{ph}}$$

$$f_{ph}^2 - 2100 \cdot f_{ph} - 2000^2 = 0 \Rightarrow f_{ph} = 3309 \quad \text{and} \quad f_{pl} = 1209$$

The specification graph is shown below:

325

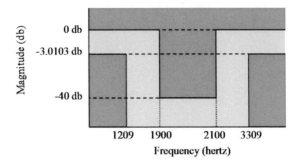

The next step is to normalize the specifications. The magnitude axis is already normalized since the pass max value is 0 db. To normalize the frequency axis divide all of the specifications by f_0 or 2000. Since this is a frequency divided by a frequency, the normalized axis can be considered either hertz or radians per second. The normalized frequency values are:

$$\omega_{pl} = \frac{1209}{2000} = 0.6045 \quad \omega_{ph} = \frac{3309}{2000} = 1.6545$$

$$\text{lower stop} = \frac{1900}{2000} = 0.95 \quad \text{upper stop} = \frac{2100}{2000} = 1.05$$

The next step is to determine the low pass equivalent. To determine the low pass equivalent number of poles, determine the pass notchwidth N and the geometric stop notchwidth:

$$N = \frac{N_f}{f_0} = \frac{2100}{2000} = 1.05 \quad \text{or} \quad N = \omega_{ph} - \omega_{pl} = 1.6545 - 0.6045 = 1.05$$

$1/(\text{lower stop}) = 1/0.95 = 1.0526$ which is greater than the upper stop

Therefore: $\omega_{sl} = 0.95$ and $\omega_{sh} = \dfrac{1}{0.95} = 1.0526$

geometric stop width $= \omega_{sh} - \omega_{sl} = 1.0526 - 0.95 = 0.1026$

Determine the number of equivalent low pass poles using Equation (6-59):

$$n = \frac{\log_{10}\left(\dfrac{10^{(0.1 \cdot A_s)} - 1}{10^{(0.1 \cdot A_p)} - 1}\right)}{2 \cdot \log_{10}\left(\dfrac{\text{notchwidth}}{\text{geometric stop width}}\right)}$$

326

$$n = \frac{\log_{10}\left(\dfrac{10^{(4)}-1}{10^{(0.30103)}-1}\right)}{2 \cdot \log_{10}\left(\dfrac{1.05}{0.1026}\right)} = 1.98$$

Rounding up to the next highest integer yields n equal to 2. Using Table 6-1 for the 2 pole entry, the normalized low pass equivalent transfer function is:

$$G(s)_{LP} = \frac{1}{s^2 + \dfrac{s}{0.7071} + 1}$$

Translating this to the pass band frequency of 1/N yields:

$$G(s)_{LP} = \frac{1}{\left((1.05 \cdot s)^2 + \dfrac{1}{0.7071} \cdot (1.05 \cdot s) + 1\right)}$$

Transforming this to a notch using Equation (6-62) yields:

$$G(s) = \frac{\left(s^2 + 1\right)^2}{\left(s^4 + \dfrac{1.05}{0.7071} \cdot s^3 + 3.1025 \cdot s^2 + \dfrac{1.05}{0.7071} \cdot s + 1\right)}$$

Factoring the 4th order denominator yields:

$$G(s) = \frac{s^2 + 1}{\left(s^2 + 0.4694 \cdot s + 0.4622\right)} \cdot \frac{s^2 + 1}{\left(s^2 + 1.0156 \cdot s + 2.1636\right)}$$

Translating this out to the specification center frequency of 2000 hertz which is $4000 \cdot \pi$ radians per second yields:

$$G(s) = \frac{\left(\dfrac{s}{4000 \cdot \pi}\right)^2 + 1}{\left(\left(\dfrac{s}{4000 \cdot \pi}\right)^2 + 0.4694 \cdot \left(\dfrac{s}{4000 \cdot \pi}\right) + 0.4622\right)} \cdot$$

$$\frac{\left(\dfrac{s}{4000 \cdot \pi}\right)^2 + 1}{\left(\left(\dfrac{s}{4000 \cdot \pi}\right)^2 + 1.0156 \cdot \left(\dfrac{s}{4000 \cdot \pi}\right) + 2.1636\right)}$$

$$G(s) = \frac{s^2 + 1.5791 \cdot 10^8}{\left(s^2 + 5899 \cdot s + 7.2988 \cdot 10^7\right)} \cdot \frac{s^2 + 1.5791 \cdot 10^8}{\left(s^2 + 12,762 \cdot s + 3.4166 \cdot 10^8\right)}$$

Below is a computer generated bode magnitude plot of the filter transfer function with the specification superimposed in red lines.

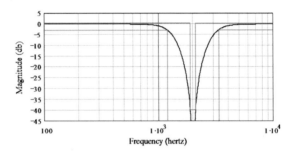

Note that the filter design meets the pass band specification and the stop band specification.

6-12.   Determine the transfer function of a notch filter with a maximum pass band magnitude of 0 db. The notch center frequency is 1000 hertz with a 10,000 hertz notchwidth at the -3.0103 db points. There shall be a maximum stop band magnitude of -30 db above 700 hertz and below 1500 hertz. Create a specification graph and a computer generated bode plot of the transfer function and compare it with the filter specification.

First determine the filter Q using Equation (6-53):

$$Q = \frac{\omega_0}{N} = \frac{2 \cdot \pi \cdot 1000}{2 \cdot \pi \cdot 10,000} = 0.1$$

Since Q is less than 0.5, use the wide band approach.

Next, create the filter specification graph. Since the center frequency and bandwidth are given, they must be normalized and converted to the upper and lower pass frequencies. The normalized center frequency and bandwidth are:

$$\omega_0 = \frac{f_0}{1000} = 1 \quad N = \frac{10,000}{1000} = 10$$

The upper and lower pass frequencies can be calculated from Equation (6-52):

$$f_0 = 1000 = \sqrt{f_{ph} \cdot f_{pl}} \Rightarrow f_{pl} = \frac{1000^2}{f_{ph}}$$

$$N_f = 10,000 = f_{ph} - f_{pl} = f_{ph} - \frac{1000^2}{f_{ph}}$$

$$f_{ph}^2 - 10,000 \cdot f_{ph} - 1000^2 = 0 \Rightarrow f_{ph} = 10,099 \quad \text{and} \quad f_{pl} = 99$$

The specification graph is shown below:

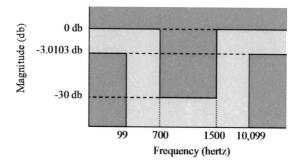

Since this is a wide band approach, it will be accomplished as separate low pass and high pass implementations. The notch specification graph must be converted to two specification graphs, one for the low pass implementation and one for the high pass implementation. The low pass implementation will be considered first. Below is the low pass specification graph:

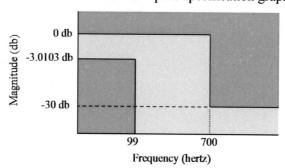

The frequencies must be converted from hertz to radians/sec to determine the filter transfer function:

$$\text{pass band frequency} = \omega_p = 2 \cdot \pi \cdot 99 = 622$$

$$\text{stop band frequency} = \omega_s = 2 \cdot \pi \cdot 700 = 4398$$

329

The magnitude axis is already normalized to pass max equal to 0 db. The pass min value (A_p) is 3.0103 db and the stop max value (A_s) is 30 db. Use Equation (6-14) to determine the number of poles:

$$n = \frac{\log_{10}\left(\frac{10^{(0.1 \cdot A_s)}-1}{10^{(0.1 \cdot A_p)}-1}\right)}{2 \cdot \log_{10}\left(\frac{\omega_s}{\omega_p}\right)} = \frac{\log_{10}\left(\frac{10^3-1}{10^{0.30103}-1}\right)}{2 \cdot \log_{10}\left(\frac{4398}{622}\right)} = 1.766$$

Rounding up to the next highest integer yields n equal to 2. Since the pass magnitude is 3.0103 db the cutoff frequency will be ω_p or 622.

Using the Table 6-1 for the 2 pole entry and ω_c equal to 622, determine the low pass transfer function:

$$G_{LP}(s) = \frac{1}{\left(\left(\frac{s}{622}\right)^2 + \frac{1}{0.7071}\cdot\left(\frac{s}{622}\right)+1\right)}$$

Next consider the high pass implementation. Below is the high pass specification graph:

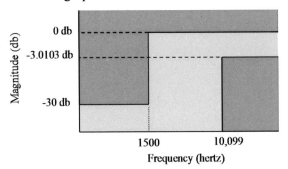

The frequencies must be converted from hertz to radians/sec to determine the filter transfer function:

pass band frequency $= \omega_p = 2 \cdot \pi \cdot 10,099 = 63,454$

stop band frequency $= \omega_s = 2 \cdot \pi \cdot 1500 = 9425$

The magnitude axis is already normalized to pass max equal to 0 db. The pass min value (A_p) is 3.0103 db and the stop max value (A_s) is 30 db. Use Equation (6-31) to determine the number of poles:

$$n = \dfrac{\log_{10}\left(\dfrac{10^{(0.1 \cdot A_s)}-1}{10^{(0.1 \cdot A_p)}-1}\right)}{2 \cdot \log_{10}\left(\dfrac{\omega_p}{\omega_s}\right)} = \dfrac{\log_{10}\left(\dfrac{10^3-1}{10^{0.30103}-1}\right)}{2 \cdot \log_{10}\left(\dfrac{63,454}{9425}\right)} = 1.811$$

Rounding up to the next highest integer yields n equal to 2. Since the pass magnitude is 3.0103 db the cutoff frequency will be ω_p or 63,454.

Using the Table 6-1 for the 2 pole entry and ω_c equal to 63,454, determine the high pass transfer function:

$$G_{HP}(s) = \dfrac{s^2}{\left(s^2 + \dfrac{63,454}{0.7071}\cdot s + 63,454^2\right)}$$

The notch transfer function is then the sum of the low pass and high pass transfer functions. For the band pass filter, the pass band is the overlap of the high pass and low pass functions, so the product is required. For the notch filter, the notch is the gap between the low pass and high pass functions, so the sum is required.

$$G(s) = G_{LP}(s) + G_{HP}(s)$$

$$= \dfrac{1}{\left(\left(\dfrac{s}{622}\right)^2 + \dfrac{1}{0.7071}\cdot\left(\dfrac{s}{622}\right)+1\right)} +$$

$$\dfrac{s^2}{\left(s^2 + \dfrac{63,454}{0.7071}\cdot s + 63,454^2\right)}$$

Below is a computer generated bode magnitude plot of the notch filter transfer function with the specification superimposed in red lines. Note that the filter design meets the pass band specification and exceeds the stop band specification. Also note that the notch does not go to zero like the narrow band approach. This is because the transfer function is a combination of a low pass and a high pass transfer function rather than a narrow band notch function.

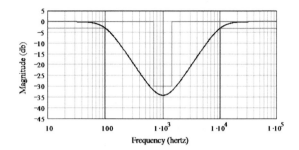

Problem Solutions Chapter 7

7-1. Design a passive Butterworth low pass filter with a maximum pass band magnitude of 0 db, a minimum pass band magnitude of -2 db out to 1 megahertz, and a maximum stop band magnitude of -30 db from 5 megahertz on. The filter load resistance will be 1000 ohm.

First create the filter specification graph shown below:

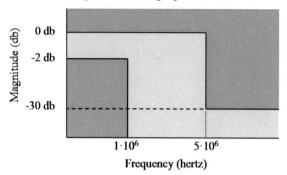

Note that the frequencies are in hertz and not radians/sec. The frequencies must be converted from hertz to radians/sec:

$$\text{pass band frequency} = \omega_p = 2 \cdot \pi \cdot 1 \cdot 10^6 = 6.283 \cdot 10^6$$

$$\text{stop band frequency} = \omega_s = 2 \cdot \pi \cdot 5 \cdot 10^6 = 31.416 \cdot 10^6$$

The magnitude axis is already normalized to pass max equal to 0 db. In order to allow for component variations, the design minimum pass band magnitude needs to be tightened by 10%. Since the magnitude values are in db, the 10% cannot be applied to the db values. Converting -2 db from db to magnitude yields:

$$-2 \text{ db} \Rightarrow 10^{(-2/20)} = 0.7943$$

Tightening this value by 10% yields 0.8738. Converting this back to db yields -1.172 db.

The pass min value (A_p) is 1.172 db and the stop max value (A_s) is 30 db. Use Equation (6-14) to determine the number of poles:

$$n = \frac{\log_{10}\left(\dfrac{10^{(0.1 \cdot A_s)} - 1}{10^{(0.1 \cdot A_p)} - 1}\right)}{2 \cdot \log_{10}\left(\dfrac{\omega_s}{\omega_p}\right)} = \frac{\log_{10}\left(\dfrac{10^3 - 1}{10^{0.1172} - 1}\right)}{2 \cdot \log_{10}\left(\dfrac{31.416 \cdot 10^6}{6.283 \cdot 10^6}\right)} = 2.51$$

Rounding n up to the next highest integer means that 3 poles will be required. Next use Equation (6-15) to determine the filter cutoff frequency ω_c:

$$\omega_c = \frac{\omega_p}{\left(10^{(0.1 \cdot A_p)} - 1\right)^{1/2 \cdot n}} = \frac{6.283 \cdot 10^6}{\left(10^{0.1172} - 1\right)^{1/6}} = 7.638 \cdot 10^6$$

Using the Table 7-1 for the 3 pole entry yields:

$$L_0 = 0.50 \quad C_1 = 1.3333 \quad L_1 = 1.50$$

Use Equation (7-9) to translate the component values:

$$\text{Translated Resistance} = K_R = 1000$$

$$\text{Translated Inductance} = \frac{K_R}{K_\omega} \cdot L = \frac{1000}{7.638 \cdot 10^6} \cdot L$$

$$\text{Translated Capacitance} = \frac{C}{K_R \cdot K_\omega} = \frac{C}{1000 \cdot 7.638 \cdot 10^6}$$

The final circuit is shown below:

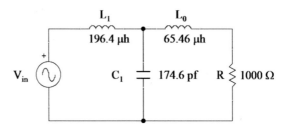

The component values on the above circuit schematic are actual calculated values. Standard values will need to be selected as close as possible to those values. If necessary, inductors can be implemented with two inductors in series and capacitors can be implemented with two capacitors in parallel (one a large value and one a small value) to get closer to the calculated value. The specification was intentionally tightened to allow for component value variances.

Below is the frequency response of the above circuit using a computer software circuit analysis program. The analysis was run with the calculated component values. The specification is superimposed in red lines. Note that the filter response exceeds both the pass band and stop band specifications.

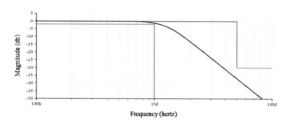

7-2. Design a passive Chebyshev low pass filter with a maximum pass band magnitude of 0 db, a minimum pass band magnitude of -2 db out to 1 megahertz, and a maximum stop band magnitude of -40 db from 3 megahertz on. The filter load resistance will be 1000 ohm.

First create the filter specification graph shown below:

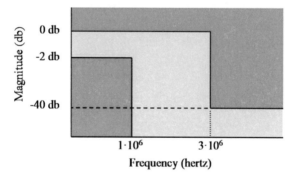

Note that the frequencies are in hertz and not radians/sec. The frequencies must be converted from hertz to radians/sec:

$$\text{pass band frequency} = \omega_p = 2 \cdot \pi \cdot 1 \cdot 10^6 = 6.283 \cdot 10^6$$

$$\text{stop band frequency} = \omega_s = 2 \cdot \pi \cdot 3 \cdot 10^6 = 18.85 \cdot 10^6$$

The magnitude axis is already normalized to pass max equal to 0 db. In order to allow for component variations, the design minimum pass band magnitude needs to be tightened by 10%. Since the magnitude values are in db, the 10% cannot be applied to the db values. Converting -2 db from db to magnitude yields:

$$-2 \text{ db} \Rightarrow 10^{(-2/20)} = 0.7943$$

Tightening this value by 10% yields 0.8738. Converting this back to db yields -1.172 db.

The pass min value (A_p) is 1.172 db and the stop max value (A_s) is 40 db. Use Equation (6-28) to determine the number of poles:

$$n = \frac{\cosh^{-1}\left(\sqrt{\dfrac{10^{(0.1 \cdot A_s)} - 1}{10^{(0.1 \cdot A_p)} - 1}}\right)}{\cosh^{-1}\left(\dfrac{\omega_s}{\omega_p}\right)} = \frac{\cosh^{-1}\left(\sqrt{\dfrac{10^4 - 1}{10^{0.1172} - 1}}\right)}{\cosh^{-1}\left(\dfrac{18.85 \cdot 10^6}{6.283 \cdot 10^6}\right)} = 3.338$$

Rounding n up to the next highest integer means that 4 poles will be required. Using Equation (6-19), determine ε:

$$\varepsilon = \sqrt{10^{(0.1 \cdot A_p)} - 1} = \sqrt{10^{0.1172} - 1} = 0.5566$$

Using the Table 6-1 for the 4 pole entry, determine the Butterworth pole values:

$$\alpha_1 = 0.3827 \quad \omega_1 = 0.9239$$
$$\alpha_2 = 0.9239 \quad \omega_2 = 0.3827$$

The Chebyshev transformation factors using Equation (6-21) are:

$$K_\alpha = \sinh\left(\frac{1}{n} \cdot \sinh^{-1}\left(\frac{1}{\varepsilon}\right)\right)$$

$$= \sinh\left(\frac{1}{4} \cdot \sinh^{-1}\left(\frac{1}{0.5566}\right)\right) = 0.3436$$

$$K_\omega = \cosh\left(\frac{1}{n} \cdot \sinh^{-1}\left(\frac{1}{\varepsilon}\right)\right)$$

$$= \cosh\left(\frac{1}{4} \cdot \sinh^{-1}\left(\frac{1}{0.5566}\right)\right) = 1.0574$$

The Chebyshev pole values are then:

$$\alpha_1 = 0.3827 \cdot 0.3436 = 0.1315$$
$$\omega_1 = 0.9239 \cdot 1.0574 = 0.9769$$
$$\alpha_2 = 0.9239 \cdot 0.3436 = 0.3175$$

$$\omega_2 = 0.3827 \cdot 1.0574 = 0.4047$$

The normalized Chebyshev poles are then:

$$\left((s+0.1315)^2 + 0.9769^2\right) \cdot \left((s+0.3175)^2 + 0.4047^2\right)$$

Converting this to unfactored polynomial form yields:

$$s^4 + 0.898 \cdot s^3 + 1.4032 \cdot s^2 + 0.6866 \cdot s + 0.2571$$

Then, using Equation (7-6):

$$Z_{TH} = \frac{0.898 \cdot s^3 + 0.6866 \cdot s}{s^4 + 1.4032 \cdot s^2 + 0.2571} = 0 +$$

$$\cfrac{1}{\cfrac{0.898 \cdot s^3 + 0.6866 \cdot s}{s^4 + 1.4032 \cdot s^2 + 0.2571}} \Rightarrow L_0 = 0$$

$$\cfrac{s^4 + 1.4032 \cdot s^2 + 0.2571}{0.898 \cdot s^3 + 0.6866 \cdot s} = 1.1136 \cdot s +$$

$$\cfrac{1}{\cfrac{0.6386 \cdot s^2 + 0.2571}{0.898 \cdot s^3 + 0.6866 \cdot s}} \Rightarrow C_1 = 1.1136$$

$$\cfrac{0.898 \cdot s^3 + 0.6866 \cdot s}{0.6386 \cdot s^2 + 0.2571} = 1.4062 \cdot s +$$

$$\cfrac{1}{\cfrac{0.3251 \cdot s}{0.6386 \cdot s^2 + 0.2571}} \Rightarrow L_1 = 1.4062$$

$$\cfrac{0.6386 \cdot s^2 + 0.2571}{0.3251 \cdot s} = 1.9643 \cdot s +$$

$$\cfrac{1}{\cfrac{0.2571}{0.3251 \cdot s}} \Rightarrow C_2 = 1.9643$$

$$\frac{0.3251 \cdot s}{0.2571} = 1.2645 \cdot s \Rightarrow L_2 = 1.2645$$

Use Equation (7-9) to translate the component values:

$$\text{Translated Re sis tan ce} = K_R = 1000$$

$$\text{Translated Induc tan ce} = \frac{K_R}{K_\omega} \cdot L = \frac{1000}{6.283 \cdot 10^6} \cdot L$$

$$\text{Translated Capaci tan ce} = \frac{C}{K_R \cdot K_\omega} = \frac{C}{1000 \cdot 6.283 \cdot 10^6}$$

Since n is even, the frequency response for low frequencies must be the pass min value rather than the pass max value. The pass min value is converted from db using Equation (6-6):

$$\text{magnitude} = \frac{1}{10^{(A_p/20)}} = \frac{1}{10^{(1.172/20)}} = 0.8738$$

This modification can be accommodated by making the load resistor a voltage divider. The sum of the two resistors will equal the load resistance of 1000 ohms. The output of the circuit will then be the voltage across the R2 resistor:

$$V_{out} = V_{R1+R2} \cdot \frac{R2}{R1 + R2} \Rightarrow \frac{R2}{R1 + R2} = 0.8738$$

$$\text{Since: } R1 + R2 = 1000 \quad R2 = 873.8 \quad \text{and} \quad R1 = 126.2$$

The final circuit is shown below:

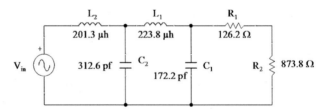

The component values on the above circuit schematic are actual calculated values. Standard values will need to be selected as close as possible to those values. If necessary, resistors can be implemented with two resistors in series, inductors can be implemented with two inductors in series, and capacitors can be implemented with two capacitors in parallel (one a large value and one a small value) to get closer to the calculated value. The specification was intentionally tightened to allow for component value variances.

Below is the frequency response of the above circuit using a computer software circuit analysis program. The analysis was run with the calculated component values. The specification is superimposed in red lines. Note that the filter response exceeds both the pass band and stop band specifications.

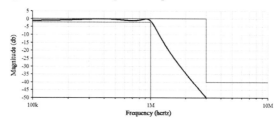

7-3. Design a passive Butterworth high pass filter with a maximum pass band magnitude of 0 db, a minimum pass band magnitude of -1 db from 1 megahertz on, and a maximum stop band magnitude of -40 db out to 200 kilohertz. The filter load resistance will be 1000 ohm.

First create the filter specification graph shown below:

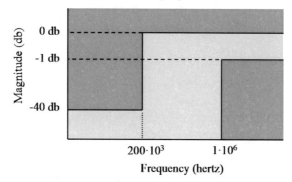

Note that the frequencies are in hertz and not radians/sec. The frequencies must be converted from hertz to radians/sec:

pass band frequency $= \omega_p = 2 \cdot \pi \cdot 1 \cdot 10^6 = 6.283 \cdot 10^6$

stop band frequency $= \omega_s = 2 \cdot \pi \cdot 200 \cdot 10^3 = 1.257 \cdot 10^6$

The magnitude axis is already normalized to pass max equal to 0 db. In order to allow for component variations, the design minimum pass band magnitude needs to be tightened by 10%. Since the magnitude values are in db, the 10% cannot be applied to the db values. Converting -1 db from db to magnitude yields:

$$-1 \text{ db} \Rightarrow 10^{(-1/20)} = 0.8913$$

Tightening this value by 10% yields 0.9804. Converting this back to db yields -0.1721 db.

The pass min value (A_p) is 0.1721 db and the stop max value (A_s) is 40 db. Use Equation (6-31) to determine the number of poles:

$$n = \frac{\log_{10}\left(\dfrac{10^{(0.1 \cdot A_s)} - 1}{10^{(0.1 \cdot A_p)} - 1}\right)}{2 \cdot \log_{10}\left(\dfrac{\omega_p}{\omega_s}\right)} = \frac{\log_{10}\left(\dfrac{10^4 - 1}{10^{0.1721} - 1}\right)}{2 \cdot \log_{10}\left(\dfrac{6.283 \cdot 10^6}{1.257 \cdot 10^6}\right)} = 3.858$$

Rounding n up to the next highest integer means that 4 poles will be required. Next use Equation (6-32) to determine the filter cutoff frequency ω_c:

$$\omega_c = \omega_p \cdot \left(10^{(0.1 \cdot A_p)} - 1\right)^{1/2 \cdot n}$$

$$= 6.283 \cdot 10^6 \cdot \left(10^{(0.1721)} - 1\right)^{1/8} = 4.207 \cdot 10^6$$

Using the Table 7-1 for the 4 pole entry yields:

$$C_{1LP} = 0.3827 \quad L_{1LP} = 1.0824 \quad C_{2LP} = 1.5772 \quad L_{2LP} = 1.5307$$

Use Equation (7-10) to transform the components to a high pass configuration.

$$L_{1HP} = \frac{1}{C_{1LP}} = \frac{1}{0.3827} = 2.613$$

$$L_{2HP} = \frac{1}{C_{2LP}} = \frac{1}{1.5772} = 0.634$$

$$C_{1HP} = \frac{1}{L_{1LP}} = \frac{1}{1.0824} = 0.9239$$

$$C_{2HP} = \frac{1}{L_{2LP}} = \frac{1}{1.5307} = 0.6533$$

Use Equation (7-9) to translate the component values:

$$\text{Translated Resistance} = K_R = 1000$$

$$\text{Translated Inductance} = \frac{K_R}{K_\omega} \cdot L = \frac{1000}{4.207 \cdot 10^6} \cdot L$$

$$\text{Translated Capacitance} = \frac{C}{K_R \cdot K_\omega} = \frac{C}{1000 \cdot 4.207 \cdot 10^6}$$

The final circuit is shown below:

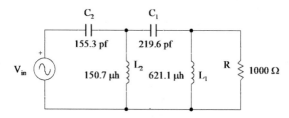

The component values on the above circuit schematic are actual calculated values. Standard values will need to be selected as close as possible to those values. If necessary, inductors can be implemented with two inductors in series and capacitors can be implemented with two capacitors in parallel (one a large value and one a small value) to get closer to the calculated value. The specification was intentionally tightened to allow for component value variances.

Below is the frequency response of the above circuit using a computer software circuit analysis program. The analysis was run with the calculated component values. The specification is superimposed in red lines. Note that the filter response exceeds both the pass band and stop band specifications.

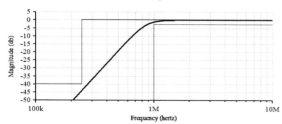

7-4. Design a passive Chebyshev high pass filter with a maximum pass band magnitude of 0 db, a minimum pass band magnitude of -2 db from 1 megahertz on, and a maximum stop band magnitude of -40 db out to 200 kilohertz. The filter load resistance will be 1000 ohm.

First create the filter specification graph shown below:

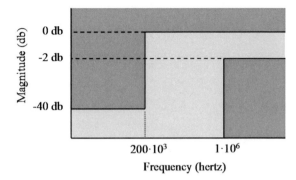

Frequency (hertz)

Note that the frequencies are in hertz and not radians/sec. The frequencies must be converted from hertz to radians/sec:

$$\text{pass band frequency} = \omega_p = 2 \cdot \pi \cdot 1 \cdot 10^6 = 6.283 \cdot 10^6$$

$$\text{stop band frequency} = \omega_s = 2 \cdot \pi \cdot 200 \cdot 10^3 = 1.257 \cdot 10^6$$

The magnitude axis is already normalized to pass max equal to 0 db. In order to allow for component variations, the design minimum pass band magnitude needs to be tightened by 10%. Since the magnitude values are in db, the 10% cannot be applied to the db values. Converting -2 db from db to magnitude yields:

$$-2 \text{ db} \Rightarrow 10^{(-2/20)} = 0.7943$$

Tightening this value by 10% yields 0.8738. Converting this back to db yields -1.172 db.

The pass min value (A_p) is 1.172 db and the stop max value (A_s) is 40 db. Use Equation (6-39) to determine the number of poles:

$$n = \frac{\cosh^{-1}\left(\sqrt{\dfrac{10^{(0.1 \cdot A_s)} - 1}{10^{(0.1 \cdot A_p)} - 1}}\right)}{\cosh^{-1}\left(\dfrac{\omega_p}{\omega_s}\right)} = \frac{\cosh^{-1}\left(\sqrt{\dfrac{10^4 - 1}{10^{0.1172} - 1}}\right)}{\cosh^{-1}\left(\dfrac{6.283 \cdot 10^6}{1.257 \cdot 10^6}\right)} = 2.567$$

Rounding n up to the next highest integer means that 3 poles will be required. Next use Equation (6-19) to determine ε:

$$\varepsilon = \sqrt{10^{(0.1 \cdot A_p)} - 1} = \sqrt{10^{0.1172} - 1} = 0.5566$$

Using the Table 6-1 for the 3 pole entry, determine the Butter-worth pole values:

$$\alpha_1 = 0.5 \qquad \omega_1 = 0.866$$
$$\alpha_2 = 1.0$$

The Chebyshev transformation factors using Equation (6-21) are:

$$K_\alpha = \sinh\left(\frac{1}{n} \cdot \sinh^{-1}\left(\frac{1}{\varepsilon}\right)\right)$$

$$= \sinh\left(\frac{1}{3} \cdot \sinh^{-1}\left(\frac{1}{0.5566}\right)\right) = 0.4649$$

$$K_\omega = \cosh\left(\frac{1}{n} \cdot \sinh^{-1}\left(\frac{1}{\varepsilon}\right)\right)$$

$$= \cosh\left(\frac{1}{3} \cdot \sinh^{-1}\left(\frac{1}{0.5566}\right)\right) = 1.1028$$

The Chebyshev pole values are then:

$$\alpha_1 = 0.5 \cdot 0.4649 = 0.2325$$
$$\omega_1 = 0.866 \cdot 1.1028 = 0.9550$$
$$\alpha_2 = 1.0 \cdot 0.4649 = 0.4649$$

The normalized Chebyshev poles are then:

$$(s + 0.4649) \cdot \left((s + 0.2325)^2 + 0.955^2\right)$$

Converting this to unfactored polynomial form yields:

$$s^3 + 0.9299 \cdot s^2 + 1.1823 \cdot s + 0.4491$$

Then, using Equation (7-6):

$$Z_{TH} = \frac{s^3 + 1.1823 \cdot s}{0.9299 \cdot s^2 + 0.4491} = 1.0754 \cdot s +$$

$$\frac{1}{\dfrac{0.6993 \cdot s}{0.9299 \cdot s^2 + 0.4491}} \Rightarrow L_0 = 1.0754$$

$$\frac{0.9299 \cdot s^2 + 0.4491}{0.6993 \cdot s} = 1.3298 \cdot s +$$

$$\frac{\dfrac{1}{0.4491}}{0.6993 \cdot s} \Rightarrow C_1 = 1.3298$$

$$\frac{0.6993 \cdot s}{0.4491} = 1.5571 \cdot s \Rightarrow L_1 = 1.5571$$

Use Equation (7-10) to transform the components to a high pass configuration.

$$L_{1HP} = \frac{1}{C_{1LP}} = \frac{1}{1.3298} = 0.752$$

$$C_{0HP} = \frac{1}{L_{0LP}} = \frac{1}{1.0754} = 0.9299$$

$$C_{1HP} = \frac{1}{L_{1LP}} = \frac{1}{1.5571} = 0.6422$$

Use Equation (7-9) to translate the component values:

$$\text{Translated Resis tan ce} = K_R = 1000$$

$$\text{Translated Induc tan ce} = \frac{K_R}{K_\omega} \cdot L = \frac{1000}{6.283 \cdot 10^6} \cdot L$$

$$\text{Translated Capaci tan ce} = \frac{C}{K_R \cdot K_\omega} = \frac{C}{1000 \cdot 6.283 \cdot 10^6}$$

The final circuit is shown below:

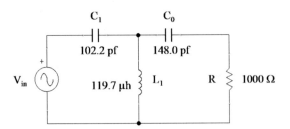

The component values on the above circuit schematic are actual calculated values. Standard values will need to be selected as close as possible to those values. If necessary, inductors can be implemented with two inductors in series and capacitors can be implemented with two capacitors in parallel (one a large value and one a small value) to get closer to the calculated value. The specification was intentionally tightened to allow for component value variances.

Below is the frequency response of the above circuit using a computer software circuit analysis program. The analysis was run with the calculated component values. The specification is superimposed in red lines. Note that the filter response exceeds both the pass band and stop band specifications.

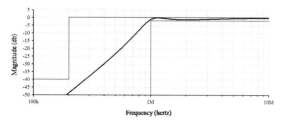

7-5. Design a passive Butterworth band pass filter with a maximum pass band magnitude of 0 db at a center frequency of 1 megahertz and a 200 kilohertz bandwidth at the -3.0103 db points. There shall be a maximum stop band magnitude of -40 db below 500 kilohertz and above 2.5 megahertz. The filter load resistance will be 1000 ohm.

First determine the filter Q using Equation (6-41):

$$Q = \frac{\omega_0}{B} = \frac{2 \cdot \pi \cdot 1 \cdot 10^6}{2 \cdot \pi \cdot 200 \cdot 10^3} = 5$$

Since Q is greater than 0.5, use the narrow band approach.

Next, create the filter specification graph. Since the center frequency and bandwidth are given, they must be normalized and converted to the upper and lower pass frequencies. The normalized center frequency and bandwidth are:

$$\omega_0 = \frac{f_0}{1 \cdot 10^6} = 1 \qquad B = \frac{200 \cdot 10^3}{1 \cdot 10^6} = 0.2$$

The upper and lower pass frequencies can be calculated from Equation (6-40):

345

$$f_0 = 1 \cdot 10^6 = \sqrt{f_{ph} \cdot f_{pl}} \Rightarrow f_{pl} = \frac{1 \cdot 10^{12}}{f_{ph}}$$

$$B_f = 200 \cdot 10^3 = f_{ph} - f_{pl} = f_{ph} - \frac{1 \cdot 10^{12}}{f_{ph}}$$

$$f_{ph}^2 - 200 \cdot 10^3 \cdot f_{ph} - 1 \cdot 10^{12} = 0$$

$$\Rightarrow f_{ph} = 1.105 \cdot 10^6 \quad \text{and} \quad f_{pl} = 905 \cdot 10^3$$

The specification graph is shown below:

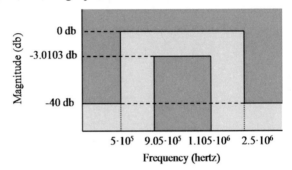

The next step is to normalize the specifications. The magnitude axis is already normalized since the pass max value is 0 db. In order to allow for component variations, the design bandwidth needs to be increased by 10%. The design bandwidth then becomes 220 kilohertz. The design upper and lower pass frequencies can be calculated from Equation (6-40):

$$f_0 = 1 \cdot 10^6 = \sqrt{f_{ph} \cdot f_{pl}} \Rightarrow f_{pl} = \frac{1 \cdot 10^{12}}{f_{ph}}$$

$$B_f = 220 \cdot 10^3 = f_{ph} - f_{pl} = f_{ph} - \frac{1 \cdot 10^{12}}{f_{ph}}$$

$$f_{ph}^2 - 220 \cdot 10^3 \cdot f_{ph} - 1 \cdot 10^{12} = 0$$

$$\Rightarrow f_{ph} = 1.116 \cdot 10^6 \quad \text{and} \quad f_{pl} = 896 \cdot 10^3$$

To normalize the frequency axis divide all of the specifications by f_0 or 1 megahertz. Since this is a frequency divided by a frequency, the normalized axis can be considered either hertz or radians per second. The normalized frequency values are:

$$\omega_{pl} = \frac{896 \cdot 10^3}{1 \cdot 10^6} = 0.896 \quad \omega_{ph} = \frac{1.116 \cdot 10^6}{1 \cdot 10^6} = 1.116$$

$$\text{lower stop} = \frac{500 \cdot 10^3}{1 \cdot 10^6} = 0.5 \quad \text{upper stop} = \frac{2.5 \cdot 10^6}{1 \cdot 10^6} = 2.5$$

The next step is to determine the low pass equivalent. To determine the low pass equivalent number of poles, determine the pass bandwidth B and the geometric stop bandwidth:

$$B = \frac{B_f}{f_0} = \frac{220 \cdot 10^3}{1 \cdot 10^6} = 0.22$$

or $B = \omega_{ph} - \omega_{pl} = 1.116 - 0.896 = 0.22$

$1/(\text{lower stop}) = 1/0.5 = 2.0$

which is less than the upper stop

Therefore: $\omega_{sl} = 0.5$ and $\omega_{sh} = \dfrac{1}{0.5} = 2.0$

geometric stop width $= \omega_{sh} - \omega_{sl} = 2.0 - 0.5 = 1.5$

Determine the number of low pass equivalent poles using Equation (6-47):

$$n = \frac{\log_{10}\left(\dfrac{10^{(0.1 A_s)} - 1}{10^{(0.1 A_p)} - 1}\right)}{2 \cdot \log_{10}\left(\dfrac{\text{geometric stop width}}{\text{bandwidth}}\right)}$$

$$n = \frac{\log_{10}\left(\dfrac{10^{(4)} - 1}{10^{(0.30103)} - 1}\right)}{2 \cdot \log_{10}\left(\dfrac{1.5}{0.22}\right)} = 2.399$$

Rounding up to the next highest integer yields n equal to 3. Using the Table 7-1 for the 3 pole entry yields:

$$L_{0LP} = 0.50 \quad C_{1LP} = 1.3333 \quad L_{1LP} = 1.50$$

Use Equation (7-11) to transform the components to a band pass configuration.

$$L_{OBP} = \frac{L_{OLP}}{B} = \frac{0.5}{0.22} = 2.2727$$

$$C_{OBP} = \frac{B}{L_{OLP}} = \frac{0.22}{0.5} = 0.440$$

$$L_{1BP} = \frac{B}{C_{1LP}} = \frac{0.22}{1.3333} = 0.1650$$

$$C_{1BP} = \frac{C_{1LP}}{B} = \frac{1.3333}{0.22} = 6.0605$$

$$L_{2BP} = \frac{L_{1LP}}{B} = \frac{1.5}{0.22} = 6.8182$$

$$C_{2BP} = \frac{B}{L_{1LP}} = \frac{0.22}{1.5} = 0.1467$$

Use Equation (7-9) to translate the component values:

$$\text{Translated Resistance} = K_R = 1000$$

$$\text{Translated Inductance} = \frac{K_R}{K_\omega} \cdot L = \frac{1000}{2 \cdot \pi \cdot 10^6} \cdot L$$

$$\text{Translated Capacitance} = \frac{C}{K_R \cdot K_\omega} = \frac{C}{1000 \cdot 2 \cdot \pi \cdot 10^6}$$

The final circuit is shown below:

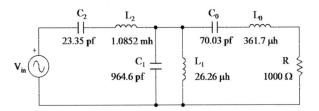

The component values on the above circuit schematic are actual calculated values. Standard values will need to be selected as close as possible to those values. If necessary, inductors can be implemented with two inductors in series and capacitors can be implemented with two capacitors in parallel (one a large value and one a small value) to get closer to the calculated value. The specification was intentionally tightened to allow for component value variances.

Below is the frequency response of the above circuit using a computer software circuit analysis program. The analysis was run with the calculated component values. The specification is superimposed in red lines. Note that the filter response exceeds both the pass band and stop band specifications.

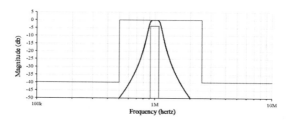

7-6. Design a passive Butterworth notch filter with a maximum pass band magnitude of 0 db. The notch center frequency is 1 mega-hertz with a 1.1 megahertz notchwidth at the -3.0103 db points. There shall be a maximum stop band magnitude of -40 db above 0.95 megahertz and below 1.04 megahertz. The filter load resistance will be 1000 ohm.

First determine the filter Q using Equation (6-53):

$$Q = \frac{\omega_0}{N} = \frac{2 \cdot \pi \cdot 1 \cdot 10^6}{2 \cdot \pi \cdot 1.1 \cdot 10^6} = 0.909$$

Since Q is greater than 0.5, use the narrow band approach.

Next, create the filter specification graph. Since the center frequency and notchwidth are given, they must be normalized and converted to the upper and lower pass frequencies. The normalized center frequency and notchwidth are:

$$\omega_0 = \frac{f_0}{1 \cdot 10^6} = 1 \quad N = \frac{1.1 \cdot 10^6}{1 \cdot 10^6} = 1.1$$

The upper and lower pass frequencies can be calculated from Equation (6-52):

$$f_0 = 1 \cdot 10^6 = \sqrt{f_{ph} \cdot f_{pl}} \Rightarrow f_{pl} = \frac{1 \cdot 10^{12}}{f_{ph}}$$

$$N_f = 1.1 \cdot 10^6 = f_{ph} - f_{pl} = f_{ph} - \frac{1 \cdot 10^{12}}{f_{ph}}$$

$$f_{ph}^2 - 1.1 \cdot 10^6 \cdot f_{ph} - 1 \cdot 10^{12} = 0$$

$$\Rightarrow f_{ph} = 1.691 \cdot 10^6 \text{ and } f_{pl} = 0.591 \cdot 10^6$$

The specification graph is shown below:

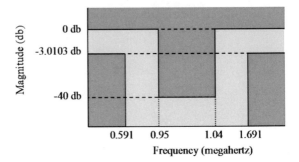

The next step is to normalize the specifications. The magnitude axis is already normalized since the pass max value is 0 db. In order to allow for component variations, the design notchwidth needs to be decreased by 10%. The design notchwidth then becomes 0.99 megahertz. The design upper and lower pass frequencies can be calculated from Equation (6-52):

$$f_0 = 1 \cdot 10^6 = \sqrt{f_{ph} \cdot f_{pl}} \Rightarrow f_{pl} = \frac{1 \cdot 10^{12}}{f_{ph}}$$

$$N_f = 0.99 \cdot 10^6 = f_{ph} - f_{pl} = f_{ph} - \frac{1 \cdot 10^{12}}{f_{ph}}$$

$$f_{ph}^2 - 0.99 \cdot 10^6 \cdot f_{ph} - 1 \cdot 10^{12} = 0$$

$$\Rightarrow f_{ph} = 1.611 \cdot 10^6 \quad \text{and} \quad f_{pl} = 0.621 \cdot 10^6$$

To normalize the frequency axis divide all of the specifications by f_0 or 1 megahertz. Since this is a frequency divided by a frequency, the normalized axis can be considered either hertz or radians per second. The normalized frequency values are:

$$\omega_{pl} = \frac{0.621 \cdot 10^6}{1 \cdot 10^6} = 0.621 \quad \omega_{ph} = \frac{1.611 \cdot 10^6}{1 \cdot 10^6} = 1.611$$

$$\text{lower stop} = \frac{0.95 \cdot 10^6}{1 \cdot 10^6} = 0.95 \quad \text{upper stop} = \frac{1.04 \cdot 10^6}{1 \cdot 10^6} = 1.04$$

The next step is to determine the low pass equivalent. To determine the equivalent low pass number of poles, determine the pass notchwidth N and the geometric stop notchwidth:

$$N = \frac{N_f}{f_0} = \frac{0.99 \cdot 10^6}{1 \cdot 10^6} = 0.99$$

or $N = \omega_{ph} - \omega_{pl} = 1.611 - 0.621 = 0.99$

$1/(\text{lower stop}) = 1/0.95 = 1.0526$ which is greater than
the upper stop

Therefore: $\omega_{sl} = 0.95$ and $\omega_{sh} = \frac{1}{0.95} = 1.0526$

geometric stop width $= \omega_{sh} - \omega_{sl} = 1.0526 - 0.95 = 0.1026$

Determine the number of equivalent low pass poles using Equation (6-59):

$$n = \frac{\log_{10}\left(\dfrac{10^{(0.1 \cdot A_s)} - 1}{10^{(0.1 \cdot A_p)} - 1}\right)}{2 \cdot \log_{10}\left(\dfrac{\text{notchwidth}}{\text{geometric stop width}}\right)}$$

$$= \frac{\log_{10}\left(\dfrac{10^{(4)} - 1}{10^{(0.30103)} - 1}\right)}{2 \cdot \log_{10}\left(\dfrac{0.99}{0.1026}\right)} = 2.031$$

Rounding up to the next highest integer yields n equal to 3. Using the Table 7-1 for the 2 pole entry yields:

$L_{0LP} = 0.50$ $C_{1LP} = 1.3333$ $L_{1LP} = 1.50$

Use Equation (7-12) to transform the components to a notch configuration.

$L_{0N} = L_{0LP} \cdot N = 0.5 \cdot 0.99 = 0.495$

$$C_{0N} = \frac{1}{L_{0LP} \cdot N} = \frac{1}{0.5 \cdot 0.99} = 2.0202$$

$$L_{1N} = \frac{1}{C_{1LP} \cdot N} = \frac{1}{1.3333 \cdot 0.99} = 0.7576$$

$C_{1N} = C_{1LP} \cdot N = 1.3333 \cdot 0.99 = 1.32$

$L_{2N} = L_{1LP} \cdot N = 1.5 \cdot 0.99 = 1.485$

$$C_{2N} = \frac{1}{L_{1LP} \cdot N} = \frac{1}{1.5 \cdot 0.99} = 0..6734$$

Use Equation (7-9) to translate the component values:

$$\text{Translated Resis} \tan \text{ce} = K_R = 1000$$

$$\text{Translated Induc} \tan \text{ce} = \frac{K_R}{K_\omega} \cdot L = \frac{1000}{2 \cdot \pi \cdot 10^6} \cdot L$$

$$\text{Translated Capaci} \tan \text{ce} = \frac{C}{K_R \cdot K_\omega} = \frac{C}{1000 \cdot 2 \cdot \pi \cdot 10^6}$$

The final circuit is shown below:

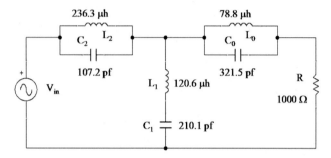

The component values on the above circuit schematic are actual calculated values. Standard values will need to be selected as close as possible to those values. If necessary, inductors can be implemented with two inductors in series and capacitors can be implemented with two capacitors in parallel (one a large value and one a small value) to get closer to the calculated value. The specification was intentionally tightened to allow for component value variances.

Below is the frequency response of the above circuit using a computer software circuit analysis program. The analysis was run with the calculated component values. The specification is superimposed in red lines. Note that the filter response exceeds both the pass band and stop band specifications.

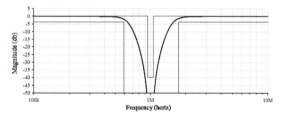

Problem Solutions Chapter 8

8-1. Design an active Butterworth low pass filter with a maximum pass band magnitude of 20 db, a minimum pass band magnitude of 18 db out to 2,000 hertz, and a maximum stop band magnitude of -20 db from 10,000 hertz on.

First create the filter specification graph shown in below:

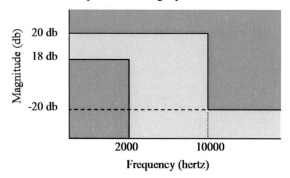

Note that the frequencies are in hertz and not radians/sec. This is typical in filter designs and the frequencies must be converted from hertz to radians/sec to determine the filter transfer function:

$$\text{pass band frequency} = \omega_p = 2 \cdot \pi \cdot 2,000 = 12,566$$
$$\text{stop band frequency} = \omega_s = 2 \cdot \pi \cdot 10,000 = 62,832$$

The magnitude axis is needs to be normalized to pass max equal to 0 db.

Pass Max $= 20 - 20 = 0$ db
Pass Min $= 18 - 20 = -2$ db
Stop Max $= -20 - 20 = -40$ db

In order to allow for component variations, the design minimum pass band magnitude needs to be tightened by 10%. Since the magnitude values are in db, the 10% cannot be applied to the db values. Converting -2 db from db to magnitude yields:

$$-2 \text{ db} \Rightarrow 10^{(-2/20)} = 0.7943$$

Tightening this value by 10% yields a gain of 0.8738. Converting this back to db yields -1.172 db.

353

The pass min value (A_p) is 1.172 db and the stop max value (A_s) is 40 db. Use Equation (6-14) to determine the number of poles:

$$n = \frac{\log_{10}\left(\frac{10^{(0.1 \cdot A_s)}-1}{10^{(0.1 \cdot A_p)}-1}\right)}{2 \cdot \log_{10}\left(\frac{\omega_s}{\omega_p}\right)} = \frac{\log_{10}\left(\frac{10^4-1}{10^{0.1172}-1}\right)}{2 \cdot \log_{10}\left(\frac{62,832}{12,566}\right)} = 3.225$$

Rounding n up to the next highest integer means that 4 poles will be required. Next use Equation (6-15) to determine the filter cutoff frequency ω_c:

$$\omega_c = \frac{\omega_p}{\left(10^{(0.1 \cdot A_p)}-1\right)^{1/2 \cdot n}} = \frac{12,566}{\left(10^{0.1172}-1\right)^{1/8}} = 14,548$$

Using the Table 6-1 for the 4 pole entry and ω_c from above determine the filter normalized translated transfer function:

$$G_T(s) = \frac{1}{\left(\left(\frac{s}{14,548}\right)^2 + \frac{1}{0.5412}\cdot\left(\frac{s}{14,548}\right)+1\right)} \cdot$$
$$\frac{1}{\left(\left(\frac{s}{14,548}\right)^2 + \frac{1}{1.3066}\cdot\left(\frac{s}{14,548}\right)+1\right)}$$

This is a 4 pole filter that can be implemented as an active filter with 2 two pole stages of the type in Figure 8-8. The value of ω_c for both stages is 14,548 rad/sec and the Q for the first stage (the input stage) is 0.5412 and the Q for the second stage (the output stage) is 1.3066.

The denormalized transfer function gain is 10 (20 db). The gain of 10 must be distributed between the two stages using Equations (8-29) with Q1 defaulting to 1:

$$K_1 \cdot K_2 = 10 \Rightarrow K_2 = \frac{10}{K_1}$$
$$K_1 \cdot Q_1 = K_2 \cdot Q_2 \Rightarrow K_1 = K_2 \cdot 1.3066$$
$$K_1^2 = 13.066 \Rightarrow K_1 = 3.615 \text{ and } K_2 = 2.766$$

The final transfer function then becomes:

$$G(s) = \cfrac{\cfrac{3.615}{\left(\left(\dfrac{s}{14,548}\right)^2 + \dfrac{1}{0.5412}\cdot\left(\dfrac{s}{14,548}\right)+1\right)}}{\cfrac{2.766}{\left(\left(\dfrac{s}{14,548}\right)^2 + \dfrac{1}{1.3066}\cdot\left(\dfrac{s}{14,548}\right)+1\right)}} \cdot$$

The stage 1 design has K_1 equal to 3.615. Choose $R2_1$ equal to 1,000 ohms and calculate $R1_1$ using Equation (8-9):

$$R1_1 = R2_1 \cdot (K_1 - 1) = 1,000 \cdot (3.615 - 1) = 2,615$$

Select $R1_1$ equal to 2,610 ohms which gives a gain of 3.61. Select CB_1 equal to 0.012 microfarads and use Equation (8-28) To determine CD_{1max}:

$$CD_{1max} = CB_1 \cdot \left(\frac{1}{4 \cdot Q_1^2} + K_1 - 1\right) = 0.0416 \cdot 10^{-6}$$

Select CD_1 equal to 0.01 microfarads. Use Equation (8-27) to calculate RC_1:

$$RC_1 = \frac{1}{2 \cdot \omega_c \cdot Q_1 \cdot CD_1} + \frac{1}{\omega_c} \cdot \sqrt{\frac{1}{4 \cdot Q_1^2 \cdot CD_1^2} - \frac{1}{CD_1}\cdot\left(\frac{1}{CB_1} + \frac{1-K_1}{CD_1}\right)}$$

$$RC_1 = 17.498 \text{ kohms}$$

Use Equation (8-25) to calculate RA_1:

$$RA_1 = \frac{1}{RC_1 \cdot CB_1 \cdot CD_1 \cdot \omega_c^2} = 2.2502 \text{ kohms}$$

Select standard values of 17.4 kohms and 2.26 kohms for RC_1 and RA_1 respectively.

The stage 2 design has K_2 equal to 2.766. Choose $R2_2$ equal to 1,130 ohms and calculate $R1_2$ using Equation (8-29):

$$R1_2 = R2_2 \cdot (K_2 - 1) = 1,130 \cdot (2.766 - 1) = 1,996$$

Select $R1_2$ equal to 2,000 ohms which gives a gain of 2.77. Select CB_2 equal to 0.015 microfarads and use Equation (8-28) To determine CD_{2max}:

$$CD_{2\,max} = CB_2 \cdot \left(\frac{1}{4 \cdot Q_2^2} + K_2 - 1\right) = 0.0287 \cdot 10^{-6}$$

Select CD_2 equal to 0.0082 microfarads. Use Equation (8-27) to calculate RC_2:

$$RC_2 = \frac{1}{2 \cdot \omega_c \cdot Q_2 \cdot CD_2} + \frac{1}{\omega_c} \cdot \sqrt{\frac{1}{4 \cdot Q_2^2 \cdot CD_2^2} - \frac{1}{CD_2}\left(\frac{1}{CB_2} + \frac{1 - K_2}{CD_2}\right)}$$

$$RC_2 = 13.02 \text{ kohms}$$

Use Equation (8-25) to calculate RA_2:

$$RA_2 = \frac{1}{RC_2 \cdot CB_2 \cdot CD_2 \cdot \omega_c^2} = 2.951 \text{ kohms}$$

Select standard values of 13.0 kohms and 2.94 kohms for RC_2 and RA_2 respectively. The final circuit is shown below:

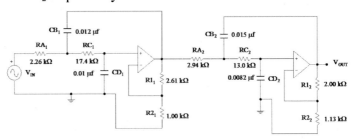

Below is the frequency response of the above circuit using a computer software circuit analysis program. The specification is superimposed in red lines. Note that the filter response exceeds both the pass band and stop band specifications.

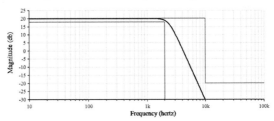

8-2. Design an active Chebyshev low pass filter with a maximum pass band magnitude of 0 db, a minimum pass band magnitude of -2 db out to 1000 hertz, and a maximum stop band magnitude of -40 db from 5000 hertz on.

First create the filter specification graph shown below:

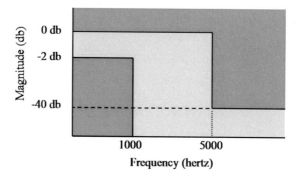

Note that the frequencies are in hertz and not radians/sec. This is typical in filter designs and the frequencies must be converted from hertz to radians/sec to determine the filter transfer function:

$$\text{pass band frequency} = \omega_p = 2 \cdot \pi \cdot 1000 = 6{,}283$$

$$\text{stop band frequency} = \omega_s = 2 \cdot \pi \cdot 5000 = 31{,}416$$

The magnitude axis is already normalized to pass max equal to 0 db.

In order to allow for component variations, the design minimum pass band magnitude needs to be tightened by 10%. Since the magnitude values are in db, the 10% cannot be applied to the db values. Converting -2 db from db to magnitude yields:

$$-2 \text{ db} \Rightarrow 10^{(-2/20)} = 0.7943$$

Tightening this value by 10% yields 0.8738. Converting this back to db yields -1.172 db.

The pass min value (A_p) is 1.172 db and the stop max value (A_s) is 40 db. Use Equation (6-28) to determine the number of poles:

$$n = \frac{\cosh^{-1}\left(\sqrt{\dfrac{10^{(0.1 \cdot A_s)} - 1}{10^{(0.1 \cdot A_p)} - 1}}\right)}{\cosh^{-1}\left(\dfrac{\omega_s}{\omega_p}\right)} = \frac{\cosh^{-1}\left(\sqrt{\dfrac{10^4 - 1}{10^{0.1172} - 1}}\right)}{\cosh^{-1}\left(\dfrac{31{,}416}{6{,}283}\right)} = 2.567$$

Rounding n up to the next highest integer means that 3 poles will be required. Using Equation (6-19), determine ε:

$$\varepsilon = \sqrt{10^{(0.1 \cdot A_p)} - 1} = \sqrt{10^{0.1172} - 1} = 0.5566$$

Using the Table 6-1 for the 3 pole entry, determine the Butterworth pole values:

$$\alpha_1 = 0.5 \quad \omega_1 = 0.866$$
$$\alpha_2 = 1.0$$

The Chebyshev transformation factors using Equation (6-21) are:

$$K_\alpha = \sinh\left(\frac{1}{n} \cdot \sinh^{-1}\left(\frac{1}{\varepsilon}\right)\right)$$
$$= \sinh\left(\frac{1}{3} \cdot \sinh^{-1}\left(\frac{1}{0.5566}\right)\right) = 0.4649$$

$$K_\omega = \cosh\left(\frac{1}{n} \cdot \sinh^{-1}\left(\frac{1}{\varepsilon}\right)\right)$$
$$= \cosh\left(\frac{1}{3} \cdot \sinh^{-1}\left(\frac{1}{0.5566}\right)\right) = 1.1028$$

The Chebyshev pole values are then:

$$\alpha_1 = 0.5 \cdot 0.4649 = 0.2325$$
$$\omega_1 = 0.866 \cdot 1.1028 = 0.955$$
$$\alpha_2 = 1.0 \cdot 0.4649 = 0.4649$$

The normalized Chebyshev poles are then:

$$(s + 0.4649) \cdot \left((s + 0.2325)^2 + 0.955^2\right)$$

Converting this to a normalized transfer function in break frequency form yields:

$$G_N(s) = \cfrac{1}{\left(\cfrac{s}{0.4649} + 1\right) \cdot \left(\left(\cfrac{s}{0.9829}\right)^2 + \cfrac{1}{2.1142} \cdot \left(\cfrac{s}{0.9829}\right) + 1\right)}$$

Translating this to the pass frequency ω_p yields:

$$G_T(s) = \cfrac{1}{\left(\cfrac{s}{2921} + 1\right) \cdot \left(\left(\cfrac{s}{6176}\right)^2 + \cfrac{1}{2.1142} \cdot \left(\cfrac{s}{6176}\right) + 1\right)}$$

The gain of a Chebyshev filter with an odd number of poles must be the pass max value. Since the pass max value is 0 db, the gain will be 1 and both stages will have a gain of 1 so the final transfer function is:

$$G(s) = \frac{1}{\left(\dfrac{s}{2921}+1\right)} \cdot \frac{1}{\left(\left(\dfrac{s}{6176}\right)^2 + \dfrac{1}{2.1142}\cdot\left(\dfrac{s}{6176}\right)+1\right)}$$

This is a 3 pole filter that can be implemented as an active filter with 1 two pole stage of the type in Figure 8-8 and 1 one pole stage of the type in Figure 8-5. The value of ω_c for stage 1 (the input stage) is 2,921 rad/sec. The value of ω_c for stage 2 (the output stage) is 6,176 rad/sec and the Q for stage 2 is 2.1142.

The stage 1 design has K_1 equal to 1 which makes $R1_1$ a short and $R2_1$ an open. Select CB_1 equal to 0.018 microfarads and use Equation (8-10) To determine RA_1:

$$RA_1 = \frac{1}{CB_1 \cdot \omega_{c1}} = \frac{1}{0.018 \cdot 10^{-6} \cdot 2921} = 19.02 \text{ kohms}$$

Select the standard value of RA_1 equal to 19.1 kohms.

The stage 2 design has K_2 equal to 1 which makes $R1_2$ a short and $R2_2$ an open. Select CB_2 equal to 0.1 microfarads and use Equation (8-28) To determine CD_{max}:

$$CD_{max} = CB_2 \cdot \left(\frac{1}{4 \cdot Q_2^2} + K_2 - 1\right) = 0.00559 \cdot 10^{-6}$$

Select CD equal to 0.0022 microfarads. Use Equation (8-27) to calculate RC:

$$RC = \frac{1}{2 \cdot \omega_{c2} \cdot Q_2 \cdot CD} + \frac{1}{\omega_{c2}} \cdot \sqrt{\frac{1}{4 \cdot Q_2^2 \cdot CD^2} - \frac{1}{CD}\left(\frac{1}{CB_2} + \frac{1-K_2}{CD}\right)}$$

RC = 30.96 kohms

Use Equation (8-25) to calculate RA_2:

$$RA_2 = \frac{1}{RC_2 \cdot CB_2 \cdot CD_2 \cdot \omega_{c2}^2} = 3.849 \text{ kohms}$$

Select standard values of 30.1 kohms and 3.83 kohms for RC_2 and RA_2 respectively. The final circuit is shown below:

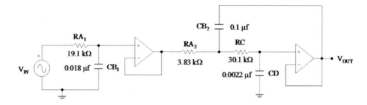

Below is the frequency response of the above circuit using a computer software circuit analysis program. The specification is superimposed in red lines. Note that the filter response exceeds both the pass band and stop band specifications.

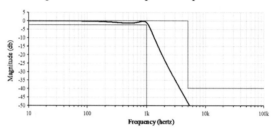

8-3. Design an active Butterworth high pass filter with a maximum pass band magnitude of 20 db, a minimum pass band magnitude of 18 db from 10000 hertz on out, and a maximum stop band magnitude of -20 db below 2000 hertz.

First create the filter specification graph shown below:

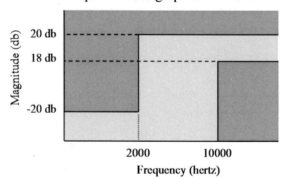

The frequencies must be converted from hertz to radians/sec to determine the filter transfer function:

$$\text{pass band frequency} = \omega_p = 2 \cdot \pi \cdot 10,000 = 62,832$$

$$\text{stop band frequency} = \omega_s = 2 \cdot \pi \cdot 2,000 = 12,566$$

The magnitude axis needs to be normalized to pass max equal to 0 db.

360

$$\text{Pass Max} = 20 - 20 = 0 \text{ db}$$
$$\text{Pass Min} = 18 - 20 = -2 \text{ db}$$
$$\text{Stop Max} = -20 - 20 = -40 \text{ db}$$

In order to allow for component variations, the design minimum pass band magnitude needs to be tightened by 10%. Since the magnitude values are in db, the 10% cannot be applied to the db values. Converting -2 db from db to magnitude yields:

$$-2 \text{ db} \Rightarrow 10^{(-2/20)} = 0.7943$$

Tightening this value by 10% yields 0.8738. Converting this back to db yields -1.172 db.

The pass min value (A_p) is 1.172 db and the stop max value (A_s) is 40 db. Use Equation (6-31) to determine the number of poles:

$$n = \frac{\log_{10}\left(\frac{10^{(0.1 \cdot A_s)} - 1}{10^{(0.1 \cdot A_p)} - 1}\right)}{2 \cdot \log_{10}\left(\frac{\omega_p}{\omega_s}\right)} = \frac{\log_{10}\left(\frac{10^4 - 1}{10^{0.1172} - 1}\right)}{2 \cdot \log_{10}\left(\frac{62,832}{12,566}\right)} = 3.225$$

Rounding n up to the next highest integer means that 4 poles will be required. Next use Equation (6-32) to determine the filter cutoff frequency ω_c:

$$\omega_c = \omega_p \cdot \left(10^{(0.1 \cdot A_p)} - 1\right)^{1/2 \cdot n} = 62,832 \cdot \left(10^{(0.1172)} - 1\right)^{1/8} = 54,270$$

Using the Table 6-1 for the 4 pole entry, Equation (6-34), and ω_c from above determine the filter transfer function:

$$G_N(s) = \frac{s^2}{\left(s^2 + \dfrac{54,270}{0.5412} \cdot s + 54,270^2\right)} \cdot \frac{s^2}{\left(s^2 + \dfrac{54,270}{1.3066} \cdot s + 54,270^2\right)}$$

The transfer function magnitudes must also be denormalized by adding a multiplying factor of 10 ($20 \cdot \log_{10} 10 = 20$ db). The total numerator gain factor of 10 must be distributed between the two stages using Equations (8-29). Note that Q values less than 1 default to 1.

$$K_1 \cdot K_2 = 10 \Rightarrow K_2 = \frac{10}{K_1}$$

$$K_1 \cdot Q_1 = K_2 \cdot Q_2 \Rightarrow K_1 \cdot 1.0 = K_2 \cdot 1.3066$$

$$K_1^2 = 13.066 \Rightarrow K_1 = 3.615 \text{ and } K_2 = 2.766$$

The transfer function then becomes:

$$G(s) = \frac{3.615 \cdot s^2}{\left(s^2 + \dfrac{54,270}{0.5412} \cdot s + 54,270^2\right)} \cdot \frac{2.766 \cdot s^2}{\left(s^2 + \dfrac{54,270}{1.3066} \cdot s + 54,270^2\right)}$$

This is a 4 pole filter that can be implemented as an active filter with 2 two pole stages of the type in Figure 8-16. The value of ω_c for both stages is 54,270 rad/sec and the Q for the first stage (the input stage) is 0.5412 and the Q for the second stage (the output stage) is 1.3066.

The stage 1 design has K_1 equal to 3.615. Choose $R2_1$ equal to 1,000 ohms and calculate $R1_1$ using Equation (8-9):

$$R1_1 = R2_1 \cdot (K_1 - 1) = 1,000 \cdot (5.723 - 1) = 2,615$$

Select $R1_1$ equal to 2,610 ohms which gives a gain of 3.61.

Select C_1 equal to 0.0018 microfarads and use Equation (8-40) to determine RB_1:

$$RB_1 = \frac{1}{4 \cdot C_1 \cdot Q_1 \cdot \omega_c} + \frac{1}{C_1 \cdot \omega_c} \cdot \sqrt{\frac{1}{16 \cdot Q_1^2} - \frac{1 - K_1}{2}}$$

$$RB_1 = 17.343 \text{ kohms}$$

Use Equation (8-37) to calculate RD_1:

$$RD_1 = \frac{1}{RB_1 \cdot C_1^2 \cdot \omega_c^2} = 6.0424 \text{ kohms}$$

Select standard values of 17.4 kohms and 6.04 kohms for RB_1 and RD_1 respectively.

The stage 2 design has K_2 equal to 2.766. Choose $R2_2$ equal to 5,760 ohms and calculate $R1_2$ using Equation (8-9):

$$R1_2 = R2_2 \cdot (K_2 - 1) = 5,760 \cdot (2.766 - 1) = 10,172$$

Select $R1_2$ equal to 10,200 ohms which gives a gain of 2.771.

Select C_2 equal to 0.0015 microfarads and use Equation (8-40) to determine RB_2:

$$RB_2 = \frac{1}{4 \cdot C_2 \cdot Q_2 \cdot \omega_c} + \frac{1}{C_2 \cdot \omega_c} \cdot \sqrt{\frac{1}{16 \cdot Q_2^2} - \frac{1-K_2}{2}}$$

$RB_2 = 14.147$ kohms

Use Equation (8-37) to calculate RD_2:

$$RD_2 = \frac{1}{RB_2 \cdot C_2^2 \cdot \omega_c^2} = 10.667 \text{ kohms}$$

Select standard values of 14.0 kohms and 10.7 kohms for RB_2 and RD_2 respectively. The final circuit is shown below:

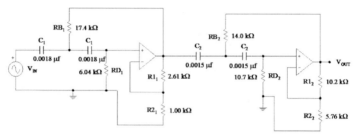

Below is the frequency response of the above circuit using a computer software circuit analysis program. The specification is superimposed in red lines. Note that the filter response exceeds both the pass band and stop band specifications.

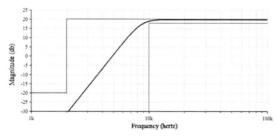

8-4. Design an active Chebyshev high pass filter with a maximum pass band magnitude of 20 db, a minimum pass band magnitude of 18 db from 3000 hertz on out, and a maximum stop band magnitude of -20 db below 1000 hertz.

First create the filter specification graph shown below:

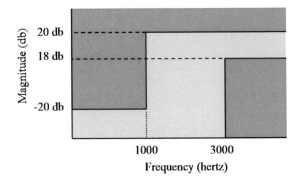

Frequency (hertz)

The frequencies must be converted from hertz to radians/sec to determine the filter transfer function:

$$\text{pass band frequency} = \omega_p = 2 \cdot \pi \cdot 3000 = 18{,}850$$

$$\text{stop band frequency} = \omega_s = 2 \cdot \pi \cdot 1000 = 6{,}283$$

The magnitude axis needs to be normalized to pass max equal to 0 db.

$$\text{Pass Max} = 20 - 20 = 0 \text{ db}$$

$$\text{Pass Min} = 18 - 20 = -2 \text{ db}$$

$$\text{Stop Max} = -20 - 20 = -40 \text{ db}$$

In order to allow for component variations, the design minimum pass band magnitude needs to be tightened by 10%. Since the magnitude values are in db, the 10% cannot be applied to the db values. Converting -2 db from db to magnitude yields:

$$-2 \text{ db} \Rightarrow 10^{(-2/20)} = 0.7943$$

Tightening this value by 10% yields 0.8738. Converting this back to db yields -1.172 db.

The pass min value (A_p) is 1.172 db and the stop max value (A_s) is 40 db. Use Equation (6-39) to determine the number of poles:

$$n = \frac{\cosh^{-1}\left(\sqrt{\dfrac{10^{(0.1 A_s)} - 1}{10^{(0.1 A_p)} - 1}}\right)}{\cosh^{-1}\left(\dfrac{\omega_p}{\omega_s}\right)} = \frac{\cosh^{-1}\left(\sqrt{\dfrac{10^4 - 1}{10^{0.1172} - 1}}\right)}{\cosh^{-1}\left(\dfrac{18{,}850}{6{,}283}\right)} = 3.338$$

Rounding n up to the next highest integer means that 4 poles will be required. Using Equation (6-19), determine ε:

$$\varepsilon = \sqrt{10^{(0.1 \cdot A_p)} - 1} = \sqrt{10^{0.1172} - 1} = 0.5566$$

Using the Table 6-1 for the 4 pole entry, determine the Butterworth pole values:

$$\alpha_1 = 0.3827 \quad \omega_1 = 0.9239$$
$$\alpha_2 = 0.9239 \quad \omega_2 = 0.3827$$

The Chebyshev transformation factors are:

$$K_\alpha = \sinh\left(\frac{1}{n} \cdot \sinh^{-1}\left(\frac{1}{\varepsilon}\right)\right)$$

$$= \sinh\left(\frac{1}{4} \cdot \sinh^{-1}\left(\frac{1}{0.5566}\right)\right) = 0.3436$$

$$K_\omega = \cosh\left(\frac{1}{n} \cdot \sinh^{-1}\left(\frac{1}{\varepsilon}\right)\right)$$

$$= \cosh\left(\frac{1}{4} \cdot \sinh^{-1}\left(\frac{1}{0.5566}\right)\right) = 1.0574$$

The Chebyshev pole values are then:

$$\alpha_1 = 0.3827 \cdot 0.3436 = 0.1315$$
$$\omega_1 = 0.9239 \cdot 1.0574 = 0.9769$$
$$\alpha_2 = 0.9239 \cdot 0.3436 = 0.3175$$
$$\omega_2 = 0.3827 \cdot 1.0574 = 0.4046$$

The normalized Chebyshev poles are then:

$$\left((s + 0.1315)^2 + 0.9769^2\right) \cdot \left((s + 0.3175)^2 + 0.4046^2\right)$$

Converting the poles to break frequency format yields:

$$\left(\left(\frac{s}{0.9857}\right)^2 + \frac{1}{3.7479} \cdot \left(\frac{s}{0.9857}\right) + 1\right) \cdot$$

$$\left(\left(\frac{s}{0.5143}\right)^2 + \frac{1}{0.81} \cdot \left(\frac{s}{0.5143}\right) + 1\right)$$

Using Equation (6-38), convert these poles to the high pass form translated to ω_p.

$$s^2 + \frac{1}{3.7479}\cdot\left(\frac{18,850}{0.9857}\right)\cdot s + \left(\frac{18,850}{0.9857}\right)^2 = s^2 + \frac{19,123}{3.7479}\cdot s + 19,123^2$$

$$s^2 + \frac{1}{0.81}\cdot\left(\frac{18,850}{0.5143}\right)\cdot s + \left(\frac{18,850}{0.5143}\right)^2 = s^2 + \frac{36,652}{0.81}\cdot s + 36,652^2$$

Since n is even the numerator constant of the transfer function must be the pass min value converted using Equation (6-6):

$$\text{numerator} = \frac{1}{10^{(A_p/20)}} = \frac{1}{10^{0.0586}} = 0.8738$$

The transfer function magnitudes must also be denormalized by adding a multiplying factor of 10 $(20\cdot\log_{10}10 = 20\text{ db})$. The total numerator gain factor of 8.738 must be distributed between the two stages using Equations (8-29). Note that Q values less than 1 default to 1.

$$K_1\cdot K_2 = 8.738 \Rightarrow K_2 = \frac{8.738}{K_1}$$

$$K_1\cdot Q_1 = K_2\cdot Q_2 \Rightarrow K_1\cdot 1.0 = K_2\cdot 3.7479$$
$$K_1^2 = 32.749 \Rightarrow K_1 = 5.723 \text{ and } K_2 = 1.527$$

The transfer function then becomes:

$$G(s) = \frac{5.723\cdot s^2}{\left(s^2 + \frac{36,652}{0.81}\cdot s + 36,652^2\right)}\cdot\frac{1.527\cdot s^2}{\left(s^2 + \frac{19,123}{3.7479}\cdot s + 19,123^2\right)}$$

This is a 4 pole filter that can be implemented as an active filter with 2 two pole stages of the type in Figure 8-16. The value of ω_p for stage 1 (the input stage) is 36,652 rad/sec and for stage 2 (the output stage) is 19,123 rad/sec. The Q for stage 1 is 0.81 and the Q for stage 2 is 3.7479.

The stage 1 design has K_1 equal to 5.723. Choose $R2_1$ equal to 2,150 ohms and calculate $R1_1$ using Equation (8-9):

$$R1_1 = R2_1\cdot(K_1 - 1) = 2,150\cdot(5.723 - 1) = 10,154$$

Select $R1_1$ equal to 10,200 ohms which gives a gain of 5.744.

Select C_1 equal to 0.0047 microfarads and use Equation (8-40) to determine RB_1:

$$RB_1 = \frac{1}{4 \cdot C_1 \cdot Q_1 \cdot \omega_{c1}} + \frac{1}{C_1 \cdot \omega_{c1}} \cdot \sqrt{\frac{1}{16 \cdot Q_1^2} - \frac{1 - K_1}{2}}$$

$RB_1 = 10.91$ kohms

Use Equation (8-37) to calculate RD_1:

$$RD_1 = \frac{1}{RB_1 \cdot C_1^2 \cdot \omega_{c1}^2} = 3.088 \text{ kohms}$$

Select standard values of 11.0 kohms and 3.09 kohms for RB_1 and RD_1 respectively.

The stage 2 design has K_2 equal to 1.527. Choose $R2_2$ equal to 2,260 ohms and calculate $R1_2$ using Equation (8-9):

$$R1_2 = R2_2 \cdot (K_2 - 1) = 2,260 \cdot (1.527 - 1) = 1,191$$

Select $R1_2$ equal to 1,180 ohms which gives a gain of 1.522.

Select C_2 equal to 0.0047 microfarads and use Equation (8-40) to determine RB_2:

$$RB_2 = \frac{1}{4 \cdot C_2 \cdot Q_2 \cdot \omega_{c2}} + \frac{1}{C_2 \cdot \omega_{c2}} \cdot \sqrt{\frac{1}{16 \cdot Q_2^2} - \frac{1 - K_2}{2}}$$

$RB_2 = 6.475$ kohms

Use Equation (8-37) to calculate RD_2:

$$RD_2 = \frac{1}{RB_2 \cdot C_2^2 \cdot \omega_{c2}^2} = 19.12 \text{ kohms}$$

Select standard values of 6.49 kohms and 19.1 kohms for RB_2 and RD_2 respectively. The final circuit is shown below:

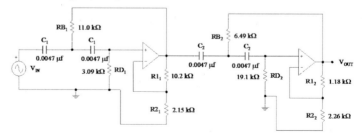

Below is the frequency response of the above circuit using a computer software circuit analysis program. The specification is superimposed in red lines. Note that the filter response exceeds both the pass band and stop band specifications.

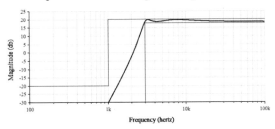

Frequency (hertz)

8-5. Design an active band pass Butterworth filter with a maximum pass band magnitude of 20 db at a center frequency of 1000 hertz and a 900 hertz bandwidth at the 16.9897 db points. There shall be a maximum stop band magnitude of -20 db below 50 hertz and above 20 khertz.

First determine the filter Q using Equation (6-41):

$$Q = \frac{\omega_0}{B} = \frac{2 \cdot \pi \cdot 1000}{2 \cdot \pi \cdot 900} = 1.111$$

Since Q is greater than 0.5, use the narrow band approach.

Next, create the filter specification graph. Since the center frequency and bandwidth are given, they must be normalized and converted to the upper and lower pass frequencies. The normalized center frequency and bandwidth are:

$$\omega_0 = \frac{f_0}{1000} = 1 \quad B = \frac{900}{1000} = 0.9$$

The upper and lower pass frequencies can be calculated from Equation (6-40):

$$f_0 = 1000 = \sqrt{f_{ph} \cdot f_{pl}} \Rightarrow f_{pl} = \frac{1000^2}{f_{ph}}$$

$$B_f = 900 = f_{ph} - f_{pl} = f_{ph} - \frac{1000^2}{f_{ph}}$$

$$f_{ph}^2 - 900 \cdot f_{ph} - 1000^2 = 0 \Rightarrow f_{ph} = 1547 \quad \text{and} \quad f_{pl} = 647$$

The specification graph is shown below:

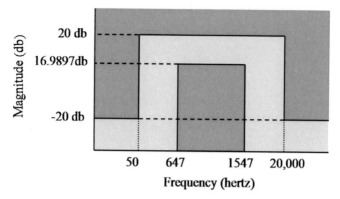

The next step is to normalize the specifications. The magnitude axis needs to be normalized to pass max equal to 0 db.

Pass Max $= 20 - 20 = 0$ db

Pass Min $= 16.9897 - 20 = -3.0103$ db

Stop Max $= -20 - 20 = -40$ db

In order to allow for component variations, the design bandwidth needs to be increased by 10%. The design bandwidth then becomes 990 hertz. The design upper and lower pass frequencies can be calculated from Equation (6-40):

$$f_0 = 1000 = \sqrt{f_{ph} \cdot f_{pl}} \Rightarrow f_{pl} = \frac{1000^2}{f_{ph}}$$

$$B_f = 990 = f_{ph} - f_{pl} = f_{ph} - \frac{1000^2}{f_{ph}}$$

$$f_{ph}^2 - 990 \cdot f_{ph} - 1000^2 = 0 \Rightarrow f_{ph} = 1611 \quad \text{and} \quad f_{pl} = 621$$

To normalize the frequency axis divide all of the specifications by f_0 or 1000. Since this is a frequency divided by a frequency, the normalized axis can be considered either hertz or radians per second. The normalized frequency values are:

$$\omega_{pl} = \frac{621}{1000} = 0.621 \quad \omega_{ph} = \frac{1611}{1000} = 1.611$$

$$\text{lower stop} = \frac{50}{1000} = 0.05 \quad \text{upper stop} = \frac{20000}{1000} = 20.0$$

The next step is to determine the low pass equivalent. To determine the low pass equivalent number of poles, determine the pass bandwidth B and the geometric stop bandwidth:

$$B = \frac{B_f}{f_0} = \frac{990}{1000} = 0.99$$

$$\text{or } B = \omega_{ph} - \omega_{pl} = 1.611 - 0.621 = 0.99$$

$$1/(\text{lower stop}) = 1/0.05 = 20.0$$

which is the same as the upper stop frequency

Therefore: $\omega_{sl} = 0.05$ and $\omega_{sh} = 20.0$

geometric stop width $= \omega_{sh} - \omega_{sl} = 20.0 - 0.05 = 19.95$

Determine the number of low pass equivalent poles using Equation (6-47):

$$n = \frac{\log_{10}\left(\frac{10^{(0.1 A_s)} - 1}{10^{(0.1 A_p)} - 1}\right)}{2 \cdot \log_{10}\left(\frac{\text{geometric stop width}}{\text{bandwidth}}\right)}$$

$$= \frac{\log_{10}\left(\frac{10^{(4)} - 1}{10^{(0.30103)} - 1}\right)}{2 \cdot \log_{10}\left(\frac{19.95}{0.99}\right)} = 1.533$$

Rounding up to the next highest integer yields n equal to 2. Using Table 6-1 for the 2 pole entry, the normalized low pass equivalent transfer function is:

$$G(s)_{LP} = \frac{1}{\left(s^2 + \dfrac{s}{0.7071} + 1\right)}$$

Translating this to the pass band frequency of B yields:

$$G(s)_{LP} = \frac{1}{\left(\dfrac{s}{0.99}\right)^2 + \dfrac{1}{0.7071} \cdot \left(\dfrac{s}{0.99}\right) + 1}$$

Transforming this to band pass using Equation (6-50) yields:

$$G(s) = \frac{B^2 \cdot s^2}{s^4 + \dfrac{B}{Q_{LP}} \cdot s^3 + (2 + B^2) \cdot s^2 + \dfrac{B}{Q_{LP}} \cdot s + 1}$$

$$G(s) = \frac{0.9801 \cdot s^2}{\left(s^4 + 1.40 \cdot s^3 + 2.9801 \cdot s^2 + 1.40 \cdot s + 1\right)}$$

Factoring the 4th order denominator yields:

$$G(s) = \frac{0.99 \cdot s}{\left(s^2 + 0.9432 \cdot s + 2.065\right)} \cdot$$

$$\frac{0.99 \cdot s}{\left(s^2 + 0.4568 \cdot s + 0.4843\right)}$$

Translating this out to the specification center frequency of 1000 hertz which is $2000 \cdot \pi$ radians per second yields:

$$G(s) = \frac{0.99 \cdot \left(\dfrac{s}{2000 \cdot \pi}\right)}{\left(\left(\dfrac{s}{2000 \cdot \pi}\right)^2 + 0.9432 \cdot \left(\dfrac{s}{2000 \cdot \pi}\right) + 2.065\right)} \cdot$$

$$\frac{0.99 \cdot \left(\dfrac{s}{2000 \cdot \pi}\right)}{\left(\left(\dfrac{s}{2000 \cdot \pi}\right)^2 + 0.4568 \cdot \left(\dfrac{s}{2000 \cdot \pi}\right) + 0.4843\right)}$$

$$G(s) = \frac{6220 \cdot s}{\left(s^2 + 5926 \cdot s + 8.1523 \cdot 10^7\right)} \cdot$$

$$\frac{6220 \cdot s}{\left(s^2 + 2870 \cdot s + 1.9118 \cdot 10^7\right)}$$

At this point the pass max value needs to be denormalized. The magnitude value of 10 (20 db) needs to be distributed between the two stages:

stage1 value $\cdot$ stage2 value $= 10$

therefore : magnitude value $= \sqrt{10} = 3.1623$

Applying this factor to the above equation yields:

$$G(s) = \frac{19,699 \cdot s}{\left(s^2 + 5926 \cdot s + 8.1523 \cdot 10^7\right)} \cdot \frac{19,699 \cdot s}{\left(s^2 + 2870 \cdot s + 1.9118 \cdot 10^7\right)}$$

This is a 4 pole filter that can be implemented as an active filter with 2 two pole stages of the type in Figure 8-24. By comparing the desired transfer function for each stage with Equation (8-41) the following requirements can be defined:

Stage 1:

$$\omega_{01} = \sqrt{8.1523 \cdot 10^7} = 9029$$

$$B_1 = 5926$$

$$K_1 = \frac{19,699}{5926} = 3.3242$$

Stage 2:

$$\omega_{02} = \sqrt{1.9118 \cdot 10^7} = 4372$$

$$B_2 = 2870$$

$$K_2 = \frac{19,699}{2870} = 6.864$$

Choose CB_1 equal to 0.0027 microfarads and calculate RA_1 using Equation (8-55).

$$RA_1 = \frac{1}{K_1 \cdot B_1 \cdot CB_1}$$

$$RA_1 = 18.80 \text{ kohms}$$

Choose RA_1 equal to 18.7 kohms. Calculate CC_{1min} using Equation (8-59).

$$CC_{1min} = \left(\frac{K_1 \cdot B_1^2}{\omega_{01}^2} - 1\right) \cdot CB_1$$

$$CC_{1min} = 0.00117 \text{ microfarads}$$

Choose CC_1 equal to 0.0039 microfarads and Calculate RD_1 using Equation (8-56).

$$RD_1 = \frac{1}{B_1} \cdot \left(\frac{1}{CC_1} + \frac{1}{CB_1} \right)$$

$$RD_1 = 105.8 \text{ kohms}$$

Choose RD_1 equal to 105 kohms. Calculate RE_1 using Equation (8-57).

$$RE_1 = \frac{1}{\dfrac{\omega_{01}^2 \cdot (CB_1 + CC_1)}{B_1} - K_1 \cdot B_1 \cdot CB_1}$$

$$RE_1 = 26.59 \text{ kohms}$$

Choose RE_1 equal to 26.7 kohms.

Choose CB_2 equal to 0.0027 microfarads and calculate RA_2 using Equation (8-55).

$$RA_2 = \frac{1}{K_2 \cdot B_2 \cdot CB_2}$$

$$RA_2 = 18.80 \text{ kohms}$$

Choose RA_2 equal to 18.7 kohms. Calculate CC_{2min} using Equation (8-59).

$$CC_{2\,min} = \left(\frac{K_2 \cdot B_2^2}{\omega_{02}^2} - 1 \right) \cdot CB_2$$

$$CC_{2\,min} = 0.00529 \text{ microfarads}$$

Choose CC_2 equal to 0.018 microfarads and Calculate RD_2 using Equation (8-56).

$$RD_2 = \frac{1}{B_2} \cdot \left(\frac{1}{CC_2} + \frac{1}{CB_2} \right)$$

$$RD_2 = 148.4 \text{ kohms}$$

Choose RD_2 equal to 147 kohms. Calculate RE_2 using Equation (8-57).

$$RE_2 = \frac{1}{\dfrac{\omega_{02}^2 \cdot (CB_2 + CC_2)}{B_2} - K_2 \cdot B_2 \cdot CB_2}$$

$$RE_2 = 11.81 \text{ kohms}$$

373

Choose RE_2 equal to 11.8 kohms. The final circuit is shown below:

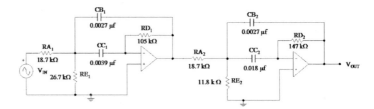

Below is the frequency response of the above circuit using a computer software circuit analysis program. The specification is superimposed in red lines. Note that the filter response exceeds both the pass band and stop band specifications.

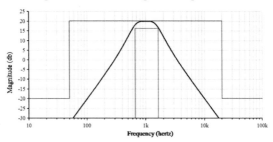

8-6. Design an active notch Butterworth filter with a maximum pass band magnitude of 20 db. The notch center frequency is 2000 hertz with a 2300 hertz notchwidth at the 16.9897 db points. There shall be a maximum stop band magnitude of -20 db above 1900 hertz and below 2100 hertz.

First determine the filter Q using Equation (6-53):

$$Q = \frac{\omega_0}{N} = \frac{2 \cdot \pi \cdot 2000}{2 \cdot \pi \cdot 2300} = 0.8696$$

Since Q is greater than 0.5, use the narrow band approach.

Next, create the filter specification graph. Since the center frequency and notchwidth are given, they must be normalized and converted to the upper and lower pass frequencies. The normalized center frequency and notchwidth are:

$$\omega_0 = \frac{f_0}{2000} = 1 \qquad N = \frac{2300}{2000} = 1.15$$

The upper and lower pass frequencies can be calculated from Equation (6-52):

$$f_0 = 2000 = \sqrt{f_{ph} \cdot f_{pl}} \Rightarrow f_{pl} = \frac{2000^2}{f_{ph}}$$

$$N_f = 2300 = f_{ph} - f_{pl} = f_{ph} - \frac{2000^2}{f_{ph}}$$

$$f_{ph}^2 - 2300 \cdot f_{ph} - 2000^2 = 0 \Rightarrow f_{ph} = 3457 \quad \text{and} \quad f_{pl} = 1157$$

The specification graph is shown below:

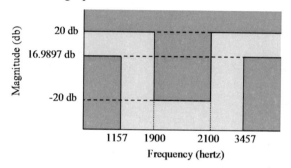

The next step is to normalize the specifications. The magnitude axis needs to be normalized to pass max equal to 0 db.

Pass Max $= 20 - 20 = 0$ db

Pass Min $= 16.9897 - 20 = -3.0103$ db

Stop Max $= -20 - 20 = -40$ db

In order to allow for component variations, the design notch-width needs to be decreased by 10%. The design notchwidth then becomes 2070 hertz. The design upper and lower pass frequencies can be calculated from Equation (6-52):

$$f_0 = 2000 = \sqrt{f_{ph} \cdot f_{pl}} \Rightarrow f_{pl} = \frac{2000^2}{f_{ph}}$$

$$N_f = 2070 = f_{ph} - f_{pl} = f_{ph} - \frac{2000^2}{f_{ph}}$$

$$f_{ph}^2 - 2070 \cdot f_{ph} - 2000^2 = 0 \Rightarrow f_{ph} = 3287 \quad \text{and} \quad f_{pl} = 1217$$

To normalize the frequency axis divide all of the specifications by f_0 or 2000. Since this is a frequency divided by a frequency, the normalized axis can be considered either hertz or radians per second. The normalized frequency values are:

$$\omega_{pl} = \frac{1217}{2000} = 0.6085 \quad \omega_{ph} = \frac{3287}{2000} = 1.6435$$

$$\text{lower stop} = \frac{1900}{2000} = 0.95 \quad \text{upper stop} = \frac{2100}{2000} = 1.05$$

The next step is to determine the low pass equivalent. To determine the equivalent low pass number of poles, determine the pass notchwidth N and the geometric stop notchwidth:

$$N = \frac{N_f}{f_0} = \frac{2070}{2000} = 1.035$$

$$\text{or } N = \omega_{ph} - \omega_{pl} = 1.6435 - 0.6085 = 1.035$$

$$1/(\text{lower stop}) = 1/0.95 = 1.0526 \quad \text{which is greater than}$$
$$\text{the upper stop}$$

$$\text{Therefore: } \omega_{sl} = 0.95 \text{ and } \omega_{sh} = \frac{1}{0.95} = 1.0526$$

$$\text{geometric stop width} = \omega_{sh} - \omega_{sl} = 1.0526 - 0.95 = 0.1026$$

Determine the number of equivalent low pass poles using Equation (6-59):

$$n = \frac{\log_{10}\left(\dfrac{10^{(0.1 \cdot A_s)} - 1}{10^{(0.1 \cdot A_p)} - 1}\right)}{2 \cdot \log_{10}\left(\dfrac{\text{notchwidth}}{\text{geometric stop width}}\right)}$$

$$= \frac{\log_{10}\left(\dfrac{10^{(4)} - 1}{10^{(0.30103)} - 1}\right)}{2 \cdot \log_{10}\left(\dfrac{1.035}{0.1026}\right)} = 1.992$$

Rounding up to the next highest integer yields n equal to 2. Using Table 6-1 for the 2 pole entry, the normalized low pass equivalent transfer function is:

$$G(s)_{LP} = \frac{1}{s^2 + \dfrac{s}{0.7071} + 1}$$

Translating this to the pass band frequency of 1/N yields:

$$G(s)_{LP} = \cfrac{1}{\left((1.035 \cdot s)^2 + \cfrac{1}{0.7071} \cdot (1.035 \cdot s) + 1 \right)}$$

Transforming this to a notch using Equations (6-60) and (6-62) yields:

$$G(s) = \cfrac{(s^2 + 1)^2}{\left(s^4 + \cfrac{1.035}{0.7071} \cdot s^3 + 3.0712 \cdot s^2 + \cfrac{1.035}{0.7071} \cdot s + 1 \right)}$$

Factoring the 4th order denominator yields:

$$G(s) = \cfrac{s^2 + 1}{(s^2 + 0.9973 \cdot s + 2.1384)} \cdot \cfrac{s^2 + 1}{(s^2 + 0.4664 \cdot s + 0.4676)}$$

Translating this out to the specification center frequency of 2000 hertz which is $4000 \cdot \pi$ radians per second yields:

$$G(s) = \cfrac{\left(\cfrac{s}{4000 \cdot \pi} \right)^2 + 1}{\left(\left(\cfrac{s}{4000 \cdot \pi} \right)^2 + 0.9973 \cdot \left(\cfrac{s}{4000 \cdot \pi} \right) + 2.1384 \right)} \cdot$$

$$\cfrac{\left(\cfrac{s}{4000 \cdot \pi} \right)^2 + 1}{\left(\left(\cfrac{s}{4000 \cdot \pi} \right)^2 + 0.4664 \cdot \left(\cfrac{s}{4000 \cdot \pi} \right) + 0.4676 \right)}$$

$$G(s) = \cfrac{s^2 + 1.5791 \cdot 10^8}{(s^2 + 12532 \cdot s + 3.3768 \cdot 10^8)} \cdot$$

$$\cfrac{s^2 + 1.5791 \cdot 10^8}{(s^2 + 5861 \cdot s + 7.384 \cdot 10^7)}$$

At this point the pass max value needs to be denormalized. The magnitude value of 10 (20 db) needs to be distributed between the two stages:

$$K_1 \cdot K_2 = 10$$

therefore: $K_1 = K_2 = \sqrt{10} = 3.1623$

Placing stage 1 in the form of Equation (8-61) yields:

$$G_1(s) = \frac{3.1623(s^2 + 1.5791 \cdot 10^8)}{(s^2 + 12532 \cdot s + 3.3768 \cdot 10^8)}$$

$$= \frac{3.1623 \cdot s^2}{(s^2 + 12532 \cdot s + 3.3768 \cdot 10^8)}$$

$$+ \frac{3.1623 \cdot 0.4676 \cdot 3.3768 \cdot 10^8}{(s^2 + 12532 \cdot s + 3.3768 \cdot 10^8)}$$

For stage 1 ω_b, ω_0, Q, and K are:

$$\omega_{b1} = \sqrt{3.3768 \cdot 10^8} = 18,376$$

$$\omega_0 = \sqrt{1.5791 \cdot 10^8} = 12,566$$

$$\frac{\omega_{b1}}{Q_1} = 12,532 \Rightarrow Q_1 = \frac{18,376}{12,532} = 1.4663$$

$$K_1 = 3.1623$$

For the stage 1 high pass section design, select CA_1 and CC_1 equal to 0.015 microfarads and use Equation (8-40) to determine RB_1:

$$RB_1 = \frac{1}{4 \cdot C_1 \cdot Q_1 \cdot \omega_{c1}} + \frac{1}{C_1 \cdot \omega_{c1}} \cdot \sqrt{\frac{1}{16 \cdot Q_1^2} - \frac{1 - K_1}{2}}$$

$$RB_1 = 1.237 \text{ kohms}$$

Use Equation (8-37) to calculate RD_1:

$$RD_1 = \frac{1}{RB_1 \cdot C_1^2 \cdot \omega_{c1}^2} = 10.64 \text{ kohms}$$

Select standard values of 1.24 kohms and 10.7 kohms for RB_1 and RD_1 respectively.

For the stage 1 low pass section design, select CB_1 equal to 0.1 microfarads and use Equation (8-28) To determine CD_{1max}:

$$CD_{1max} = CB_1 \cdot \left(\frac{1}{4 \cdot Q_1^2} + K_1 - 1 \right) = 0.0116 \cdot 10^{-6}$$

Select CD_1 equal to 0.01 microfarads. Use Equation (8-27) to calculate RC_1:

$$RC_1 = \frac{1}{2 \cdot \omega_{c1} \cdot Q_1 \cdot CD_1} + \frac{1}{\omega_{c1}} \cdot \sqrt{\frac{1}{4 \cdot Q_1^2 \cdot CD_1^2} - \frac{1}{CD_1} \cdot \left(\frac{1}{CB_1} + \frac{1-K_1}{CD_1} \right)}$$

$$RC_1 = 2.5499 \text{ kohms}$$

Use Equation (8-25) to calculate RA_1:

$$RA_1 = \frac{1}{RC_1 \cdot CB_1 \cdot CD_1 \cdot \omega_{c1}^2} = 1.161 \text{ kohms}$$

Select standard values of 2.55 kohms and 1.15 kohms for RC_1 and RA_1 respectively.

For the stage 1 summing op-amp section design, select R_1 equal to 4750 ohms and use Equations (8-62) to calculate the remaining resistors:

$$\text{Resistor } (K_1 \cdot R_1) = K_1 \cdot R_1 = 3.1623 \cdot 4750 = 15{,}021 \text{ ohms}$$

$$\text{Resistor } \left(R_1 / \left(\omega_0^2 / \omega_{b1}^2 \right) \right) = \frac{R_1}{\omega_0^2 / \omega_{b1}^2} = \frac{4750}{0.4676} = 10{,}158 \text{ ohms}$$

Select K_1R_1 equal to 15.0 kohms and $R_1 / \left(\omega_0^2 / \omega_{b1}^2 \right)$ equal to 10.2 kohms.

Placing stage 2 in the form of Equation (8-61) yields:

$$G_2(s) = \frac{3.1623 \cdot \left(s^2 + 1.5791 \cdot 10^8 \right)}{\left(s^2 + 5861 \cdot s + 7.384 \cdot 10^7 \right)}$$

$$G_2(s) = \frac{3.1623 \cdot s^2}{\left(s^2 + 5861 \cdot s + 7.384 \cdot 10^7 \right)}$$
$$+ \frac{3.1623 \cdot 2.1384 \cdot 7.384 \cdot 10^7}{\left(s^2 + 5861 \cdot s + 7.384 \cdot 10^7 \right)}$$

For stage 2 ω_b, ω_0, Q, and K are:

$$\omega_{b2} = \sqrt{7.384 \cdot 10^7} = 8{,}593$$

$$\omega_0 = \sqrt{1.5791 \cdot 10^8} = 12{,}566$$

$$\frac{\omega_{b2}}{Q_2} = 5{,}861 \Rightarrow Q_2 = \frac{8{,}593}{5{,}861} = 1.4661$$

$$K_2 = 3.1623$$

For the stage 2 high pass section design, select CA_2 and CC_2 equal to 0.015 microfarads and use Equation (8-40) to determine RB_2:

$$RB_2 = \frac{1}{4 \cdot C_2 \cdot Q_2 \cdot \omega_{c2}} + \frac{1}{C_2 \cdot \omega_{c2}} \cdot \sqrt{\frac{1}{16 \cdot Q_2^2} - \frac{1-K_2}{2}}$$

$$RB_2 = 2.6459 \text{ kohms}$$

Use Equation (8-37) to calculate RD_2:

$$RD_2 = \frac{1}{RB_2 \cdot C_2^2 \cdot \omega_{c2}^2} = 22.749 \text{ kohms}$$

Select standard values of 2.67 kohms and 22.6 kohms for RB_2 and RD_2 respectively.

For the stage 2 low pass section design, select CB_2 equal to 0.1 microfarads and use Equation (8-28) To determine CD_{2max}:

$$CD_{2max} = CB_2 \cdot \left(\frac{1}{4 \cdot Q_2^2} + K_2 - 1 \right) = 0.01163 \cdot 10^{-6}$$

Select CD_2 equal to 0.0047 microfarads. Use Equation (8-27) to calculate RC_2:

$$RC_2 = \frac{1}{2 \cdot \omega_{c2} \cdot Q_2 \cdot CD_2} + \frac{1}{\omega_{c2}} \cdot \sqrt{\frac{1}{4 \cdot Q_2^2 \cdot CD_2^2} - \frac{1}{CD_2} \cdot \left(\frac{1}{CB_2} + \frac{1-K_2}{CD_2} \right)}$$

$$RC_2 = 14.963 \text{ kohms}$$

Use Equation (8-25) to calculate RA_2:

$$RA_2 = \frac{1}{RC_2 \cdot CB_2 \cdot CD_2 \cdot \omega_{c2}^2} = 1.9257 \text{ kohms}$$

Select standard values of 15.0 kohms and 1.91 kohms for RC_2 and RA_2 respectively.

For the stage 2 summing op-amp section design, select R_2 equal to 4750 ohms and use Equations (8-62) to calculate the remaining resistors:

$$\text{Resistor } (K_2 \cdot R_2) = K_2 \cdot R_2 = 3.1623 \cdot 4750 = 15,021 \text{ ohms}$$

$$\text{Resistor } (R_2 / (\omega_0^2 / \omega_{b2}^2)) = \frac{R_2}{\omega_0^2 / \omega_{b2}^2} = \frac{4750}{2.1384} = 2,221 \text{ ohms}$$

Select K_2R_2 equal to 15.0 kohms and $R_2/\left(\omega_0^2/\omega_{b2}^2\right)$ equal to 2.21 kohms. The final circuit is shown below:

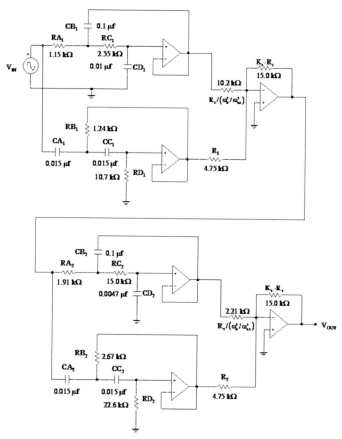

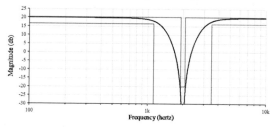

Below is the frequency response of the above circuit using a computer software circuit analysis program. The specification is superimposed in red lines. Note that the filter response exceeds both the pass band and stop band specifications.

8-7. Implement the low pass filter of Problem 8-1 using the state variable filter realization.

From Problem 8-1 solution, the transfer function equation is:

$$G(s) = \frac{3.615}{\left(\left(\frac{s}{14,548}\right)^2 + \frac{1}{0.5412} \cdot \left(\frac{s}{14,548}\right) + 1\right)} \cdot$$

$$\frac{2.766}{\left(\left(\frac{s}{14,548}\right)^2 + \frac{1}{1.3066} \cdot \left(\frac{s}{14,548}\right) + 1\right)}$$

The filter can be implemented with 2 two pole stages using the filter circuit of Figure 8-37. The stage 1 requirements are:

$$K_1 = 3.615$$
$$\omega_{c1} = 14,548$$
$$Q_1 = 0.5412$$

The stage 2 requirements are:

$$K_2 = 2.766$$
$$\omega_{c2} = 14,548$$
$$Q_2 = 1.3066$$

For stage 1, select a standard value for CA_1 of 0.0022 microfarads and calculate RB_1 from Equation (8-82).

$$RB_1 = \frac{Q_1}{CA_1 \cdot \omega_{c1}} = 16.91 \text{ kohms}$$

Select a standard value of 16.9 kohms for RB_1. Next, select a standard value for CB_1 of 0.01 microfarads and calculate RC_1 from Equation (8-83).

$$RC_1 = \frac{1}{CB_1 \cdot Q_1 \cdot \omega_{c1}} = 12.701 \text{ kohms}$$

Select a standard value of 12.7 kohms for RC_1. Finally, select a standard value for RA_1 of 49.9 kohms and calculate $RA_1/(K_1-1)$ and RA_1/K_1 from Equations (8-84).

$$\text{Resistor } RA_1/(K_1-1) = \frac{RA_1}{K_1-1} = 19.08 \text{ kohms}$$

$$\text{Resistor } RA_1/K_1 = \frac{RA_1}{K_1} = 13.80 \text{ kohms}$$

Select a standard values of 19.1 kohms for $RA_1/(K_1-1)$ and 13.7 kohms for RA_1/K_1.

For stage 2, select a standard value for CA_2 of 0.0082 microfarads and calculate RB_2 from Equation (8-82).

$$RB_2 = \frac{Q_2}{CA_2 \cdot \omega_{c2}} = 10.95 \text{ kohms}$$

Select a standard value of 11.0 kohms for RB_2. Next, select a standard value for CB_2 of 0.0047 microfarads and calculate RC_2 from Equation (8-83).

$$RC_2 = \frac{1}{CB_2 \cdot Q_2 \cdot \omega_{c2}} = 11.19 \text{ kohms}$$

Select a standard value of 11.3 kohms for RC_2. Finally, select a standard value for RA_2 of 28.0 kohms and calculate $RA_2/(K_2-1)$ and RA_2/K_2 from Equations (8-84).

$$\text{Resistor } RA_2/(K_2-1) = \frac{RA_2}{K_2-1} = 15.86 \text{ kohms}$$

$$\text{Resistor } RA_2/K_2 = \frac{RA_2}{K_2} = 10.12 \text{ kohms}$$

Select a standard value of 15.8 kohms for $RA_2/(K_2-1)$ and 10.2 kohms for RA_2/K_2. The final circuit is shown below:

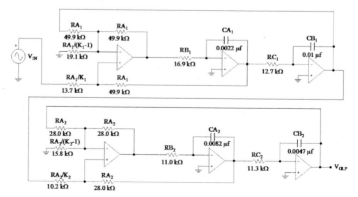

Below is the frequency response of the above circuit using a computer software circuit analysis program. The specification is superimposed in red lines. Note that the filter response exceeds both the pass band and stop band specifications.

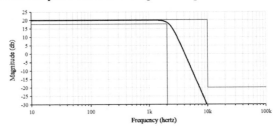

8-8. Implement the high pass filter of Problem 8-3 using the state variable filter realization.

From Problem 8-3 solution, the transfer function equation is:

$$G(s) = \frac{3.615 \cdot s^2}{\left(s^2 + \dfrac{54,270}{0.5412} \cdot s + 54,270^2\right)}$$

$$\cdot \frac{2.766 \cdot s^2}{\left(s^2 + \dfrac{54,270}{1.3066} \cdot s + 54,270^2\right)}$$

The filter can be implemented with 2 two pole stages using the filter circuit of Figure 8-37. The stage 1 requirements are:

$$K_1 = 3.615$$
$$\omega_{c1} = 54,270$$
$$Q_1 = 0.5412$$

The stage 2 requirements are:

$$K_2 = 2.766$$
$$\omega_{c2} = 54,270$$
$$Q_2 = 1.3066$$

For stage 1, select a standard value for CA_1 of 820 picofarads and calculate RB_1 from Equation (8-82).

$$RB_1 = \frac{Q_1}{CA_1 \cdot \omega_{c1}} = 12.16 \text{ kohms}$$

Select a standard value of 12.1 kohms for RB_1. Next, select a standard value for CB_1 of 0.0022 microfarads and calculate RC_1 from Equation (8-83).

$$RC_1 = \frac{1}{CB_1 \cdot Q_1 \cdot \omega_{c1}} = 15.48 \text{ kohms}$$

Select a standard value of 15.4 kohms for RC_1. Finally, select a standard value for RA_1 of 49.9 kohms and calculate $RA_1/(K_1-1)$ and RA_1/K_1 from Equations (8-84).

$$\text{Resistor } RA_1/(K_1-1) = \frac{RA_1}{K_1 - 1} = 19.08 \text{ kohms}$$

$$\text{Resistor } RA_1/K_1 = \frac{RA_1}{K_1} = 13.80 \text{ kohms}$$

Select a standard values of 19.1 kohms for $RA_1/(K_1-1)$ and 13.7 kohms for RA_1/K_1.

For stage 2, select a standard value for CA_2 of 0.0018 microfarads and calculate RB_2 from Equation (8-82).

$$RB_2 = \frac{Q_2}{CA_2 \cdot \omega_{c2}} = 13.38 \text{ kohms}$$

Select a standard value of 13.3 kohms for RB_2. Next, select a standard value for CB_2 of 0.0012 microfarads and calculate RC_2 from Equation (8-83).

$$RC_2 = \frac{1}{CB_2 \cdot Q_2 \cdot \omega_{c2}} = 11.75 \text{ kohms}$$

Select a standard value of 11.8 kohms for RC_2. Finally, select a standard value for RA_2 of 28.0 kohms and calculate $RA_2/(K_2-1)$ and RA_2/K_2 from Equations (8-84).

$$\text{Resistor } RA_2/(K_2-1) = \frac{RA_2}{K_2 - 1} = 15.86 \text{ kohms}$$

$$\text{Resistor } RA_2/K_2 = \frac{RA_2}{K_2} = 10.12 \text{ kohms}$$

Select a standard value of 15.8 kohms for $RA_2/(K_2-1)$ and 10.2 kohms for RA_2/K_2. The final circuit is shown below:

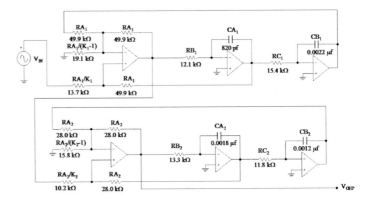

Below is the frequency response of the above circuit using a computer software circuit analysis program. The specification is superimposed in red lines. Note that the filter response exceeds both the pass band and stop band specifications.

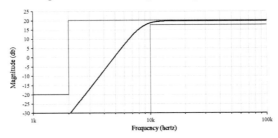

8-9. Implement the band pass filter of Problem 8-5 using the state variable filter realization.

From Problem 8-5 solution, the transfer function equation is:

$$G(s) = \frac{19,699 \cdot s}{\left(s^2 + 5926 \cdot s + 8.1523 \cdot 10^7\right)} \cdot$$

$$\frac{19,699 \cdot s}{\left(s^2 + 2870 \cdot s + 1.9118 \cdot 10^7\right)}$$

$$= \frac{3.3242 \cdot 5926 \cdot s}{\left(s^2 + 5926 \cdot s + 9029^2\right)} \cdot$$

$$\frac{6.864 \cdot 2870 \cdot s}{\left(s^2 + 2870 \cdot s + 4372^2\right)}$$

The filter can be implemented with 2 two pole stages using the filter circuit of Figure 8-37. The stage 1 requirements are:

$$K_1 = 3.3242$$

386

$$\omega_{c1} = 9029$$

$$B_1 = 5926$$

$$Q_1 = \frac{\omega_{c1}}{B_1} = 1.5236$$

The stage 2 requirements are:

$$K_2 = 6.864$$

$$\omega_{c2} = 4372$$

$$B_2 = 2870$$

$$Q_2 = \frac{\omega_{c2}}{B_2} = 1.5233$$

For stage 1, select a standard value for CA_1 of 0.01 microfarads and calculate RB_1 from Equation (8-82).

$$RB_1 = \frac{Q_1}{CA_1 \cdot \omega_{c1}} = 16.87 \text{ kohms}$$

Select a standard value of 16.9 kohms for RB_1. Next, select a standard value for CB_1 of 0.0068 microfarads and calculate RC_1 from Equation (8-83).

$$RC_1 = \frac{1}{CB_1 \cdot Q_1 \cdot \omega_{c1}} = 10.69 \text{ kohms}$$

Select a standard value of 10.7 kohms for RC_1. Finally, select a standard value for RA_1 of 34.0 kohms and calculate $RA_1/(K_1-1)$ and RA_1/K_1 from Equations (8-84).

$$\text{Resistor } RA_1/(K_1-1) = \frac{RA_1}{K_1 - 1} = 14.63 \text{ kohms}$$

$$\text{Resistor } RA_1/K_1 = \frac{RA_1}{K_1} = 10.23 \text{ kohms}$$

Select a standard values of 14.7 kohms for $RA_1/(K_1-1)$ and 10.2 kohms for RA_1/K_1.

For stage 2, select a standard value for CA_2 of 0.022 micro-farads and calculate RB_2 from Equation (8-82).

$$RB_2 = \frac{Q_2}{CA_2 \cdot \omega_{c2}} = 15.84 \text{ kohms}$$

Select a standard value of 15.8 kohms for RB_2. Next, select a standard value for CB_2 of 0.01 microfarads and calculate RC_2 from Equation (8-83).

$$RC_2 = \frac{1}{CB_2 \cdot Q_2 \cdot \omega_{c2}} = 15.02 \text{ kohms}$$

Select a standard value of 15.0 kohms for RC_2. Finally, select a standard value for RA_2 of 73.2 kohms and calculate $RA_2/(K_2-1)$ and RA_2/K_2 from Equations (8-84).

$$\text{Resistor } RA_2/(K_2-1) = \frac{RA_2}{K_2-1} = 12.48 \text{ kohms}$$

$$\text{Resistor } RA_2/K_2 = \frac{RA_2}{K_2} = 10.66 \text{ kohms}$$

Select a standard value of 12.4 kohms for $RA_2/(K_2-1)$ and 10.7 kohms for RA_2/K_2. The final circuit is shown below:

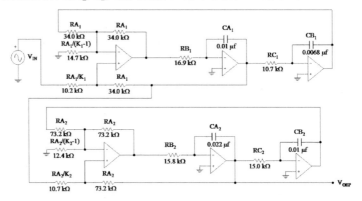

Below is the frequency response of the above circuit using a computer software circuit analysis program. The specification is superimposed in red lines. Note that the filter response exceeds both the pass band and stop band specifications.

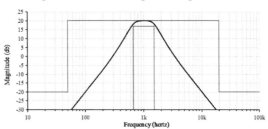

388

8-10. Implement the notch filter of Problem 8-6 using the state variable filter realization.

From Problem 8-6, the transfer function equation is:

$$G(s) = \frac{3.1623(s^2 + 1.5791 \cdot 10^8)}{(s^2 + 12532 \cdot s + 3.3768 \cdot 10^8)} \cdot$$

$$\frac{3.1623 \cdot (s^2 + 1.5791 \cdot 10^8)}{(s^2 + 5861 \cdot s + 7.384 \cdot 10^7)}$$

The filter can be implemented with 2 two pole stages using the filter circuit of Figure 8-44.

Splitting the first stage into high pass and low pass components and placing in the form of Equation (8-85) yields:

$$G_1(s) = \frac{3.1623(s^2 + 1.5791 \cdot 10^8)}{(s^2 + 12532 \cdot s + 3.3768 \cdot 10^8)}$$

$$= \frac{3.1623(s^2 + 12566^2)}{(s^2 + 12532 \cdot s + 18376^2)}$$

$$= \frac{3.1623 \cdot s^2}{(s^2 + 12532 \cdot s + 18376^2)}$$

$$+ \frac{3.1623 \cdot 0.4676 \cdot 18376^2}{(s^2 + 12532 \cdot s + 18376^2)}$$

The stage 1 requirements are:

$$\omega_0 = 12{,}566$$

$$K_{1N} = 3.1623$$

$$K_{1HP} = 1$$

$$K_{1LP} = 0.4676$$

$$\omega_{c1} = 18{,}376$$

$$\frac{\omega_{c1}}{Q_1} = 12{,}532$$

$$Q_1 = 1.466$$

For stage 1, select a standard value for CA_1 of 0.0068 micro-farads and calculate RB_1 from Equation (8-82).

389

$$RB_1 = \frac{Q_1}{CA_1 \cdot \omega_{c1}} = 11.73 \text{ kohms}$$

Select a standard value of 11.8 kohms for RB_1. Next, select a standard value for CB_1 of 0.0033 microfarads and calculate RC_1 from Equation (8-83).

$$RC_1 = \frac{1}{CB_1 \cdot Q_1 \cdot \omega_{c1}} = 11.25 \text{ kohms}$$

Select a standard value of 11.3 kohms for RC_1. Next, select a standard value for RA_1 of 10.0 kohms.

Select a standard value of 4.99 kohms for RD_1 and use Equation (8-86) to calculate the remaining resistors.

Resistor $(K_{1N} \cdot RD_1) = 15.78$ kohms

Resistor $(RD_1 / (\omega_0^2 / \omega_{c1}^2)) = 10.67$ kohms

Select standard values of 15.8 kohms for $K_{1N} \cdot RD_1$ and 10.7 kohms for $RD_1 / (\omega_0^2 / \omega_{c1}^2)$.

Splitting the second stage into high pass and low pass components and placing in the form of Equation (8-85) yields:

$$G_2(s) = \frac{3.1623 \cdot (s^2 + 1.5791 \cdot 10^8)}{(s^2 + 5861 \cdot s + 7.384 \cdot 10^7)}$$

$$= \frac{3.1623 \cdot (s^2 + 12566^2)}{(s^2 + 5861 \cdot s + 8593^2)}$$

$$= \frac{3.1623 \cdot s^2}{(s^2 + 5861 \cdot s + 8593^2)}$$

$$+ \frac{3.1623 \cdot 2.1384 \cdot 8593^2}{(s^2 + 5861 \cdot s + 8593^2)}$$

The stage 2 requirements are:

$$\omega_0 = 12,566$$
$$K_{2N} = 3.1623$$
$$K_{2HP} = 1$$
$$K_{2LP} = 2.1384$$
$$\omega_{c2} = 8593$$

$$\frac{\omega_{c2}}{Q_2} = 5861$$

$$Q_2 = 1.466$$

For stage 2, select a standard value for CA_2 of 0.012 microfarads and calculate RB_2 from Equation (8-82).

$$RB_2 = \frac{Q_2}{CA_2 \cdot \omega_{c2}} = 14.22 \text{ kohms}$$

Select a standard value of 14.3 kohms for RB_2. Next, select a standard value for CB_2 of 0.0047 microfarads and calculate RC_2 from Equation (8-83).

$$RC_2 = \frac{1}{CB_2 \cdot Q_2 \cdot \omega_{c2}} = 16.89 \text{ kohms}$$

Select a standard value of 16.9 kohms for RC_2. Next, select a standard value for RA_2 of 10.0 kohms.

Select a standard value of 12.7 kohms for RD_2 and use Equation (8-86) to calculate the remaining resistors.

Resistor $\left(K_{2N} \cdot RD_2 \right) = 40.16$ kohms

Resistor $\left(RD_2 / \left(\omega_0^2 / \omega_{c2}^2 \right) \right) = 5.939$ kohms

Select standard values of 40.2 kohms for $K_{2N} \cdot RD_2$ and 5.90 kohms for $RD_2 / \left(\omega_0^2 / \omega_{c2}^2 \right)$.

The final circuit is shown below.

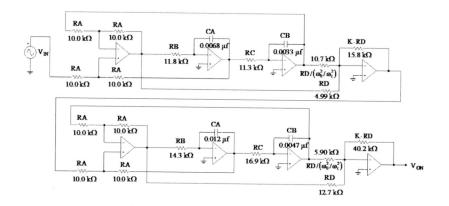

Below is the frequency response of the above circuit using a computer software circuit analysis program. The specification is superimposed in red lines. Note that the filter response exceeds both the pass band and stop band specifications.

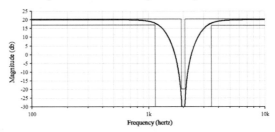

8-11. Verify that the notch filter circuit of Figure 8-28 implements the transfer function of Equation (8-61).

The notch filter circuit of Figure 8-28 is shown below:

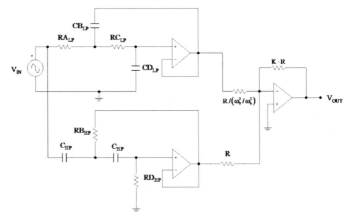

The circuit consists of a two pole low pass section, a two pole high pass section, and a summer circuit. First consider the summer circuit shown below:

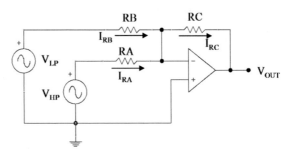

Since the op-amp input current is zero, summing the currents at the negative input node yields:

$$I_{RA} + I_{RB} = I_{RC}$$

Since the positive input node is grounded and the difference between the input node voltages must be zero for the output to be a finite value, converting the node current equation to node voltages yields:

$$\frac{V_{HP}}{RA} + \frac{V_{LP}}{RB} = \frac{-V_{OUT}}{RC}$$

Solving for V_{OUT} yields:

$$V_{OUT} = -\left(\frac{RC}{RA} \cdot V_{HP} + \frac{RC}{RB} \cdot V_{LP}\right)$$

Letting RA equal R, RB equal $R/\left(\omega_0^2/\omega_{b1}^2\right)$, and RC equal to $K \cdot R$ in the above equation yields:

$$V_{OUT} = -\left(\frac{K \cdot R}{R} \cdot V_{HP} + \frac{K \cdot R}{R/\left(\omega_0^2/\omega_{b1}^2\right)} \cdot V_{LP}\right)$$

$$= -\left(K \cdot V_{HP} + K \cdot \left(\omega_0^2/\omega_{b1}^2\right) \cdot V_{LP}\right)$$

In the above equation V_{HP} is the output of a two pole high pass filter stage with a transfer function (see Equation (8-30)) of:

$$G(s) = \frac{V_{OUT}}{V_{IN}} = \frac{K \cdot s^2}{s^2 + \frac{\omega_c}{Q} \cdot s + \omega_c^2}$$

where: $V_{OUT} = V_{HP}$, $K = 1$, $\omega_c = \omega_b$, and $\frac{\omega_c}{Q} = N$

$$\Rightarrow V_{HP} = V_{IN} \cdot \left(\frac{s^2}{s^2 + N \cdot s + \omega_b^2}\right)$$

Also V_{LP} is the output of a two pole low pass filter stage with a transfer function (see Equation (8-21)) of:

$$G(s) = \frac{V_{OUT}}{V_{IN}} = \frac{K}{\frac{s^2}{\omega_c^2} + \frac{1}{Q} \cdot \frac{s}{\omega_c} + 1} = \frac{K \cdot \omega_c^2}{s^2 + \frac{\omega_c}{Q} \cdot s + \omega_c^2}$$

where: $V_{OUT} = V_{LP}$, $K = 1$, $\omega_c = \omega_b$, and $\dfrac{\omega_c}{Q} = N$

$$\Rightarrow V_{LP} = V_{IN} \cdot \left(\frac{\omega_b^2}{s^2 + N \cdot s + \omega_b^2} \right)$$

Substituting V_{HP} and V_{LP} into the above equation for V_{OUT} yields:

$$V_{OUT} = -\left(K \cdot V_{HP} + K \cdot \left(\frac{\omega_0^2}{\omega_b^2} \right) \cdot V_{LP} \right)$$

$$= -\left(V_{IN} \cdot \left(\frac{K \cdot s^2}{s^2 + N \cdot s + \omega_b^2} \right) + V_{IN} \cdot \left(\frac{K \cdot \left(\frac{\omega_0^2}{\omega_b^2} \right) \cdot \omega_b^2}{s^2 + N \cdot s + \omega_b^2} \right) \right)$$

$$\Rightarrow G(s) = \frac{V_{OUT}}{V_{IN}} = -\left(\left(\frac{K \cdot s^2}{s^2 + N \cdot s + \omega_b^2} \right) + \left(\frac{K \cdot \left(\frac{\omega_0^2}{\omega_b^2} \right) \cdot \omega_b^2}{s^2 + N \cdot s + \omega_b^2} \right) \right)$$

Which agrees with Equation (8-61).

INDEX

ε (pass band ripple), 93, 140

Active filter realizations, 157
Analog waveforms, 31

Band pass active filter realization, 186, 187
Band pass filter, 82
Band pass number of poles, 111, 112
Band pass two pole active filter circuit, 187, 188, 189, 190
Band reject filter, 82
Band stop filter, 82
Bandwidth, 109
Bode, Hendrik, 61
Bode approximation plot, 62, 63, 64, 65, 66, 68, 70, 71
Bode magnitude plot, 62, 64, 67, 69
Bode phase plot, 63, 65, 68, 70
Bode plot, 61
Break frequency format, 86
Butterworth approximation, 85, 86, 87, 88, 89, 90
Butterworth circle, 86
Butterworth cutoff frequency, 89, 90
Butterworth passive low pass filter component values, 135
Butterworth number of poles, 89, 90
Butterworth pole values, 87
Butterworth, Stephen, 85

Butterworth to Chebyshev transformation, 94
Butterworth transfer function, 90

Capacitor charge, 22
Capacitor current, 22, 23, 26, 27
Capacitor electric field, 22
Capacitor initial voltage, 47
Capacitor Laplace impedance, 46
Capacitor voltage, 22, 23, 25
Center frequency, 109, 117
Chebyshev approximation, 92, 93, 94, 95, 96, 97
Chebyshev ellipse, 94
Chebyshev number of poles, 96, 97
Chebyshev numerator, 95
Chebyshev, Pafnuty, 92
Chebyshev polynomial, 93
Chebyshev transfer function, 97
Complex pole approximation, 68, 69
Complex poles, 58, 67, 68
Complex zero approximation, 70, 71
Complex zeros, 69, 70
Continued fraction expansion, 132
Continuous time waveforms, 31
Converting Laplace to frequency domain, 61
Cubic function, 9
Cutoff frequency, 87, 137, 144

Damping factor, 11
Decaying sinusoid, 15, 16, 37
Digital waveforms, 31
Dirac delta function, 10
Dirac, Paul, 10
Discrete time waveforms, 31

Euler exponential form of Sine and Cosine, 32
Exponential function, 11, 36

Filters, 4, 5, 81
Filter specification, 82, 83, 84
Fourier, Joseph, 31
Fourier series, 31, 32, 33
Fourier transform, 31, 33
Frequency and Impedance translation, 135, 136
Frequency domain, 3, 4, 5, 6, 31, 32, 60

Geffe, P. R., 113
Geometric stop width, 111

Heavyside, Oliver, 8, 31
Heavyside step function, 8
High pass active filter realization, 175, 176
High pass cutoff frequency, 102
High pass filter, 82
High pass number of poles, 102, 106
High pass one pole active filter circuit, 176, 177
High pass two pole active filter circuit, 177, 178, 179

Impulse function, 10, 36
Inductor current, 23, 24, 26, 27
Inductor initial current, 46
Inductor Laplace impedance, 46
Inductor voltage, 23, 24, 25

Inductor magnetic field, 23
Inductor magnetic flux, 23, 24
Integral-differential equations, 24, 25, 26, 27, 41
Inverse Laplace transform, 42, 43, 44, 45

Kirchhoff's current law, 2, 25, 26
Kirchhoff's voltage law, 22, 23, 25, 41

l'Hospital's rule, 57
Laplace, Pierre Simon De, 31
Laplace complex differentiation property, 39
Laplace complex integration property, 40
Laplace complex translation property, 36, 37
Laplace domain, 34
Laplace impedance, 45
Laplace linearity property, 34, 35
Laplace real differentiation property, 39
Laplace real integration property, 40
Laplace real translation property, 38
Laplace transform, 1, 21, 27, 31, 33
Laplace transform of a circuit, 45, 46, 47
Laplace transform of a derivative, 38, 39
Laplace transform of an integral, 39, 40
Laplace transform of integral-differential equations, 41
Low pass active filter realization, 160
Low pass filter, 81
Low pass ladder circuit, 130

Low pass one pole active filter circuit, 160, 161, 162
Low pass passive filter realization, 129, 130, 131, 132
Low pass to band pass transformation, 109, 110, 111, 112, 113, 146
Low pass to high pass transformation, 101, 102, 103, 142, 143
Low pass to notch transformation, 117, 118, 119, 120, 121, 122, 150, 151
Low pass two pole active filter circuit, 162, 163, 164, 165, 166, 167

Multiple pole/zero plots, 72, 73, 74, 75, 76, 77, 78, 79

Nodal voltage analysis, 1
Non-periodic functions, 31, 33, 34
Notch active filter realization, 197, 198
Notch filter, 82
Notch number of poles, 120
Notch two pole active filter circuit, 198
Notchwidth, 117

Ohm, Georg, 21
Ohm's law, 2, 21
Operational amplifier, 157, 158, 159, 160
Optical frequency spectrum, 4

Partial Fraction Expansion, 51, 56
Pass band, 81, 82
Pass band ripple, 92
Passive filter realizations, 129

Periodic functions, 7, 31, 32, 33
Pole zero format, 86
Pole zero plot, 56, 57
Prism, 4
Pulse source, 1

Q, 67, 68, 109, 117, 118

Ramp function, 9, 35
Ramp source, 1
Real pole approximation plot, 64
Real poles, 57, 62, 63
Real zero approximation plot, 66
Real zeros, 64, 65
Rectangular wave source, 7
Repeating pulse, 16, 17
Resistor current, 21, 22, 26, 27
Resistor Laplace impedance, 46
Resistor voltage, 21, 25

Sallen-Key active filter circuit, 162
Saw tooth wave source, 7
Sinusoid pulse, 15
Sinusoidal function, 11, 12, 37
Sinusoidal wave source, 7
Square function, 9
Square wave source, 1, 7
Stability, 55, 56, 57, 58, 59
State variable active filter circuit, 213, 214
State variable active filter realization, 206, 207, 208, 209, 210
State variable active filter integrator circuit, 211
State variable active filter summer circuit, 211, 212, 213
State variable active notch filter, 221, 222
State variable active notch filter circuit, 222

Steady state response, 1, 3
Steady state solution, 7
Step function, 8, 35
Stop band, 81, 82
System, 55

t^n functions, 7, 8
Time constant, 11
Time domain, 3
Time shifted pulse, 13
Time shifting, 12, 13, 38
Transfer function, 55, 56
Transient response, 1, 3
Transient solution, 7
Transition band, 82
Trapezoid Pulse, 14
Triangle pulse, 14
Triangle wave source, 7

Unit step function, 8
Unit impulse function, 10

Voltage divider rule, 48, 49

Made in the USA
San Bernardino, CA
02 May 2016